ELECTRONICS, INFORMATION TECHNOLOGY AND INTELLECTUALIZATION

PROCEEDINGS OF THE INTERNATIONAL CONFERENCE EITI 2014, SHENZHEN, CHINA, 16–17 AUGUST 2014

Electronics, Information Technology and Intellectualization

Editors

Young Min Song
Pusan National University, Busan, Korea

Kyung Sup Kwak
Inha University, Incheon, Korea

CRC Press
Taylor & Francis Group
Boca Raton London New York

CRC Press is an imprint of the
Taylor & Francis Group, an **informa** business

A BALKEMA BOOK

Published by:
CRC Press/Balkema
P.O. Box 447, 2300 AK Leiden, The Netherlands
e-mail: Pub.NL@taylorandfrancis.com
www.crcpress.com – www.taylorandfrancis.com

First issued in paperback 2020

Typeset by MPS Limited, Chennai, India

ISBN 13: 978-0-367-73856-3 (pbk)
ISBN 13: 978-1-138-02741-1 (hbk)

Visit the Taylor & Francis Web site at
http://www.taylorandfrancis.com

and the CRC Press Web site at
http://www.crcpress.com

Table of contents

Electronics, Information Technology and Intellectualization – Song & Kwak (Eds)
© 2015 Taylor & Francis Group, London, ISBN 978-1-138-02741-1

Organizing committee

CONFERENCE ORGANIZATION

Co-Chairmen
Prof. Kyung Sup, Kwak Inha University
Prof. Young Min Song, Pusan National University
Dr. Jason Lee, Hong Kong Information Science and Engineering Research Center

International Scientific Committee
Prof. Mu Chun Su, National Central University
Prof. Yu En Wu, National Kaohsiung First University of Science and Technology
Prof. Chien-Yue Chen, National Yunlin University of Science and Technology
Prof. Yeong-Chau Kuo, National Kaohsiung First University of Science and Technology
Prof. Jung Chuan Chou, National Yunlin University of Science and Technology
Prof. Jiyun Kim, Kyung Hee University

Local Organizing Chairman
Dr. Jason Lee, Hong Kong Information Science and Engineering Research Center

Electronics, Information Technology and Intellectualization – Song & Kwak (Eds)
© 2015 Taylor & Francis Group, London, ISBN 978-1-138-02741-1

Preface

The 2014 International Conference on Electronics, Information Technology and Intellectualization (ICEITI 2014) provides an academic platform for leading experts and researchers in these fields for sharing and exchanging the latest research results and problem solving solutions. ICEITI has collected advanced results and trends in the field of Electronics, Information Technology and Intellectualization.

This book is a collection of quality papers. All the papers accepted are reviewed and edited by 2–3 expert referees from our academic committee board. This book aims to collect the advanced and practical result of research developments on electric machines, electric apparatus, power systems and automation, high voltage techniques, microelectronic devices, computer and image processing technologies, wireless networks, mobile networks, etc.

The Scientific Committee of the Conference has made sure that this book will provide the readers with the latest advanced knowledge in these fields, also providing a valuable summary and reference.

We express our sincere appreciations to the authors for their contribution to this book. We would also like to express our sincere gratitude to all the experts and referees for their valuable comments and the editing of the papers. Thanks also to CRC Press/Balkema (Taylor & Francis Group).

Electronics, Information Technology and Intellectualization – Song & Kwak (Eds)
© *2015 Taylor & Francis Group, London, ISBN 978-1-138-02741-1*

Approximation properties of Kantorovich type q-Szász-Mirakjan-Schurer operators

Qing-Bo Cai

School of Mathematics and Computer Science, Quanzhou Normal University, Quanzhou, China

ABSTRACT: In this paper, we investigate the moments and the central moments of Kantorovich type q-Szász-Mirakjan-Schurer operators by computation and obtain a weighted approximation theorem and a Voronovskaya type asymptotic formula.

Keywords: q-integer; q-Szász-Mirakjan-Schurer operators; moments

1   INTRODUCTION

Mahmudov (2010) mentioned the Kantorovich type q-Szász-Mirakjan-Schurer operators as follows

$$K_{n,q}(f;x) = e_q\left(-[n+l]_q x\right)\sum_{k=0}^{\infty} q^{\frac{k(k-1)}{2}} \frac{[n+l]_q^k x^k}{[k]_q!} \tag{1}$$
$$\cdot \int_0^1 f\left(\frac{[k]_q + q^k t}{q^{k-1}[n]_q}\right) d_q t$$

for every $l \in N \cup \{0\}$, $q \in (0,1)$, $x \in [0,\infty)$ and for every real valued continuous and bounded function f on $[0,\infty)$. Cai (2011) obtained the weighted statistical approximation theorem and investigated local approximation properties of the operators (1).

In this paper, we compute the third and fourth moments and fourth central moments of operators (1) and obtain a weighted approximation theorem and a Voronvskaya type asymptotic formula.

Firstly, we recall some concepts of q-calculus. All of the results can be found in (Kac & Cheung 2002). For any fixed real number $0 < q \le 1$ and each nonnegative integer k, we denote q-integers by $[k]_q$, where

$$[k]_q = \begin{cases} \dfrac{1-q^k}{1-q}, & q \neq 1; \\ k, & q = 1. \end{cases}$$

Also, q-factorial and q-binomial coefficients are defined as follows:

$$[k]_q! = \begin{cases} [k]_q [k-1]_q \dots [1]_q, & k = 1,2,\dots; \\ 1, & k = 0, \end{cases}$$

and

$$\begin{bmatrix} n \\ k \end{bmatrix}_q = \frac{[n]_q!}{[k]_q![n-k]_q!}, \quad (n \ge k \ge 0).$$

The q-Jackson integral is defined as

$$\int_0^a f(x) d_q x = (1-q)a \sum_{n=0}^{\infty} f\left(aq^n\right)q^n, \qquad a > 0,$$

Provided the sums converge absolutely.

The q-analogue $e_q(x)$ and $E_q(x)$ of the exponential function are given as

$$e_q(x) = \sum_{k=0}^{\infty} \frac{x^k}{[k]_q!} = \frac{1}{\left(1-(1-q)x\right)_q^{\infty}}, \ |x| < \frac{1}{1-q},$$

$$E_q(x) = \sum_{k=0}^{\infty} q^{k(k-1)/2} \frac{x^k}{[k]_q!} = \left(1+(1-q)x\right)_q^{\infty},$$

where $(1-x)_q^{\infty} = \prod_{j=0}^{\infty}(1-q^j x)$ and $|q| < 1$. Obviously,
$$e_q(x) \cdot E_q(x) = 1.$$

2   SOME PRELIMINARY RESULTS

In this section, we give the following lemmas, which are needed to prove our theorem.

Lemma 2.1 (see (Cai 2011)). *Let $m = 0, 1, 2, \dots$, $0 < q < 1$, we have*

$$\int_0^1 \left(\frac{[k]_q + q^k t}{q^{k-1}[n]_q}\right)^m d_q t$$

$$= \sum_{j=0}^{m} \left(\frac{q}{[n]_q}\right)^{m-j} \frac{1}{[m-j+1]_q} \left(\frac{[k]_q}{q^{k-1}[n]_q}\right)^j. \tag{2}$$

Lemma 2.2 (see (Cai 2011)). *Let $0 < q < 1, l \in N \cup \{0\}$, we have*

$$K_{n,q}(1;x) = 1, \tag{3}$$

$$K_{n,q}(t;x) = \frac{[n+l]_q}{[n]_q}x + \frac{q}{[2]_q[n]_q}, \tag{4}$$

$$K_{n,q}\left(t^2;x\right)=\frac{[n+l]_q^2}{q[n]_q^2}x^2+\frac{(1+3q)[n+l]_q}{[2]_q[n]_q^2}x$$

$$+\frac{q^2}{[2]_q[n]_q^2}, \tag{5}$$

$$K_{n,q}(t-x;x)=\frac{q^n[l]_q}{[n]_q}x+\frac{q}{[2]_q[n]_q}, \tag{6}$$

$$K_{n,q}\left((t-x)^2;x\right)$$
$$=\frac{(1-q)[n]_q^2+2q^n(1-q)[n]_q[l]_q+q^{2n}[l]_q^2}{q[n]_q^2}x^2$$
$$+\frac{[2]_q[n]_q+(1+3q)q^n[l]_q}{[2]_q[n]_q^2}x+\frac{q^2}{[3]_q[n]_q^2}. \tag{7}$$

Lemma 2.3. *Let* $0<q<1, l\in N\cup\{0\}$, *we have*

$$K_{n,q}\left(t^3;x\right)=\frac{[n+l]_q^3}{q^3[n]_q^3}x^3+\left[\frac{[n+l]_q^2}{[2]_q[n]_q^3}+\right.$$
$$\left.\frac{(1+2q)[n+l]_q^2}{q^2[n]_q^3}\right]x^2+\left(\frac{q^2[n+l]_q}{[3]_q[n]_q^3}+\right.$$
$$\left.\frac{q[n+l]_q}{[2]_q[n]_q^3}+\frac{[n+l]_q}{[n]_q^3}\right)x+\frac{q^3}{[4]_q[n]_q^3}. \tag{8}$$

Proof. From Lemma 2.1, we have

$$\int_0^1\left(\frac{[k]_q+q^kt}{q^{k-1}[n]_q}\right)^3d_qt$$
$$=\sum_{j=0}^3\left(\frac{q}{[n]_q}\right)^{3-j}\frac{1}{[4-j]_q}\left(\frac{[k]_q}{q^{k-1}[n]_q}\right)^j$$
$$=\left(\frac{q}{[n]_q}\right)^3\frac{1}{[4]_q}+\left(\frac{q}{[n]_q}\right)^2\frac{1}{[3]_q}\frac{[k]_q}{q^{k-1}[n]_q}+\frac{q}{[n]_q}$$
$$\cdot\frac{1}{[2]_q}\left(\frac{[k]_q}{q^{k-1}[n]_q}\right)^2+\left(\frac{[k]_q}{q^{k-1}[n]_q}\right)^3,$$

then,

$$K_{n,q}\left(t^3;x\right)$$
$$=e_q\left(-[n+l]_qx\right)\sum_{k=0}^\infty q^{\frac{k(k-1)}{2}}\frac{[n+l]_q^kx^k}{[k]_q!}$$
$$\cdot\int_0^1\left(\frac{[k]_q+q^kt}{q^{k-1}[n]_q}\right)^3d_qt$$
$$=e_q\left(-[n+l]_qx\right)\sum_{k=0}^\infty q^{\frac{k(k-1)}{2}}\frac{[n+l]_q^kx^k}{[k]_q!}\frac{q^3}{[4]_q[n]_q^3}$$
$$+e_q\left(-[n+l]_qx\right)\sum_{k=0}^\infty q^{\frac{k(k-1)}{2}}\frac{[n+l]_q^kx^k}{[k]_q!}\frac{q^2}{[3]_q[n]_q^2}$$
$$\cdot\frac{[k]_q}{q^{k-1}[n]_q}+e_q\left(-[n+l]_qx\right)\sum_{k=0}^\infty q^{\frac{k(k-1)}{2}}\frac{[n+l]_q^k}{[k]_q!}$$
$$\cdot\frac{qx^k}{[2]_q[n]_q}\frac{[k]_q^2}{q^{2k-2}[n]_q^2}+e_q\left(-[n+l]_qx\right)\sum_{k=0}^\infty q^{\frac{k(k-1)}{2}}$$
$$\cdot\frac{[n+l]_q^kx^k}{[k]_q!}\frac{[k]_q^3}{q^{3k-3}[n]_q^3},$$

since

$$[k]_q^2=[k]_q[k-1]_q+q^{k-1}[k]_q, \tag{9}$$

$$[k]_q^3=[k]_q[k-1]_q[k-2]_q+(1+2q)q^{k-2}$$
$$\{[k]_q[k-1]_q+q^{2k-2}[k]_q, \tag{10}$$

and from (3) and some simple computations, we have the desired result. Lemma 2.3 is proved.

Lemma 2.4. *Let* $0<q<1, l\in N\cup\{0\}$, *we have*

$$K_{n,q}\left(t^4;x\right)=$$
$$\frac{[n+l]_q^4x^4}{q^6[n]_q^4}+\left[\frac{1}{q^2[2]_q}+\frac{\left(1+2q+3q^2\right)}{q^5}\right]\frac{[n+l]_q^3}{[n]_q^4}$$
$$\cdot x^3+\left(\frac{q}{[3]_q}+\frac{1+2q}{q[2]_q}+\frac{1+3q+3q^2}{q^3}\right)\frac{[n+l]_q^2x^2}{[n]_q^4}$$
$$+\left(\frac{q^3}{[4]_q}+\frac{q^2}{[3]_q}+\frac{q}{[2]_q}+1\right)\frac{[n+l]_q}{[n]_q^4}x+\frac{q^4}{[5]_q[n]_q^4}. \tag{11}$$

Proof. From Lemma 2.1, we have

$$\int_0^1\left(\frac{[k]_q+q^kt}{q^{k-1}[n]_q}\right)^4d_qt$$
$$=\sum_{j=0}^4\left(\frac{q}{[n]_q}\right)^{4-j}\frac{1}{[5-j]_q}\left(\frac{[k]_q}{q^{k-1}[n]_q}\right)^j$$
$$=\frac{q^4}{[5]_q[n]_q^4}+\frac{q^3}{[n]_q^3}\frac{[k]_q}{[4]_qq^{k-1}[n]_q}+\frac{q^2}{[n]_q^2}\frac{[k]_q^2}{q^{2k-2}[n]_q^2}$$
$$+\frac{q}{[n]_q}\frac{[k]_q^3}{[2]_qq^{3k-3}[n]_q^3}+\frac{[k]_q^4}{q^{4k-4}[n]_q^4},$$

since

$$[k]_q^4=[k]_q[k-1]_q[k-2]_q[k-3]_q+\left(1+2q+3q^2\right)$$
$$\cdot q^{k-3}[k]_q[k-1]_q[k-2]_q+\left(1+3q+3q^2\right)$$
$$\cdot q^{2k-4}[k]_q[k-1]_q+q3^{k-3}[k]_q,$$

and from (3), (9) and (10), we have

$$K_{n,q}\left(t^4;x\right)$$
$$=e_q\left(-[n+l]_qx\right)\sum_{k=0}^\infty q^{\frac{k(k-1)}{2}}\frac{[n+l]_q^kx^k}{[k]_q!}$$
$$\cdot\int_0^1\left(\frac{[k]_q+q^kt}{q^{k-1}[n]_q}\right)^4d_qt$$
$$=\frac{[n+l]_q^4}{q^6[n]_q^4}x^4+\left[\frac{1}{q^2[2]_q}+\frac{(1+2q+3q^2)}{q^5}\right]$$
$$\cdot\frac{[n+l]_q^3}{[n]_q^4}x^3+\left(\frac{q}{[3]_q}+\frac{1+2q}{q[2]_q}+\frac{1+3q+3q^2}{q^3}\right)$$
$$\cdot\frac{[n+l]_q^2}{[n]_q^4}x^2+\left(\frac{q^3}{[4]_q}+\frac{q^2}{[3]_q}+\frac{q}{[2]_q}+1\right)$$
$$\cdot\frac{[n+l]_q}{[n]_q^4}x+\frac{q^4}{[5]_q[n]_q^4}.$$

Lemma 2.4 is proved.

Lemma 2.5. *Let $q = \{q_n\}$ be a sequence satisfying $q_n \in (0,1)$, $\lim_{n\to\infty} q_n = 1$, $\lim_{n\to\infty} q_n^n = 1$ and $l \in N \cup \{0\}$. For fix $x \in [0,\infty)$, we have*

$$
K_{n,q}\left((t-x)^4;x\right) = \left(\frac{[n+l]_q^4}{q^6[n]_q^4} - \frac{4[n+l]_q^3}{q^3[n]_q^3}\right.
$$
$$
\left. + \frac{6[n+l]_q^2}{q[n]_q^2} - \frac{4[n+l]_q}{[n]_q} + 1\right)x^4 + \left(\frac{[n+l]_q^3}{q^2[2]_q[n]_q^4}\right.
$$
$$
+ \frac{(1+2q+3q^2)[n+l]_q^3}{q^5[n]_q^4} - \frac{4[n+l]_q^2}{[2]_q[n]_q^3} - \frac{4q}{[2]_q[n]_q}
$$
$$
\left. - \frac{4(1+2q)[n+l]_q^2}{q^2[n]_q^3} + \frac{6(1+3q)[n+l]_q}{[2]_q[n]_q^2}\right)x^3
$$
$$
+ \left(\frac{q[n+l]_q^2}{[3]_q[n]_q^3} + \frac{(1+2q)[n+l]_q^2}{q[2]_q[n]_q^4} + \frac{(1+3q+3q^2)}{q^3[n]_q^4}\right.
$$
$$
\{n+l]_q^2 - \frac{4q^2[n+l]_q}{[3]_q[n]_q^2} - \frac{4q[n+l]_q}{[2]_q[n]_q^3} - \frac{4[n+l]_q}{[n]_q^3}
$$
$$
\left. + \frac{6q^2}{[3]_q[n]_q^2}\right)x^2 + \left(\frac{q^3[n+l]_q}{[4]_q[n]_q^4} + \frac{q^2[n+l]_q}{[3]_q[n]_q^4} + \right.
$$
$$
\frac{q[n+l]_q}{[2]_q[n]_q^4} + \frac{[n+l]_q}{[n]_q^4} - \frac{4q^3}{[4]_q[n]_q^3}\right)x + \frac{q^4}{[5]_q[n]_q^4}.
$$

Proof. From $K_{n,q}\left((t-x)^4;x\right) = K_{n,q}\left(t^4;x\right) +$

$$
4xK_{n,q}\left(t^3;x\right) + 6x^2K_{n,q}\left(t^2;x\right) + 4x^3K_{n,q}(t;x) + x^4
$$

and Lemma 2.2, Lemma 2.3 and Lemma 2.4, we obtain it easily by computation.

3 WEIGHTED APPROXIMATION AND VORONOVSKAYA TYPE ASYMPTOTIC FORMULA

Let $B_{x^2}[0,\infty)$ be the set of all functions f defined on $[0,\infty)$ satisfying the condition $|f(x)| \leq M_f(1+x^2)$, where M_f is a constant depending only on f. We denote the subspace of all continuous functions belonging to $B_{x^2}[0,\infty)$ by $C_{x^2}[0,\infty)$. Also, let $C_{x^2}^*[0,\infty)$ be the subspace of all functions $f \in C_{x^2}[0,\infty)$, for which $\lim_{x\to\infty} \frac{f(x)}{1+x^2}$ is finite. The norm on $C_{x^2}^*[0,\infty)$ is $\|f\|_{x^2} = \sup_{x\in[0,\infty)} \frac{|f(x)|}{1+x^2}$.

Now we will discuss the weighted approximation theorem.

Theorem 3.1. *Let $q = \{q_n\}$ be a sequence satisfying $q_n \in (0,1)$, $\lim_{n\to\infty} q_n = 1$, $\lim_{n\to\infty} q_n^n = 1$, $l \in N \cup \{0\}$. For $f \in C_{x^2}^*[0,\infty)$, we have*

$$
\lim_{n\to\infty}\left\|K_{n,q_n}(f) - f\right\|_{x^2} = 0. \tag{12}
$$

Proof. By using the Korovkin theorem in (Gadjiev 1976), we see that it is sufficient to verify the following three conditions

$$
\lim_{n\to\infty}\left\|K_{n,q_n}\left(t^v;x\right) - x^v\right\|_{x^2} = 0, \qquad v = 0,1,2. \tag{13}
$$

Since $K_{n,q_n}(1;x) = 1$, (13) holds true for $v = 0$. For $v = 1$, using (4), we have

$$
\left\|K_{n,q_n}(t;x) - x\right\|_{x^2}
$$
$$
= \sup_{x\in[0,\infty)} \frac{\left|K_{n,q_n}(t;x) - x\right|}{1+x^2} = \frac{q^n[l]_q}{[n]_q}\sup_{x\in[0,\infty)}\frac{|x|}{1+x^2}
$$
$$
+ \frac{q}{[2]_q[n]_q}\sup_{x\in[0,\infty)}\frac{1}{1+x^2}
$$
$$
\leq \frac{q^n[l]_q}{[n]_q} + \frac{q}{[2]_q[n]_q},
$$

since $\lim_{n\to\infty} q_n = 1$, $\lim_{n\to\infty} q_n^n = 1$, we get $\lim_{n\to\infty}\frac{q_n^n[l]_{q_n}}{[n]_{q_n}} = 0$ and $\lim_{n\to\infty}\frac{q_n}{[2]_{q_n}[n]_{q_n}} = 0$, so the second condition of (13) holds for $v = 1$ as $n \to \infty$.

Finally, for $v = 2$, using (5), we have

$$
\left\|K_{n,q_n}\left(t^2;x\right) - x^2\right\|_{x^2}
$$
$$
= \sup_{x\in[0,\infty)}\frac{\left|K_{n,q_n}\left(t^2;x\right) - x^2\right|}{1+x^2}
$$
$$
= \left(\frac{1}{q} - 1 + \frac{2q^{n-1}[l]_q}{[n]_q} + \frac{2q^{2n-1}[l]_q^2}{[n]_q^2}\right)\sup_{x\in[0,\infty)}\frac{x^2}{1+x^2}
$$
$$
+ \frac{(1+3q)[n+l]_q}{[2]_q[n]_q^2}\sup_{x\in[0,\infty)}\frac{x}{1+x^2} + \frac{q^2}{[3]_q[n]_q^2}
$$
$$
\cdot \sup_{x\in[0,\infty)}\frac{1}{1+x^2}
$$
$$
\leq \frac{1}{q} - 1 + \frac{2q^{n-1}[l]_q}{[n]_q} + \frac{2q^{2n-1}[l]_q^2}{[n]_q^2}
$$
$$
+ \frac{(1+3q)[n+l]_q}{[2]_q[n]_q^2} + \frac{q^2}{[3]_q[n]_q^2},
$$

since $\lim_{n\to\infty} q_n = 1$, $\lim_{n\to\infty} q_n^n = 1$, we get $\lim_{n\to\infty}\frac{1}{q_n} - 1 = 0$, $\lim_{n\to\infty}\frac{2q_n^{n-1}[l]_{q_n}}{[n]_{q_n}} = 0$, $\lim_{n\to\infty}\frac{2q_n^{2n-1}[l]_{q_n}^2}{[n]_{q_n}^2} = 0$, $\lim_{n\to\infty}\frac{(1+3q_n)[n+l]_{q_n}}{[2]_{q_n}[n]_{q_n}^2} = 0$ and $\lim_{n\to\infty}\frac{q_n^2}{[3]_{q_n}[n]_{q_n}^2} = 0$, so the third condition of (13) holds for $v = 2$ as $n \to \infty$, then the proof of Theorem 3.1 is completed.

Theorem 3.2. *Let $q = \{q_n\}$ be a sequence satisfying $q_n \in (0,1)$, $\lim_{n\to\infty} q_n = 1$, $\lim_{n\to\infty} q_n^n = 1$, $l \in N \cup \{0\}$ and $f \in C_{x^2}^2[0,\infty)$. Then for fix $x \in [0,\infty)$, the following equality holds*

$$
\lim_{n\to\infty}[n]_q\left(K_{n,q}(f;x) - f(x)\right)
$$
$$
= f'(x)\left(lx + \frac{1}{2}\right) + \frac{x}{2}f''(x). \tag{14}
$$

Proof. Let $x \in [0,\infty)$ be fixed. By the Taylor formula, we may write

$$
f(t) = f(x) + f'(x)(t-x) + \frac{1}{2}f''(x)(t-x)^2
$$
$$
r(t;x)(t-x)^2, \tag{15}
$$

3

where $r(t; x)$ is the Peano form of the remainder, $r(t; x) \in C_{x^2}[0, \infty)$, using L'Hopital's rule, we have

$$\lim_{t \to x} r(t; x)$$

$$= \lim_{t \to x} \frac{f(t) - f(x) - f'(x)(t - x) - \frac{1}{2}f''(x)(t - x)^2}{(t - x)^2}$$

$$= \lim_{t \to x} \frac{f'(t) - f'(x) - f''(x)(t - x)}{2(t - x)}$$

$$\lim_{t \to x} \frac{f''(t) - f''(x)}{2} = 0.$$

Applying $K_{n,q}(f; x)$ to (15), we obtain

$$[n]_q \left(K_{n,q}(f; x) - f(x) \right)$$

$$= f'(x)[n]_q K_{n,q}(t - x; x) + \frac{f''(x)}{2}[n]_q$$

$$\cdot K_{n,q} \left((t - x)^2; x \right) + [n]_q K_{n,q} \left(r(t; x)(t - x)^2; x \right).$$

By the Cauchy-Schwarz inequality, we have

$$K_{n,q} \left(r(t; x)(t - x)^2; x \right)$$

$$\leq \sqrt{K_{n,q} \left(r(t; x); x \right)} \cdot \sqrt{K_{n,q} \left((t - x)^4; x \right)}. \tag{16}$$

Since $r^2(x; x) = 0$, then it follows from Theorem 3.1 that

$$\lim_{n \to \infty} K_{n,q} \left(r^2(t; x); x \right) = r^2(x; x) = 0. \tag{17}$$

Now, from (16), (17) and Lemma 2.5, we get $\lim_{n \to \infty} [n]_q \sqrt{K_{n,q} \left((t - x)^4; x \right)} = O(1)$, then $\lim_{n \to \infty} [n]_q K_{n,q} \left(r(t; x)(t - x)^2; x \right) = 0$. Since $\lim_{n \to \infty} [n]_q K_{n,q}(t - x; x) = lx + \frac{1}{2}$ and $\lim_{n \to \infty} [n]_q K_{n,q} \left((t - x)^2; x \right) = x$, we get the desired result.

ACKNOWLEDGMENT

This work is supported by the Educational Office of Fujian Province of China (Grant No. JA13269), the Startup Project of Doctor Scientific Research of Quanzhou Normal University, Fujian Provincial Key Laboratory of Data Intensive Computing and Key Laboratory of Intelligent Computing and Information Processing, Fujian Province University.

REFERENCES

Cai, Q. B. (2011). Weighted statistical approximation properties of the Kantorovich type q-Szász-Mirakjan-Schurer operators. *J. Sanming Univ. 28(5)*, 8–12.

Gadjiev, A. D. (1976). Theorems of the type of P. P. Korovkin type theorems. *Math. Zametki 20(5)*, 781–786.

Kac, V. G. & P. Cheung (2002). Quantum Calculus. *Universitext, Springer-Verlag, New York*.

Mahmudov, N. I. (2010). Statistical approximation of Baskakov and Baskakov-Kantorovich operators based on the q-integers. *Central European J. Math. 8(4)*, 816–826.

Electronics, Information Technology and Intellectualization – Song & Kwak (Eds)
© *2015 Taylor & Francis Group, London, ISBN 978-1-138-02741-1*

The tracking error analysis of a weld seam tracking follower system

X.G. Liu
Gui Lin College of Aerospace Technology, Gui Lin, China

R. Wu
Guang Xi University of Technology, Liu Zhou, China

ABSTRACT: In the course of robot welding operations, the CCD camera, in a Seam Tracking servo system, with its position orientated accurately in the area, directly determines the data accuracy of image processing. It becomes the decisive factor affecting the welding process. The purpose of this paper is to study and analyze the impact factors of the errors resulting from the CCD camera's position and orientation in the welding area.

Keywords: CCD camera; visual servo; tracking error

1 INTRODUCTION

Currently, in robot visual servo systems, there are two common servo methods: one is a location-based control[1], the other is an image-based control. The latter does not need to estimate the target in the Cartesian coordinate system posture, to reduce the computational delay, and can overcome the CCD camera calibration errors on positioning accuracy[2]. However, the reason why the visual approach also failed to produce the actual welding commonly being applied, is that the image was of low quality. Image processing algorithms, resulting in a relatively low degree of intelligence, a time-consuming operation, anti-interference ability and adaptability, are poor. Currently the main research method is most focused on improving the efficiency of the algorithm[3], and the hardware processing speed[4]. But owing to the fact that in the welding zone the arc light strength is too strong, the resulting weld shorter distances can be detected by means of the arc light. Therefore, based on the existing vision processing hardware and algorithms, it is difficult to improve on the welding speed, especially, in the seam tracking without an auxiliary light source[5]. The CCD camera in the seam tracking servo system ensures the accuracy of the position orientation in the space, which directly determines the accuracy of the data processed by the image. It has become the determinants of whether to implement the welding process of a welding project[6].

2 BASED ON THE ESTABLISHMENT OF A MATHEMATICAL MODEL OF ROBOT IRB1400

Figure 1 is a schematic diagram of IRB1400 robot mechanism, $x_c y_c z_c$ represents the CCD camera coordinate system shown in Figure 2, $x_8 y_8 z_8$ represents

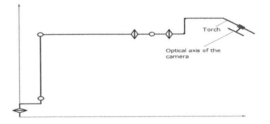

Figure 1. IRB1400 simple robot mechanism.

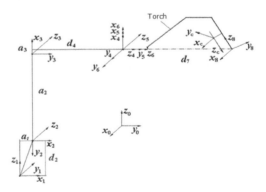

Figure 2. Institutional linkage rod coordinate system.

welding torch coordinate system, $x_c y_c z_c$ can rotate table about $x_8 y_8 z_8$. In robot welding process, CCD camera in which the spatial position orientation can offset due to various factors. $o_8 - x_8 y_8 z_8$ represents the ideal location of the weld torch coordinates shown in Figure 3. $o_8' - x_8' y_8' z_8'$ represents the actual position of the welding torch coordinate system. $o_c - x_c y_c z_c$ represents the ideal position of the CCD camera coordinate system and indicates the actual position of the CCD

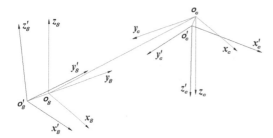

Figure 3.   Space CCD camera pose error.

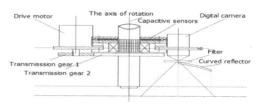

b) Principle schematic diagram

c) Real figure

Figure 4.   Seam tracking servo system.

camera coordinate system. Cause of this situation is the $o'_c - x'_c y'_c z'_c$ transmission error of the dynamic systems, welding torch attitude, and the turn error of the Seam tracking servo system rotating mechanism.

3   CCD CAMERA INSTALLATION POSE ERROR IMPACT ANALYSES

Weld torch coordinate system $o_8 - x_8 y_8 z_8$ is the coordinate system of the reference, the ideal posture of CCD camera coordinate system is $o_c - x_c y_c z_c$ coordinates. However, because of the factors of processes, installation errors and so on that resulting in the actual position of the CCD camera coordinate system and the ideal posture in the welding coordinate system. Therefore, the changes have little effect on the accuracy of weld centre line after the image, which are negligible[7]. A change in the vector direction of the light axis, causes the centre of the image to be offset, which is formed by a deviation between the ideal centres. The amount of deviation for any two of times to obtained image is invariant. So deviation is constant as a system error. Thus, the causes of error can be eliminated by calibration.

4   ANGULAR ERROR SLEWING MECHANISM OF IMPACT ANALYSIS

The basic structure of rotating seam tracking mechanism is shown in Figure 4. With the motor and drive gear, the driving gear 2 is fixed to the welding torch. The motor rotation, the drive gear 1 and a digital CCD camera and an optical filter, a curved reflector, while rotating around the axis of rotation, and a capacity sensor, all are composed of a dynamic gate and a static gate, a dynamic gate rotating with the speed of the motor around the axis of rotation. The angle precision of capacity sensors is $10'$, the follower mechanism is driven by a CCD camera rotating around the torch. Work angle and walk angle is zero. If the CCD camera coordinate system exists on a pixel position $[X_c, Y_c, Z_c]$, we have:

$$\begin{bmatrix} \Delta x_s \\ \Delta y_s \\ \Delta z_s \\ 1 \end{bmatrix} = \begin{bmatrix} -X_c \sin\varphi - Y_c \cos\varphi - \sin\varphi \cdot d_{7x} \\ -X_c \cos\varphi + Y_c \sin\varphi - \cos\varphi \cdot d_{7x} \\ 0 \\ 1 \end{bmatrix} d\varphi \qquad (1)$$

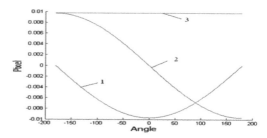

Figure 5.   Position error curves.

When $X_c = Y_c = 25\,\text{mm}$, $d_{7x} = 40\,\text{mm}$, $d\varphi = 10'$, $\varphi = [-180°, 180°]$, pixel distance relationship is 10 pixels/mm, as the error curve shows in Figure 5. Error curve 1 is the change of Δx_s-error, error curve 2 is the change of Δy_s-error, and error curve 3 is a variation of the error $\sqrt{\Delta x_s^2 + \Delta y_s^2}$.

As can be seen from Figure 5, seam tracking servo system rotation angle error affects the position error of weld image which is 0.01 pixels, and too far to meet the requirements of the accuracy of welding. Even at work angle and walk angle was twenty degrees, seam tracking servo system errors on the rotation angle of the welds image pixel position error may not be considered.

5   WELD TORCH ATTITUDE ERROR ANALYSES

Since the output angle of the joint position sensor of the robot has a positional deviation, so that there is an error at the end of the weld torch. According to the technical parameters of the robot IRB1400, the known position repeat ability is 0.05 mm, the axis of the weld torch angle error is 0.001 rad. According to Figure 2, in the work-piece coordinate system, because

Table 1. impact error value of walk angle and work angle. Unit: mm.

	α, β Maximum angle	15	20	25	30	35	40
dx_s	α Maximum error generated	0.1916	0.2705	0.3622	0.4592	0.5935	0.7665
	β Maximum error generated	0	0	0	0	0	0
	α, β Maximum error generated	0.1916	0.2705	0.3622	0.4592	0.5935	0.7665
dv_s	α Maximum error generated	0 1924	0.2714	0.3633	0.4606	0.5954	0.7691
	β Maximum error generated	0.1859	0.2548	0.3285	0.3994	0.4875	0.5874
	α, β Maximum error generated	0.4248	0.6538	0.9829	1.4194	2.1983	3.5501

the robot's each rotational joint has a motor, the position of each motor has a reading error. Therefore, the actual and theoretical calculation of the position of the weld torch coordinate system exhibits errors. According to the extraction established mathematical model of the robot IRB1400, it is known that the transformation coordinates between the weld torch coordinate system and the weldment is:

$$
d\begin{bmatrix} x_s \\ y_s \\ z_s \\ 1 \end{bmatrix} = \begin{bmatrix} x_8\dfrac{\tan\alpha\,\sec^2\alpha}{\sqrt{\tan^2\alpha+1}}\,d\alpha \\ (a_{11}+a_{12})\,d\alpha + (b_{11}+b_{12})\,d\beta \\ 0 \\ 0 \end{bmatrix} \tag{2}
$$

It can be obtain a position error of weld information due to attitude error.

6 SIMULATIONS

6.1 Affect the work angle and walk angle error of pixel position error

According to the welding process shows that α, β is generally in the range of the angle between $0°C\sim20°C$. To illustrate the impact of the work angle and walk angle at different angles to get weld center information. Let simulation conditions are as follows:

1. 0.25 degrees angle error, namely $d\alpha = 0.25°$, $d\beta = 0.25°$.
2. In the CCD camera coordinate system, the farthest position of the reflector reflex arc light is $[Y_c, X_c] = [20, 20]$. The origin of CCD camera coordinate system in the original position is [40 0 50], namely $d_{7x} = 40$, $d_{7z} = 50$, do the error simulation calculate in ranges $0\sim15°$, $0\sim20°$, $0\sim25°$, $0\sim30°$, $0\sim35°$, $0\sim40°$. It can be obtained table 1.

From table 1 we can conclude (α is the walk angle, β is the work angle).

1. Along the direction of the weld torch's movement, the error of image pixel positions are obtained, when the travel angle increases. The work angle's change has no effect on the error of the image pixel's positions in this direction.
2. Along the vertical direction of the welding torch movement, the error in the image pixel's position is

obtained by the walk angle and work angle being increased. Changes of the generated error by walk angle along the movement are almost the same. When work angle and walk angle have errors at the same time, with the walk angle and work angle increases. When work angle and walk angle are less than 20 degrees, the maximum error of the system is 0.6538 mm.

6.2 Comprehensive error influence the image pixel position error

In order to track the weld seam when weld seam is arc curves, the follower system need a rotate angle. So that the arc light reflected by the reflector is irradiated to the weld seam. When the work angle and walk angle of the weld torch includes errors, and when the angle of rotation of the follower is not the same, it must be verified whether the position accuracy of the image weld seam meets the operational requirements. For this reason, considering the limits selected, let the work angle and walk angle be $20°$, the limit value of the reflective irradiation area is: $[Y_c, X_c] = [20, 20]$, $d_{7x} = 40$, $d_{7z} = 50$. In this case, the weld seam plane in weld torch coordinates is: $z_8 = 0.3847x_8 + 0.3421y_8$. Therefore, the angle between the plane x_8y_8 of weld torch coordinates and the plane x_sy_s of the work-piece coordinates is $42.7879°$.

The rotation angle of the follower system, which is the direction of welding torch movement 0 degrees, clockwise direction is a positive direction, and counterclockwise direction is a negative direction. The angle values of simulation test is: $\varphi = [-\frac{\pi}{2}, -\frac{3\pi}{8}, -\frac{\pi}{4}, -\frac{\pi}{8}, 0, \frac{\pi}{8}, \frac{\pi}{4}, \frac{3\pi}{8}, \frac{\pi}{2}]$. It can be obtained table 2.

From table 2 can concluded (α is the walk angle, work angle β):

1. Along the movement direction of the weld torch, the error position of image pixels, range from $-90°$ to $0°$, with the the increase of the angle α increase. Range from $0°$ to $90°$, with the increase of the angle α decrease. The walk angle's change has no effect on the error of image pixel position along this direction.
2. Along the vertical direction of the welding torch move, the error position of image pixel ranges from $-90°$ to $0°$, with the the increase of the angle α increase. Range from $0°$ to $90°$, with the the increase of the angle α decrease. The change in the

Table 2. When $\alpha = 20°$, $\beta = 20°$, the impact error value of φ.

	φ Maximum error generated	$-\pi/2\cdot$	$-3\pi/8$	$-\pi/4$	$-\pi/8$	0	$\pi/8$	$\pi/4$	$3\pi/8$	$\pi/2$
dx_s	α Maximum error generated	0	0.1035	0.1912	0.2499	0.2705	0.2499	0.1912	0.1035	0
	β Maximum error generated	0	0	0	0	0	0	0	0	0
	α, β Maximum error generated	0	0.1035	0.1912	0.2499	0.2705	0.2499	0.1912	0.1035	0
dv_s	α Maximum error generated	0	0.1044	0.1923	0.2505	0.2714	0.2510	0.1923	0.1044	0
	β Maximum error generated	0.2704	0.3473	0.3714	0.1320	0.2548	0.3389	0.3714	0.3473	0.2877
	α, β Maximum error generated	0.1758	0.4126	0.5866	0.5368	0.6538	0.6713	0.5866	0.4126	0.2216

generated error caused by the walk angle along the movement is almost same.

3. When the error of angle β between in CCD camera laid angle φ is 45° and the direction of the welding torch movement is −45°, the image position error generated by image is maximum.

4. When work angle and the walk angle's error, at the same time, range from −90° to 0°, the angle α increases and the angle β increases, with the angle α and the travel of the operating angle β increases the range from 0° to 90°, with the increase of the angle α decreasing.

5. When the error of angle α and angle β in the angle between the CCD cameras laid out angle φ and the direction of weld torch movement is 22.5°, the maximum error of the system is 0.6713 mm. By converting, the calculation of the farthest point is 80 mm.

In the embodiment of the robot welding operations, the weld gap of sheet is generally more than 1 mm. After welding its width is more than 4 mm and meets the requirements only from the impact of attitude errors. From table 1 and 2, weld torch attitude error is the main influencing factor,which influence on obtained the image pixels position. Which, especially,when angle α and angle β a little big. The corresponding error includes the maximum error generated by the rotation angle in a graph.

7 CONCLUSIONS

This paper begins with discussion and analysis. Based on the walk angle and the work angle of welding craft,

in-depth research to weld torch attitude errors, results in producing a welding torch attitude error from mathematical models and simulation tests. The simulation test shows that when the pixel positions are calculated as being less than 80 mm, the accuracy of the welding seam tracking follower system meets the requirements of welding seam tracking.

REFERENCES

Lin Jingshan & Yang Cunli et al., Auxiliary Source Image Method TIG Welding Seam Tracking Sensor System, Welding Learn Journal (2012).
Liu Xiaogang & Xie Cunxi et al., Base on Weld Seam to Get Image of Reflected by Arc Light and Image Processing, Welding Learn Journal (2008).
Liu Xiaoyu & Fang Kangling, Image-based on Robot Visual Servo Control (2006), Wuhan University of Science and Technology.
Li XiaoRan & Wu Zengyin, Positioning Accuracy Analysis Of CCD Remote Sensing Images Without Control, Chinese Society of Astronautics Deep Space Exploration Technical Committee of the Ninth Annual Meeting Proceedings (2012).
Wang Zhuo, Study Car Body Welding Image Recognition Algorithm (2010). Wuhan University of Technology.
Wang Hongwen & Li Yang et al, Based on LQR Shear — Flash Light on Welder Position Servo Control Modeling and Simulation. Welding Learn Journal. (2010).
Xu Min & Zhao Mingyang, The Research Based Laser Welding Seam Stripe Centerline of Local Threshold Detecte Rapidly Method. Instrument Technique and Sensor (2013).

Electronics, Information Technology and Intellectualization – Song & Kwak (Eds)
© 2015 Taylor & Francis Group, London, ISBN 978-1-138-02741-1

A recommendation algorithm and a system research of mobile platforms

ZhuFeng Qiao, JianXin Guo & JiChun Zhao
*Institute of Information on Science and Technology of Agriculture, Beijing Academy of Agriculture
and Forestry Sciences, Beijing, China*
The Research Center of Beijing Engineering Technology for Rural Remote Information Services, Beijing, China

ABSTRACT: The text of the mobile communication network field of personalized recommendation, the mobile user context information into collaborative filtering recommendation process. This paper proposes a collaborative filtering recommendation algorithm based on user context similarity. The context of 'pre filtering recommendation method of mobile users – Mobile Services – Context' three dimension model, combined with the traditional collaborative filtering algorithm for preference prediction and recommendation, and the application system design and development, the application effect is good.

1 GENERAL INSTRUCTIONS

1.1 Introduction

With the rapid development of mobile communication technology, more and more users use mobile devices to access mobile network services and information content. Because a mobile device has low processing ability, poor battery capacity, and input disadvantages, output capacity is limited; the ability to move the user's access to information is severely affected. Moreover, a dynamic mobile user is changeable, so he has mobile information needs in different context conditions. For example, some users like 'in the subway' to browse news, in 'the office' and rest' is like playing a game. Secondly, because he can access the information explosion there is information overload. Mobile applications need not only to provide a wealth of information to the user, but also need to provide the function of automatic screening, getting the right information to help the user. Therefore, how to help mobile users find the real interest, consistent with his context conditions (such as time, place, activity state, and network conditions) mobile network services, and information content, is of particularly importance.

Recommendation systems which combine information filtering technology and decision support technology can solve the problem of information overload. Identifying and predicting user preferences, providing personalized service, making the appropriate information available to different users, is one of the important means to alleviate the problem of 'information overload'. A recommendation system is the mainstream oriented PC Web platform, and the recommendation technology transfer to the mobile platform, as the context environment, with little consideration to the context of the user. Relying on the 'only user – item' the two-dimensional model cannot generate accurately recommended information in many application scenarios. For example, some users like in 'morning' rather than 'noon' recommended suitable news information, some users in the 'travel' or in 'a geographical position' want to recommend some appropriate surrounding restaurants, shopping malls and other information. Therefore, the question of how to help mobile users find their real interests, consistent with the context conditions of mobile Internet services, and to further improve the recommendation accuracy and user satisfaction, is particularly important.

1.2 Mobile platform characteristics

(1) The opacity in the application environment: the constraints of mobile devices, wireless network constraints, the influence of the external environment, and the behaviour characteristic of mobile users.
(2) Location: a specific time system can provide the user's position information.
(3) General: in any place, at any time the information and service delivery to the user's ability.

2 TRADITIONAL RECOMMENDATION TECHNIQUES

2.1 Content-based recommendation

The basic principles of content-based recommendation methods are, according to the user's previous items, to choose other similar items as the recommended results. For example, if there is now a new film and the user is used to see a movie starring the same actors or filming similar subjects, users can find an item just like the

Table 1. Mobile Internet recommended and traditional Internet recommended comparison.

	Mobile recommendation	Traditional recommendation
Mobility	Intelligent mobile phone, tablet computer with mobile users	No
Network access	The mobile 3G/4G network, WIFI network	The wired network, WIFI network
Data sources	The mobile user behaviour records, context information	Record the Internet behaviour
Universal	Real-time aware mobile user context information	Less
Real time	High real-time requirement	Low real-time requirement
Interactive interface	List, map, based on the evaluation mode	List, based on the evaluation mode

new film. We use the preference model (User Profile) of the feature vector to describe the user's interests, the same for each item for feature extraction, as goods model (Item Profile) content features. Then the vector features both vector and the candidate's object model to calculate the user preference model of the matching degree, and a high degree of matching candidate items can be used as a recommendation to the target user.

2.2 Collaborative filtering recommendation

Unlike content-based filtering, collaborative filtering does not have quite the similar resources and user model, but through the user evaluation of resource discovery the similarity between users and the use of similarity to recommend information can be achieved. It is based on such a premise: evaluations on some items of different users are similar, so some other resource item evaluation are similar to them. With similar interested users don't as a user, when the user is interested in a resource, the resource can be recommended to other users the same. Recommended resources are the latest service for target customers, but they must be the target customers of the nearest neighbour which have been evaluated. Collaborative filtering recommendation is essentially a recommendation methods often used in real life, such as two of the same friends recommending music, and books they love to read.

2.3 Hybrid recommendation

A hybrid recommendation refers to both the similarity comparison of resources and each user model for content based recommendation. Moreover, similar interest user groups use collaborative recommendation methods as recommended. Because the hybrid recommendation will take advantage of the two recommended methods, offset by two recommended methods, so it has better recommendation performance.

3 RECOMMENDATION ALGORITHM

3.1 Data model

Definition 1: using vector model is expressed by N context properties (such as time, place, activity) consisting of context $C = \{C_1, C_2, \ldots, C_i, \ldots, C_n\}$; a context instance is expressed as $C = (C_1, C_2, \ldots C_n)$, where C_i ($1 \leq i \leq n$) for a specific attribute context value for C_i, If C_i is the time context, C_i may have the morning, afternoon, evening value.

Definition 2: 'mobile users and mobile services – Context', three dimension model denoted as U-S-C model, which is a three dimensional vector space {User, Service, Context}, the three dimensions are: mobile users, mobile services, and context Each dimension is separate with their respective attributes and values consisting of a vector to represent them. Among them, are mobile users, mobile services, context, preference values of a record together, called a preference record. Set preferences are called preference data set.

3.2 The algorithm design

Using similar context set N(c) instead of the mobile user's current context C as the filter condition, to reduce the dimension of 'mobile users and mobile services-Context' three dimension model, alleviates the data sparsity problem, while also using the traditional two dimension recommendation algorithm of user preference prediction.

Therefore, we need to design a kind of calculation to construct N(c). Formula (1) is calculated, based on the user's context similarity, used to measure the same user different context similarity, according to the user in different context of different scores on the same object value calculation.

$$\text{sim}_u(u, x, y) =$$

$$\frac{\sum_{\theta \in S}(r_{(u,\theta,x)} - \bar{r}_{u,x})(r_{(u,\theta,y)} - \bar{r}_{u,y})}{\sqrt{\sum_{\theta \in S}(r_{(u,\theta,x)} - \bar{r}_{u,x})^2 \times \sum_{\theta \in S}(r_{(u,\theta,y)} - \bar{r}_{u,y})^2}} \quad (1)$$

The S represents the set of the user u having score to mobile service value set in the context of x and context y; R (u, s, x) represents score of the user u to service s in context x; $\bar{r}_{u,y}$ represents the average score of the user u in the context of x. In practice, if the number of mobile service mobile user data centralized common scores in different context and with little or no, it will be difficult to similar context object current user context C collection. In order to solve this problem, we can use the context feature similarity calculation formula instead of formula (1). Here, the feature similarity is using a Cosine similarity computing context instance, with X and Y, formula

$$\text{sim}(x, y) = \frac{x \cdot y}{\|x\| \cdot \|y\|} \qquad (2)$$

3.3 The algorithm process

Input: 'mobile users and mobile services – Context' three dimension model; target user u and the current context C; mobile service to recommend a set Service (R).

Output: the user u in the current context of C, the Service (R) in TOP-N mobile service's biggest preference.

The algorithm steps:

Step 1: Take out all the preference data out from the target user u, construct 2 dimensional preference matrix of context – Mobile Services.

Step 2: Using similarity formula to calculate the similarity between C and the rest of the current context Construct the current context similarity set N(c), because the matrix is the target user 2 dimensional preference matrix of context – Mobile Services, so generate the target user's current context generation based on the context of the set N(c).

Step 3: To judge whether step 2 due to data sparsity is difficult to construct similar context collection, if there is not a problem, the implementation of article step 4. Otherwise, calculate the similarity between C and the rest of the current context, and use a similar neighbour's current context C feature similarity context to fill the step 2 N(c).

Step 4: the 'mobile users and mobile services – context' three dimensional model reduction to generate U-S 2 dimensional model. Dimension reduction method is as follows: obtain similar context object the user's current context based on the set N(c), and take out context belonging to all preference record set N(c) from mobile users and mobile services – Context' models. If a user has different preference values to the same mobile in different contexts, only preserve the similarity of the preference's biggest record of context and target current user context C.

Step 5: for the two dimensional U-SS model, using a collaborative filtering algorithm, based on user calculated preference. Using collaborative filtering determine M neighbour users of target user u. Predicting user preference value to every mobile service of mobile service set Service (R), the preference TOP-N

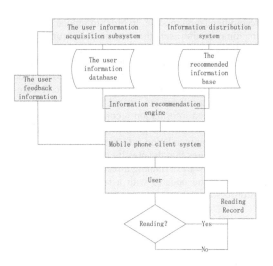

Figure 1. Mobile personalized recommendation model.

mobile service maximum recommended to the target user U.

4 SYSTEM DESIGN OF MOBILE

4.1 Architecture

This system architecture is a hierarchical model. It includes android personalized information recommendation system and server manager system based. This system is mainly divided into client data access layer, web layer, data manipulation layer, and data persistence layer. The client is mainly responsible for display and access to the user recommendation information's explicit input information and user information to the terminal user's implicit input; Web layer is mainly composed of Servlet, JSP, and is responsible for adding recommend information to the database in the system, and provides Servlet access interface for the android client's access by http/https. The data manipulation layer encapsulates data manipulation API and provides a uniform interface to the outside world. The data storage layer mainly uses Oracle 10 g database management software.

4.2 Android client application

The android client provides recommendation information display, information collection, information retrieval, user's personal preference feedback modules, and a procedure flow chart of the main function as shown in figure 1. A first use of client applications requires the registered user name, password and other information. The application can directly use the stored personal information in the user management system to login, after the successful verification of the user in the operation.

4.3 Server management system

The main module of the server program includes an information collection module and a personalized

11

recommendation module. In order to reduce the waiting time of the user, the two modules run by run in offline mode execution. That is set to run time fixed or manual operation to update the spatial feature words, probability distribution, and the classification of users, recommendation set information etc. Recommended data information stored in the database and a personalized information recommendation module are responsible for the generation and maintenance of user recommendation information user sets. This process ensures that the novelty of the recommended information starts a running thread in the background, with real-time updates with a recommended set of information.

5 SUMMARY

In this paper, on the basis of traditional collaborative recommendations, in view of the characteristics of the mobile network environment, use the context similarity of mobile users design the collaborative filtering recommendation algorithm based on user context similarity. The design of the mobile recommendation platform includes the server management system and the android client application. The application shows that, this algorithm is suitable for a personalized service in a mobile network recommended field, both to alleviate the data sparsity problem in the traditional context of pre filtering method, and can make full use of the context information, so as to improve the quality of recommendation.

REFERENCES

Adomavicius, G., Tuzhilin, A. Toward the next generation of recommender systems: A survey of the state-of-the-art and possible extensions [J]. IEEE Transactions on Knowledge and Data Engineering, 2005, 17(6):734–749.

Guo Ping, Liu Bo, Shen Yue. The Summary on Key Technology of Big Data's Self-organized Regional Push based on Agriculture Cloud. Software.

Hu Jiming, Hu Changping, Deng Shengli. Review of Research of the Information Recommendation in the Social Network Environment. Information and documentation services. 2013, No. 2.

Liu Jia, Du Xingzhong, Chen Zhenyu, He Tie-ke, Zhu Qinghua, Wu Qing, A Survey of Research on Mobile Recommendation Information Science October, 2012, Vol. 30, No. 10.

Meng Xiangwu. Mobile Recommender Systems and Their Applications. Journal of Software. January 2013, Vol. 24, No. 1.

Song Leyi, Xiong Hu, Zhang Rong. Towards the next generation of mobile recommender systems. Journal of East China Normal University (Natural Science). May 2013. No. 13.

Wang Licai, Meng Xianwu, Zhangyujie. Context-Aware recommender systems. Journal of Software, 2012, 23(1): 1–20.

Xu Hailing, Wu Xiao, Li Xiaodong. Comparison study of internet recommendation system. Journal of Software, 2009, 20(2):350–362.

Electronics, Information Technology and Intellectualization – Song & Kwak (Eds)
© 2015 Taylor & Francis Group, London, ISBN 978-1-138-02741-1

Optimization of HF-SCF on CPU-GPU Hybrid Parallel Computing infrastructure

S.R. Xu, S.Z. Ahmad & B. Li
Beihang University, Beijing, China

ABSTRACT: In molecular orbital ab initio calculations, the two-electron integral calculation is the most computationally expensive part. With the increase in the size of the molecules, calculation of integrals will require a large processing power. Improvement in the calculation time of the two-electron integrals has always been a hot spot in this research field. In this paper, we have suggested a novel implementation of algorithm on a heterogeneous system of CPU and GPU. Trace based results show that the proposed implementation can gain significant performance benefits for the calculation.

1 INTRODUCTION

The past decade has seen a remarkable increase in the High-Performance Computing (HPC), especially the improvement of the Graphics Processing Unit (GPU). Driven by the insatiable market demand for life-like real-time graphics, the GPU has developed into a processor, with unprecedented floating-point performance and programmability. Today's GPUs greatly outpace CPUs in arithmetic throughput and memory bandwidth, making them the ideal processor to accelerate a variety of data parallel applications.

Hartree-Fock (HF) is the basis of Molecular Orbital (MO) theory, which posits that the motion of each electron can be described by a single-particle function which does not depend explicitly on the instantaneous motions of the other electrons. It often provides a good starting point for more elaborate theoretical methods which are better approximations to the electronic Schrödinger equation, so we only consider HF theory in this paper.

The rest of this paper is organized as follows. Section 2 presents a brief overview of GPU and CUDA. Section 3 introduces some main integrals for our calculation. Section 4 discusses the design and implementation of the integral computation algorithms. Section 5 evaluates the result of our work and Section 6 summarizes our main work.

2 OVERVIEW OF GPU AND CUDA

The GPU architecture is built around a scalable array of multiprocessor units called Streaming Multi Processors (SMs). Each SM consists of a set of Streaming Processors (SPs). The multiprocessor executes threads in groups of thirty-two parallel threads called warps

(Nvidia, 2009). When a CUDA program on the host CPU launches a kernel grid, the blocks of the grid are distributed to SMs with available execution capacity. The threads of a thread block execute concurrently on one SM, and multiple thread blocks can execute concurrently on one SM. As thread blocks terminate, new blocks are launched on the available multiprocessors.

3 MAIN INTEGRALS

The HF equations can be written like this

$$F(C)C = SC\varepsilon \tag{1}$$

To solve the HF equations, we need to first introduce some important integrals as described by Szabo & Ostlund (2012).

3.1 *Overlap integral*

The simplest are the overlap integrals, stored in the form of the overlap matrix S. Its elements are defined as:

$$S_{\mu\nu} = \int \varphi_\mu(1)\varphi_\nu(1)\mathrm{d}\tau \tag{2}$$

3.2 *Kinetic energy integral*

The kinetic energy integrals are collected as the matrix T, whose elements are given by:

$$T_{\mu\nu} = \int \varphi_\mu(1)\left[-\frac{1}{2}\nabla^2\right]\varphi_\nu(1)\mathrm{d}\tau_1 \tag{3}$$

3.3 Nuclear attraction integrals

The nuclear-attraction integrals are collected as the matrix V^{nucl}, whose elements are given by:

$$V_{\mu\nu}^{nucl} = \int \varphi_\mu(1) \frac{Z_A}{r_{A1}} \varphi_\nu(1) d\tau_1 \qquad (4)$$

The above two integrals form the core-Hamiltonian matrix:

$$H_{\mu\nu}^{core} = T_{\mu\nu} + V_{\mu\nu}^{nucl} \qquad (5)$$

3.4 Two-electron integral

$$(\mu\nu \mid \lambda\sigma) = \iint \varphi_\mu(1)\varphi_\nu(1)\frac{1}{r_{12}}\varphi_\lambda(2)\varphi_\sigma(2)d\tau_1 d\tau_2 \qquad (6)$$

In practice, the basis functions are typically linear combinations of primitive atom-centred Gaussian basis functions, thus the two-electron integrals in the contracted basis are evaluated as

$$(\mu\nu \mid \lambda\sigma) = \sum_{p=1}^{N_\mu}\sum_{q=1}^{N_\nu}\sum_{r=1}^{N_\lambda}\sum_{s=1}^{N_\sigma} d_{\mu p} d_{\nu q} d_{\lambda r} d_{\sigma s} [pq \mid rs] \qquad (7)$$

where $d_{\mu i}$ refers to the contraction coefficients. Moreover, the number of primitives in a contracted basis function, N_μ, is the contraction length.

After giving the definitions of the above four integrals, we can finally get the expression of the Fock matrix:

$$F_{\mu\nu} = H_{\mu\nu}^{core} + \sum_{\lambda\sigma} P_{\lambda\sigma}\left[(\mu\nu \mid \sigma\lambda) - \frac{1}{2}(\mu\lambda \mid \sigma\nu)\right] \qquad (8)$$

4 CALCULATION DETAILS

The method to solve the HF equations is called Self-Consistent Field method (SCF). The SCF procedure can be defined as follows:

1. Specify a molecule and a basis set.
2. Obtain a guess at the density matrix P.
3. Calculate all required molecular integrals, $S_{\mu\nu}$, $H_{\mu\nu}$, and $(\mu\nu \mid \lambda\sigma)$.
4. Obtain the Fock matrix.
5. Solve the HF equations.
6. Form a new density matrix P from C.
7. Determine whether the procedure has converged. If not, return to step 3 with the new density matrix.
8. If the procedure has converged, we can end the calculation and get the expected values.

4.1 Division of calculation tasks

We measured the computation time of the main parts for four test molecules, C6H6, C10H18, H64 and C60. The STO-3G (Feller 1996, Schuchardt et al. 2007) basis set was used. Table 1 shows the user time of the

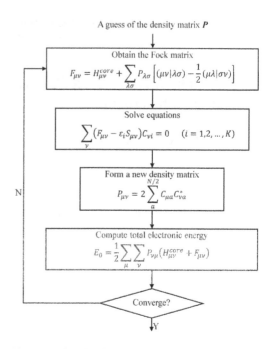

Figure 1. Flowchart for the SCF procedure. If the procedure has converged, then use the resultant solution to calculate expected values and other quantities of interest.

first SCF iteration. We can see that the evaluation of the two-electron integrals consumed the most time. Thus, we should first of all consider executing the evaluation of the two-electron integrals on GPU.

Given a particular basis set $\{\phi_\mu\}$, the integrals of T and V^{nucl} need to be evaluated and the core-Hamiltonian matrix formed. Unlike the full Fock matrix, the core-Hamiltonian matrix and the overlap integrals need only to be evaluated once as they remain constant during the iterative calculation. Thus, we execute the evaluation of the core-Hamiltonian matrix and the overlap integrals on CPU and store them on the host memory.

4.2 GPU algorithm for two-electron integrals

4.2.1 Strategy for GPU mapping

The first and most challenging problem is how to map the integrals to the GPU threads, with the load balancing. Simply, we can let each thread calculate one contracted integral by directly looping over all primitive integrals and sum up the results according to Equation 7. Because two different contracted integrals may consist of different numbers of primitive integrals, misbalancing could arise when they happened to be in the same wrap. Of course we can minimize the impact of this problem by letting each thread calculate one primitive integral. However, it may bring in many idle threads and cause misbalancing between SMs (Ufimtsev & Martinez 2008).

Figure 2 shows the mapping strategy that we proposed. By using the new feature imported in CUDA

Table 1.   Computation time of main parts.

| Molecule | Basis functions | User time (ms) (%) | | | | |
		Integral screening	H and S*	Two-electron integral	Other	Total
C6H6	36	42.9 (2.8)	38.1 (2.5)	1450.1 (94.5)	3.8 (0.2)	1534.9
C10H18	68	150.8 (1.5)	280.1 (2.9)	9341.6 (95.5)	9.9 (0.1)	9782.4
H64	64	86.2 (1.3)	429.5 (6.5)	6058.7 (92.0)	10.8 (0.2)	6585.2
C60	300	3523.0 (0.5)	11899.3 (1.8)	661020.9 (97.7)	459.5 (0.1)	676902.7

*Core-Hamiltonian matrix and overlap integrals.

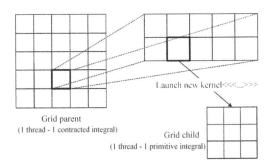

Figure 2.   GPU mapping algorithm for evaluating two-electron integrals.

Version (a)	Version (b)
Kernel parent:	Kernel A:
Inner each CI thread	Inner each thread
calculate up bound ub	calculate up bound ub
if ($ub < \lambda_{GPU}$*)	if ($ub < \lambda_{GPU}$*)
return	add to task queue
else	Kernel B:
compute two-electron integral	Inner each CI thread
	compute two-electron integral
*λ_{GPU} is the integral screening threshold we set for GPU calculation	

Figure 3.   Pseudocode of the integral screening.

5.0 – Dynamic Parallelism (Adinetz 2014a, b, c), we can dynamically and evenly assign all contracted integrals to the GPU threads.

Similarly, we still let each thread of the parent grid calculate one contracted integral. Instead of directly looping over all primitive integrals, we chose to launch a child kernel with suitable number of threads, depending on the size of each contracted integral. Each thread of the child kernel just calculates one primitive integral.

4.2.2   Integral screening procedure re-optimized

Rather than directly copying the CPU version implementation of the Schwarz (Strout & Scuseria 1995) integral screening, as shown in Figure 3a (Yasuda 2008), we found that we could re-optimize the integral screening by taking advantage of the GPU architecture. For version (a), it will experience load-balancing

Table 2.   Experiment platforms.

	CPU	GPU
Model	Intel Xeon E5620	GeForce GTX 780
Cores	4	2304
Frequency	2.40 GHz	0.94 GHz
Memory	23 GB	3 GB

issue in the following case. Consider two contracted integrals: one that meets the condition $ub < \lambda_{GPU}$ and the other does not. It will take different times to finish the calculation of those two integrals. If they happened to be in the same wrap, the misbalancing problem appeared again. To further use the ability of GPU, we broke the integral screening procedure into two parts, as shown in Figure 3b. First, we collected all contracted integrals that should not be neglected. Then we started a new kernel to compute the collected integrals.

5   EXPERIMENTS

In this section, we used our server consists of one Intel Xeon(R) CPU E5620 and one GeForce GTX 780 GPU card. Table 2 shows the details of the hardware architecture.

We performed 5 tests on the following molecules (Fig. 4) using the STO-3G basis set: Benzene (C6H6), Uridine (C9H12N2O6), Coronene (C24H12), Buckyball (C60) and a typical test system consisting of 64 H atoms organized as a $4 \times 4 \times 4$ cube. Table 3 shows the results we got.

Overall, results show that using CUDA on GPU platform is effective in evaluating the two-electron integrals. The GPU is up to forty-five times faster than a single CPU.

6   CONCLUSION

In this paper, we proposed a hybrid strategy evaluating some of the integrals on the GPU and others on the CPU. Then we performed several tests on different scale molecules and presented the corresponding results. Our results show that the GPU implementation

Table 3. Speedup for two-electron integral on GPU.

Molecule	Time spent on two-electron integrals per SCF iteration (ms)		Electronic energy (atomic units)	Speedup
	GPU	CPU		
Benzene	420.4	1451.2	−430.9208	3.5
Hydrogen cube	500.5	6070.2	−632.7700	12.1
Uridine	939.7	24047.2	−2229.4746	25.6
Coronene	1756.7	59902.2	−2737.4949	34.1
Buckyball	14687.3	658993.1	−10614.8145	44.9

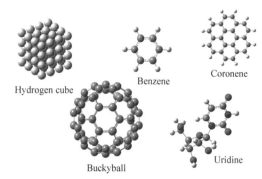

Figure 4. Test molecules. The size range from 12 to 60 atoms.

for two-electron integrals achieved nearly 45x speedup as compared to the CPU one.

However, for even more large-scale molecules, we may run out of memory on both host and GPU card. We can add multiple nodes having multicore CPUs and GPUs in a cluster. Data rate among these nodes can be optimized as mentioned by Ahmad et al. (2014b). MPI can be used to parallelize the task and distribute it among multiple nodes. This infrastructure can be made more efficient and secure as mentioned by Ahmad et al. (2014a).

ACKNOWLEDGEMENT

This research is supported by the National High Technology Research and Development Program of China (Grant No. 2012AA01A305).

REFERENCES

Adinetz, A. V. 2014a. *Adaptive Parallel Computation with CUDA Dynamic Parallelism*. [Online] Available from: http://devblogs.nvidia.com/parallelforall/introduction-cuda-dynamic-parallelism/ [Accessed 3rd July 2014].

Adinetz, A. V. 2014b. *CUDA Dynamic Parallelism API and Principles*. [Online] Available from: http://devblogs.nvidia.com/parallelforall/cuda-dynamic-parallelism-api-principles/ [Accessed 3rd July 2014].

Adinetz, A. V. 2014c. *A CUDA Dynamic Parallelism Case Study: PANDA*. [Online] Available from: http://devblogs.nvidia.com/parallelforall/a-cuda-dynamic-parallelism-case-study-panda/ [Accessed 3rd July 2014].

Ahmad, S.F., Ahmad, S.Z. & Li, B. 2014a. Honeypot, Network Forensic Technique and Framework. *The 2014 International Conference on Computer Science and Network Security, 12–13 April 2014, Xi'an, China*. Pennsylvania, DEStech Publications, Inc. pp. 407–411.

Ahmad, S.Z., Ahmad, S.F., Li, B. & Ali, A. 2014b. Novel Approach for Bandwidth Optimization in Grid Computing Infrastructure. *The 2014 International Conference on Computer Science and Network Security, 12–13 April 2014, Xi'an, China*. Pennsylvania, DEStech Publications, Inc. pp. 212–218.

Feller, D. 1996. The Role of Databases in Support of Computational Chemistry Calculations. *J Comput Chem*, 17: 1571–1586.

Nvidia, C. 2009. *NVIDIA's Next Generation CUDA Compute Architecture: FERMI*. [Online] Available from: http://www.nvidia.com/content/PDF/fermi_white_papers/NVIDIA_Fermi_Compute_Architecture_Whitepaper.pdf [Accessed 3rd July 2014].

Schuchardt, K.L., Didier, B.T., Elsethagen, T., Sun, L., Gurumoorthi, V., Chase, J., Li, J. & Windus, T.L. 2007. Basis Set Exchange: A Community Database for Computational Sciences. *J Chem Inf Model*, 47: 1045–1052.

Strout, D.L. & Scuseria, G.E. 1995. A quantitative study of the scaling properties of the Hartree–Fock method. *J Chem Phys*, 102: 8448–8452.

Szabo, A. & Ostlund, N.S. 1996. *Modern quantum chemistry: introduction to advanced electronic structure theory*. New York: Courier Dover Publications.

Ufimtsev, I.S. & Martínez, T.J. 2008. Quantum chemistry on graphical processing units. 1. Strategies for two-electron integral evaluation. *J Chem Theory Comput*, 4: 222–231.

Yasuda, K. 2008. Two-\electron integral evaluation on the graphics processor unit. *J Comput Chem*, 29: 334–342.

Next generation malware analysis techniques and tools

S.F. Ahmad
Department of Electrical and Computer Engineering, CASE, Islamabad, Pakistan

S.Z. Ahmad, S.R. Xu & B. Li
Beihang University, Beijing, China

ABSTRACT: Internet is the easiest medium to proliferate the malware into the personal computer systems and the server machines. Daily we hear about the announcement of zero day attacks from world renowned organization like ISC-Internet Strom Center. This leads the attackers to exploit those systems which remained unpatched on the internet or results in the compromising of the normal user. Researchers are working in this area of malware analysis which was started from the signature based static analysis to the most modern techniques of malware detection which could be able to detect or nearly guess the zero-day attacks. This paper reviews the most modern available research in the domain of malware analysis.

1 INTRODUCTION

Malware is a piece of software which causes any malicious activity on a computer system, which is not desired by the user. Malware analysis corresponds to the process which includes the series of steps to find out the internal structure of that malware and how it works in its different phases (Egele et al., 2012).

To avoid real users from these threats, antivirus companies offer such tools and techniques that focus on the detection of malicious software and their components. Such tools identify known threats using signature matching. It is also required that the signatures database and their updates are provided by the vendors. After having a sample, malware analysts first determine whether that piece of software can cause any harm to the users. As a second step they identify the malicious activities they performed on the computer system (Egele et al., 2012).

Hackers use a complex attack vector so that effectives could not track them. For example, in a scenario of malware distribution, first a computer system on the internet is infected with a bot which is a malicious piece of software that allows remote Command and Control (C&C) server to remotely control this bot. The network of the computer systems which has become the bots is called botnet. Spammers used these botnets to send spam emails containing links to the fake web pages which caused the installation of spyware on the visitor computer, which collects the personal information, credit card numbers and send back this information to the Command & Control Server online (Egele et al., 2012).

2 TYPES OF MALWARE

2.1 *Worm*

It is a widespread terminology in networked environments (Egele et al., 2012). Worm is defined as a program that can run independently and can propagate a fully working version of itself to other machines (Kaur & Nagpal, 2012).

2.2 *Virus*

A virus is a piece of code that adds itself to other programs, including operating systems (Kaur and Nagpal, 2012). It cannot run independently but requires its host program to be executed to activate it. As with worms, viruses usually propagate themselves by infecting every vulnerable host they can find (Egele et al., 2012).

2.3 *Trojan horse*

Software that seems to be useful and legitimate but a malicious program resides in it and performs malicious actions (Kaur and Nagpal, 2012).

2.4 *Spyware*

Software steals sensitive information from a compromised computer system and transfers this information to the attacker's server (Egele et al., 2012).

2.5 Bot

A bot is a malicious program that allows a hacker to remotely access and to control the system via C&C Server (Egele et al., 2012). Theses C&C servers are used to instruct the bots to send the spam emails or further expand their botnet network by compromising further systems (Willems et al., 2007).

2.6 Rootkit

Rootkit is a software which has the capability to hide its presence from the user of the computer system (Kaur & Nagpal 2012). Rootkit uses the techniques like hooking with API calls in user mode or kernel mode or tempering with operating system structures (Egele et al., 2012).

3 TYPES OF MALWARE ANALYSIS

Malware analysis is a multipronged process which provides us to look into the malware structure and functionality. Another important step of malware analysis is to observe its functionality and behaviour with the help of dynamic executable analysis at low level.

3.1 Static analysis

Malware forensic analysis can be done with two methodologies: static analysis and dynamic analysis using several on the shelf available tools. Software disassemblers and debuggers like Hex-Rays and OllyDBg are used to analyse malware and view its internal view regarding functionality. This is referred to as static analysis (Verma et al,. 2013).

3.2 Dynamic analysis

Dynamic analysis tries to observe the malware behavior by running it in an environment which logs every activity it performs while executing on the real system. Several plugins that extend the functionality of IDA Pro and OllyDBg include IDA-Stealth and Olly-Advanced. It enables us to use them for dynamic analysis and protects us from anti-analysis techniques (Verma et al., 2013).

4 TOOLS FOR MALWARE ANALYSIS

4.1 TTAnalyze

- TTAnalyze runs unknown binary with an operating system in an emulated environment. Therefore the malware does not actually execute on a real processor unlike the virtual machines, debuggers and API hooking. TTAnalyze is undetectable by the malware (Bayer et al., 2006).
- TTAnalyze system monitors the calls to the native kernel functions also the calls to Windows API functions. It also supports the function call arguments that contain pointers to other objects (Bayer et al. 2006).

- TTAnalyze function call injection allows the injection of our own code during execution of the program to perform more precise analysis on binary (Bayer et al. 2006).

4.2 DSD-Tracer

DSD-Tracer is a malware analysis framework that integrates dynamic and static analysis. It is expected to identify the families of obfuscation packers using VM-aware detection techniques. Detection of other non-obfuscated virtualization aware malware was implemented using a set of static analysis rules and dynamic rules applied to the output of Sophos virus engine built-in emulator (Lau & Svajcer, 2010).

In its dynamic state, DSD-Tracer provides a detailed trace of the executable. The information includes data like instructions decoded before its execution, all CPU registers, reads/writes to virtual/physical memory, interrupts/exceptions generated (Lau & Svajcer, 2010).

4.3 CWSandbox

CWSandbox system analyses a malware sample automatically. It is used for a behaviour-based analysis. In such a procedure, the malware binary is executed in a controlled environment so that all relevant routine calls to the Windows API can be monitored. In the end it generates a high-level summarized report from the analysed API calls. This generated report contains data for each process and its associated actions for example there is one subsection for file system accesses and another one for the network operations (Willems et al., 2007).

After the analysis of the API calls' parameters, system function calls are directed towards original API function calls by the sandbox. In this way, sandbox allows the malware to integrate itself into the target operating system, by copying itself into the Windows system directory or adding new registry keys. To achieve fast automatic analysis, CWSandbox runs in the virtual environment so the system returns to its clean state after doing analysis process. This technique has some shortcomings like detectability and slower execution, but using CWSandbox in a native environment with automated procedures and automated procedure bring back the sandbox into its original state and helps in reduction of this system's drawbacks (Willems et al., 2007).

4.4 Some commonly used tools

These tools are used in the investigation (Kaur and Nagpal, 2012):

- Grep: It is a command line tool which is written for UNIX system for searching of text.
- AVG Antivirus: The antivirus software is use for detecting and removing malware and trojan from the system.

- Whois: This tool is used for querying online RIPE database to determine registration information for IP addresses and domains.
- IDA Pro: Commercial disassembling and debugging software.
- Hexedit: This is useful to edit and view the raw data file in hex format.
- VMWare: Virtualization software which allows to install different operating systems in a sandbox environment.
- File Analyser: File analysis tool that allows basic analysis of files like showing the contents of file and properties of file in hex dump format.
- Helix: This is a live Linux distribution for live forensic analysis and passwords recovery from RAM and Internet history.
- Sysinternals: This is a suite of tools designed to help managing, troubleshooting and diagnosing Windows environment and the softwares running over it.

5 CODE NORMALIZATION TO FIND SELF-MUTATING MALWARE

Self-Mutating is a technique of code obfuscation in which a code changes itself during execution. Some famous self-mutating programs are METAPHOR, ZMIST and EVOL. It has been observed that some commercial virus scanning tools are also circumvented by these kind of viruses. Automatic mutation programs have to be able to analyse their own code and mutate into next generation after extracting information from their code. This type of mutation technique can be easily calculated. By using this technique, it is also possible to reverse the process of mutation and find out the actual code or it's binary (Bruschi et al., 2006).

6 AUTOMATIC ANALYSIS OF MALWARE BEHAVIOR

Malicious programs are characterized by complex behaviours in system resources and advanced network activity. Different malware types of the same family share common behavioural patterns such as usage of mutant or modification of a system file (Rieck et al., 2009).

- This framework provides binary analysis by first execrating and monitoring the malwares in the sandbox environment. Based upon the operations and actions performed by the malware binaries, a sequential report of monitored behaviour is generated. This report listed each systems call with its arguments and each network activity.
- A high-dimensional vector space is populated with the sequential reports where each dimension represents a behavioural pattern, a small series of analysed instructions. Using this vectorial representation, the similarity of behaviour can be assessed geometrically.
- Clustering and classification techniques of machine learning are applied to dig out the novel and known classes of malware. Efficient computation techniques are realized using those prototype vectors, which comprise larger groups of reports with similar behaviour, and so this way helps us to find an effective approximation for exact analysis.
- By using clustering and classification techniques alternatively, the embedded behaviour of malware can be more easily analysed. Known malware classes are identified using behaviour matching and prototype vectors technique discussed earlier.
- By alternating between clustering and classification steps, the embedded behaviour of malware can be analysed incrementally, for example, on a daily basis. First, behaviour matching known malware classes is identified using prototype vectors of previously discovered clusters. Then reports with unidentified behaviour are clustered for discovery of novel malware classes.

7 MALICIOUS EXECUTABLE CLASSIFICATION SYSTEM (MECS)

A combined technique has been introduced from machine learning and data mining. MECS detects malicious binaries which are currently not classified or not detected on the system, without going through preprocessing or removing any obfuscation which was applied on the malicious binary (Kolter and Maloof, 2006).

8 MALWARE ANALYSIS USING PHYLOGENY GENERATION AND PERMUTATIONS OF CODE

Malicious programs like viruses and worms related to the previously developed malicious programs through evolutionary relationships. Identifying those relationships and developing a phylogeny model is expected to be helpful for identifying new malware and establishing a naming scheme. Matching code permutation helps in building better models of malware evolution. It describes the method to develop phylogeny models. It uses feature names as n-perms to match and find the permutated code (Karim et al., 2005).

9 MALWARE ANALYSIS USING GENETIC ALGORITHMS

Mehdi et al. (2009) presents an architecture named IMAD-In-Execution Malware Analysis and Detection system. It first logs the system calls of the binary on Linux machine then the sequence of system calls will placed in a n-gram fixed size window, now the n-gram analyser analyses each n-gram and categorized it as malicious or normal. A goodness evaluator assigns a goodness value to each n-gram as the friendly binary. To develop an in-execution classification system, ngc786 classifier was developed to classify in execution binary as malware or friendly after putting the results in genetic optimizer.

Williams et al. (2014) uses Genetic and Evolutionary Feature Selection (GEFeS) and Genetic and Evolutionary Feature Selection and Weighting (GEFeWS) algorithms to identify the malwares which are associated with HTML. They identified that GEFeS performance was better in terms of less features selection while GEFeWS was better in high accuracy. They used X-TOOLSS (Tinker et al., 2010) which is a suite of genetic and evolutionary computations (GECs) to identify the optimal result.

10 CONCLUSION

This paper presents a comprehensive comparison of different malware analysis schemes. Some of these are found detectable by the malware. So, to counter this issue, techniques like malware clustering and classification and phylogeny models have been introduced. A proposed future work could be the implementation of biological genetic algorithms for the classification and identification of new malware, in the framework proposed by Ahmad et al. (2014). In this way we achieve a better framework with enhanced capabilities of detecting a new malware.

ACKNOWLEDGEMENTS

This research is supported by the National High Technology Research and Development Program of China (Grant No. 2012AA01A305).

REFERENCES

Ahmad, S.F., Ahmad, S.Z. & Li, B. 2014. Honeypot, Network Forensic Technique and Framework. *The 2014 International Conference on Computer Science and Network Security, 12–13 April 2014, Xi'an, China*. Pennsylvania, DEStech Publications, Inc. pp. 407–411.

Bayer, U., Moser, A., Kruegel, C. & Kirda, E. 2006. Dynamic analysis of malicious code. *Journal in Computer Virology*, 2(1): 67–77.

Bruschi, D., Martignoni, L. & Monga, M. 2006. Using code normalization for fighting self-mutating malware. *In Proceedings of the International Symposium on Secure Software Engineering*. pp. 37–44.

Egele, M., Scholte, T., Kirda, E. & Kruegel, C. 2012. A survey on automated dynamic malware-analysis techniques and tools. *ACM Comput Surv*, 44: 6.

Karim, M.E., Walenstein, A., Lakhotia, A. & Parida, L. 2005. Malware phylogeny generation using permutations of code. *Journal in Computer Virology*, 1(1–2): 13–23.

Kaur, G. & Nagpal, B. 2012. Malware Analysis & its Application to Digital Forensic. *International Journal on Computer Science and Engineering (IJCSE)*, 4(04): 622–626.

Kolter, J.Z. & Maloof, M.A. 2006. Learning to detect and classify malicious executables in the wild. *J Mach Learn Res*, 7: 2721–2744.

Lau, B. & Svajcer, V. 2010. Measuring virtual machine detection in malware using DSD tracer. *Journal in Computer Virology*, 6(3): 181–195.

Mehdi, S.B., Tanwani, A.K. & Farooq, M. 2009. IMAD: in-execution malware analysis and detection. *In Proceedings of the 11th Annual conference on Genetic and evolutionary computation*. ACM. pp. 1553–1560.

Rieck, K., Trinius, P., Willems, C. & Holz, T. 2011. Automatic analysis of malware behavior using machine learning. *Journal of Computer Security*, 19(4): 639–668.

Tinker, M.L., Dozier, G. & Garrett, A. 2010. The exploratory toolset for the optimization of launch and space systems (x-toolss).

Verma, A., Rao, M.S., Gupta, A.K., Jeberson, W. & Singh, V. 2013. A Literature review on malware and its analysis. *International Journal of Current Research and Review*, 5(16): 71–82.

Willems, C., Holz, T. & Freiling, F. 2007. CWSandbox: Towards automated dynamic binary analysis. *IEEE Secur Priv*, 5(2): 32–39.

Williams, H.C., Carter, J.N., Campbell, W.L., Roy, K. & Dozier, G.V. 2014. Genetic & Evolutionary Feature Selection for Author Identification of HTML Associated with Malware. *International Journal of Machine Learning & Computing*, 4(3): 250–255.

Research and implementation of Hybrid Parallel Computing for force field calculation

S.Z. Ahmad, S.R. Xu & B. Li
Beihang University, Beijing, China

S.F. Ahmad
Department of Electrical and Computer Engineering, CASE, Islamabad, Pakistan

ABSTRACT: Hybrid Parallel Computing (HPC) is a technique that facilitates us to run computationally expensive scientific jobs on desktop machines. Assisted Model Building with Energy Refinement (AMBER) and Force Fields calculations of Molecular Dynamics (MD) are computationally expensive jobs that require supercomputers for execution. The objective of this research is to provide researchers with desktop machines, capable of doing massive processing. A novel implementation of AMBER force field empirical formula, using MPI infrastructure on HPC, is suggested. We suggest a heterogeneous platform consisting of Central Processing Units (CPUs) and Intel Phi co-processors along with dynamic parallelism, in which we chose number of processes at run time to distribute the task among the processing nodes. Message Passing Interface (MPI) is used to achieve task-based and data-based parallelism. Trace based results and graphs shows that the proposed method can gain significant performance benefits for the processing intensive application.

1 INTRODUCTION

The HPC helps us to exploit the parallelism for computationally expensive scientific jobs on desktop machines having heterogeneity in nature. We are using MPI to calculate AMBER Force Fields energy on HPC infrastructure. Intel Phi co-processors can be used to meet the requirements of processor intensive jobs. It is also called Many Integrated Cores (MIC). HPC includes running programs that execute across heterogeneous platforms on multiple nodes consisting of Central Processing Units (CPUs) and MICs (Nguyen 2013). MPI provides parallel computing software infrastructure using task-based and data based parallelism. It has been adopted by Intel® Xeon Phi and many scientific organizations.

The rest of this paper is organized as follows. Section 2 presents a brief overview of AMBER Force Fields energy calculation. Section 3 introduces some techniques for HPC. Section 4 discusses the design and implementation of AMBER force field energy calculation in parallel program for Intel MPI. Section 4 also evaluates the result of our work and Section 5 concludes our main work.

2 AMBER FORCE FIELDS

In this paper, energy calculations of AMBER Force Fields of molecular dynamics have been chosen as

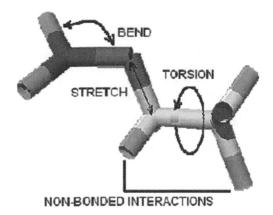

Figure 1. Force Fields bonded and non-bonded interactions.

an idea to prove the concept. The Force Fields calculation is an empirical calculation method intended to give estimates of structures and energies for conformations of molecules. The method is based on the assumption of natural bond lengths and angles, deviation from which leads to strain, and the existence of torsional interactions and attractive and/or repulsive van der Waals and dipolar forces between non-bonded atoms. The method is also called empirical force-field calculations (Muller 1994).

The actual coordinates of a model combined with the force field data create the energy expression (or target function) for the model. This energy expression is the equation that describes the potential energy surface of a particular model as a function of its atomic coordinates. The potential energy of a system can be expressed as a sum of valence (or bond), and non-bonded interactions. The energy of valence interactions is generally accounted for by diagonal terms, named as bond stretching (E_{bond}), valence angle bending (E_{angle}), dihedral angle torsion ($E_{torsion}$), and inversion ($E_{inversion}$) terms, which are part of nearly all force fields for covalent systems.

$$E_{val} = E_{bond} + E_{angle} + E_{torsion} + E_{inv} \qquad (1)$$

The energy of interactions between non-bonded atoms is accounted for by van der Waals (E_{vdW}), electrostatic ($E_{Coulomb}$), and (in some older force fields) hydrogen bond (E_{hbond}) terms.

$$E_{nonbond} = E_{vdW} + E_{Coulomb} + E_{hbond} \qquad (2)$$

The functional form of the AMBER force field is (Cornell 1995):

$$V(r^N) = \sum_{bonds} k_b (l - l_0)^2 + \sum_{angles} k_a (\theta - \theta_0)^2$$

$$+ \sum_{torsions} \frac{1}{2} V_n [1 + \cos(n\omega - \gamma)] \qquad (3)$$

$$+ \sum_{j=1}^{N-1} \sum_{i=j+1}^{N} \left\{ \varepsilon_{i,j} \left[\left(\frac{r_{0ij}}{r_{ij}} \right)^{12} - 2 \left(\frac{r_{0ij}}{r_{ij}} \right)^6 \right] + \frac{q_i q_j}{4\pi\varepsilon_0 r_{ij}} \right\}$$

where k_a = Angle Constant; k_b = Bond Constant; l_o = natural bond length; l = bond length; V_n = Torsion Constant; θ = Angle Bend; ω = Torsion Angle; γ = Phase factor; ε = well depth; q_j = Charge on j and q_i = Charge on i.

Many scientific research applications about the dynamics of proteins and DNA use data from simulations spanning nanoseconds (10^{-9} s) to microseconds (10^{-6} s). To obtain results of simulations, several CPU-days to CPU-years are needed (Tuckerman et al. 1991). Parallel algorithms allow the load to be distributed among CPUs/GPUs or MICs.

During a classical MD simulation, the most CPU intensive task is the evaluation of the potential (force field energy) as a function of the particles internal coordinates. Within that energy evaluation, the most computationally expensive one is non-bonded or non-covalent part. Which is of the order of big O notation, common MD simulations scale by $O(n^2)$ if all pairwise electrostatic and van der Waals interactions are accounted (Tuckerman et al. 1991). This computational cost can be meet by exploiting the parallelism in HPC. Similarly, the Ab-Initio Calculations Methods using the Hartree-Fock or Density functional theory (DFT) to calculate the electronic structure and associated properties are called ab-initio or first principles calculations (ATĐLHAN 2007). Ab-initio calculations are computationally expensive. We have tried

using GPU to help the improvement of this calculation (Xu et al. 2014). More practically, the system is constrained to several hundred atoms. Obviously, it is thus impossible to use ab-initio calculations to exactly simulate a large number of molecules. It also becomes very computationally expensive to calculate the energy.

3 HYBRID PARALLEL COMPUTING TECHNIQUES

3.1 HPC as a solution to MD problems

Hybrid parallel Computing HPC is one way to meet the needs of processor hungry MD simulation. Its pillars relies on an underlying heterogeneous computing infrastructure of MPI, weather Processor/Co-processor combination or Intel's Xeon Phi Co-processors are used as multiple nodes. Heterogeneous computing refers to systems that use more than one kind of processor. These are multi-core systems that gain performance not just by adding cores, but also by incorporating specialized processing capabilities to handle particular tasks. Heterogeneous System Architecture (HSA) utilizes multiple processor types (typically CPUs and co-processors), to give you the best of both worlds. Intel Phi coprocessors can also perform mathematically intensive computations on very large data sets, while CPUs can run the operating system and perform traditional serial tasks.

By the end of 2010, nearly all new desktop computers had multicore processors. Dual-core and even quad core processors enter the mainstream of affordable computer industry. Still, multicore processing posed some challenges of its own. The extra cores and cache memory required to fuel their instruction pipelines came at a cost of both increased processor size and, again, of high power consumption.

Meanwhile, the multi-core era also saw some interesting developments in processors, which were growing in sophistication and complexity, spurred on by advances in semiconductor technology. Coprocessors have vector processing capabilities that enable them to perform parallel operations on very large sets of data and to do it at much lower power consumption relative to the serial processing of similar data sets on CPUs. This is what allows co-processors to drive capabilities at such as incredibly realistic, performance by offloading processing/graphics from the CPU. They thus become increasingly attractive for more general purpose processing, such as addressing data parallel programming tasks (AMD 2014).

Advanced Micro Devices (AMD) and Nvidia Corp., which make General Purpose Graphics-Processing Units (GPGPUs) that are well-suited for this sort of application, have previously dominated the accelerator market. Nevertheless, Intel is making a strong position for the space, with a different kind of accelerator called as the Xeon Phi coprocessor.

The systems on the Top500 list were already using the Intel's chip by the time it made its official announcement in November 2012. Among top super-computers, Intel's accelerator share is still small. The machines using Nvidia's GPUs outnumber those with Intel's coprocessors 31 to 11. However, Tianhe-2 has had a big impact on the landscape. If we compare the petaflops on the list, the aggregated performance delivered by Intel Phi coprocessors is now bigger than the performance obtained by GPGPUs.

Intel's chips contain up to sixty-one cores and are built using the company's 22-nanometer manufac-turing process, which is a generation ahead of the competition. The company says its coprocessors have a few advantages over GPGPUs. They can operate inde-pendently of CPUs and they do not require special code to program (Courtland 2013). A separate Linux kernel runs in the Intel Phi to manage its own resources. We access its command prompt by secure shell.

3.2 *MPI for Hybrid Parallel Computing*

MPI is a communication protocol for parallel program-ming. MPI is specifically used to allow applications to run in parallel across a number of separate computers connected by a network. Message passing programs generally run the same code on multiple processors, which then communicate with one an-other via library calls. It is a generalized architecture for parallel pro-gramming. Intel MPI (IMPI) is used to meet the requirements of Hybrid Parallel Computing on both suggested platforms CPU and MIC. MPI Latency of IMPI Library 4.1 Update 1 is up to 10 times faster than alternative MPI libraries. Compiler Performance is also increased by using the underlying C, C++ & FORTRAN compilers by Intel. MPI works as a wrap-per to compiler (Slavova 2012).

3.3 *Intel® Cluster Studio XE 2013*

We are using the Intel Cluster Studio XE 2013. It meets the challenges of HPC developers by pro-viding IMPI, Intel compilers, parallel models and libraries with advanced performance optimizations for today's multi-core and many-core processors in HPC clusters (Slavova 2012). It provides Intel's superior shared, distributed, or hybrid application performance monitoring applications as well.

4 IMPLEMENTATION DETAILS

We are using Intel Phi 3120A (6 GB, 1.100 GHz, 57 core) on a computer with 2 CPUs (6Core, 2.1 GHz). We have used Intel MPI library (4.1.3.048) on CentOS (6.5) with Linux Kernel (2.6.32–431). We have used IMPI library to develop application that can run on multiple nodes of cluster chosen by the user at run-time. As per the requirement of MPI we have to write a single program to run as multiple processes on differ-ent cores of processor and coprocessors. Each process works on different subset of parameters and do energy calculation. In our scenario size of each subset is

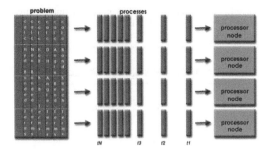

Figure 2. Each process calculates all six functions at its own subset of parameters by running it on respective processor nodes.

```
Start
    Number of processes=user input
    Parameter subset=Total parameters/Number
    of processes
    If (process=root)
        Send (Parameter subset to child)
    Else if (process=child)
        Receive (Parameter subset from root)
        Calculate 6 functions on Parameter subset
        Send (partial result to root)
    If (process=root)
        Receive (partial results)
        Summarise results to calculate energy
End
```

Figure 3. Pseudocode for parallel program to calculate energy.

calculated, keeping in view the number of processes, by the user input at starting time of root process, then sub-list of parameters is distributed to each child pro-cess buy the root process. Each child process dose calculation at its own subset of parameters and submit it back to root process.

Keeping in view the number of processors and its processing capabilities, the number of processes sub-mitted to it can lead us to better performance results. The pseudocode of this is mentioned in Figure 3. In our experiment we are using Biomer/B software by White (2014), written in Java. We have added complex-ity parameter and processing time calculation function in its source code and found the following improve-ments as mentioned in Table 1. The processing time that we obtained by running the same energy cal-culation/complexity by Biomer software is 5073 sec, which is reduced to 1850 sec by using a combination of processor and coprocessor.

The reason that Biomer software is written in Java is that it runs in a single process without exploiting the parallelism. According to our experiment as depicted in Table 1, it can be further optimized to 18 sec only if we chose fifty-seven processes on the processors and the coprocessors.

Table 1. Comparative analysis of processing time required to calculate energy by CPU, MIC, combined and CPU only with Biomer.

Number of Processes	CPU$_{(second)}$	MIC$_{(second)}$	CPU + MIC*$_{(second)}$	Biomer CPU$_{(second)}$
1	–	–	–	5073
2	676	5458	1850	–
12	78	505	250	–
57	23	103	18**	–
58	23	132	23	–

*The Number of Processes [x] in CPU+MIC [x+x] are double to the number of processes in case of CPU only [x] or MIC only [x].
**Time required for processing is minimum.

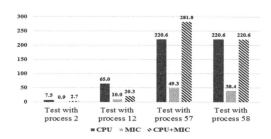

Figure 4. Speedup as compared to Biomer. Test results with different number of processes running on CPU, MIC and a combination of CPU + MIC.

5 CONCLUSION

It is realized that molecular dynamics simulations are bounded by time constraints, due to available computational power, as these computations are computationally expensive. Dynamic parallelism technique with MPI is suggested to exploit the parallelism provided by HPC infrastructures, having Intel® Xeon Phi™ coprocessors at its core. Intel MPI infrastructure can be used for energy calculation for a larger number of molecules on multiple processing nodes consisting of processors and coprocessors. We can add multiple nodes having multicore CPUs and MICs in a cluster. Data rate among these nodes can be optimized, as mentioned by Ahmad et al. (2014b). MPI can be used to parallelize the task and distribute it among multiple nodes. This multimode infrastructure can be made more efficient and secure, as mentioned by Ahmad et al. (2014a).

ACKNOWLEDGEMENT

This research is supported by the National High Technology Research and Development Program of China (Grant No. 2012AA01A305).

REFERENCES

Ahmad, S.F., Ahmad, S.Z. & Li, B. 2014a. Honeypot, Network Forensic Technique and Framework. *The 2014 International Conference on Computer Science and Network Security, 12–13 April 2014, Xi'an, China.* Pennsylvania, DEStech Publications, Inc. pp. 407–411

Ahmad, S.Z., Ahmad, S.F., Li, B. & Ali, A. 2014b. Novel Approach for Bandwidth Optimization in Grid Computing Infrastructure. *The 2014 International Conference on Computer Science and Network Security, 12–13 April 2014, Xi'an, China.* Pennsylvania, DEStech Publications, Inc. pp. 212–218

AMD. 2014. *What is Heterogeneous Computing?* [Online] Available from: http://developer.amd.com/resources/heterogeneous-computing/what-is-heterogeneous-computing/ [Accessed 10th July 2014].

ATĐLHAN, S. 2007. Molecular dynamics simulation of montmorillonite and mechanical and thermodynamic properties calculations (Doctoral dissertation, Texas A&M University).

Cornell, W.D., Cieplak, P., Bayly, C.I., Gould, I.R., Merz, K.M., Ferguson, D.M., Spellmeyer, D.C., Fox, T., Caldwell, J.W. & Kollman, P.A. 1995. A second generation force field for the simulation of proteins, nucleic acids, and organic molecules. *Journal of the American Chemical Society*, 117(19): 5179–5197.

Courtland, R. 2013. What Intels xeon phi coprocessor means for the future of supercomputing. *IEEE Spectrum.*

Muller, P. 1994. Glossary of terms used in physical organic chemistry (IUPAC Recommendations 1994). *Pure and applied chemistry*, 66(5): 1077–1184.

Nguyen, L.Q. 2013. *Intel Xeon Phi Coprocessor Developer's Quick Start Guide for Microsoft Windows Host.* [Online] Available from: https://software.intel.com/sites/default/files/article/335818/intel-xeon-phi-coprocessor-quick-start-developers-guide-windows-v1-2.pdf [Accessed 10th July 2014].

Slavova, G. 2012. Intel® MPI Library for Linux OS. [Online] Available from: https://software.intel.com/sites/default/files/managed/38/27/Reference_Manual.pdf [Accessed 10th July 2014].

Tuckerman, M.E., Berne, B.J. & Martyna, G.J. 1991. Molecular dynamics algorithm for multiple time scales: Systems with long range forces. *The Journal of chemical physics*, 94(10): 6811–6815.

White, N. 2014. *B (formerly Biomer).* [Online] Available from: http://casegroup.rutgers.edu/Biomer/index.html [Accessed 10th July 2014].

Xu, S., Ahmad, S.Z. & Li, B. 2014. *Optimization of HF-SCF on CPU-GPU Hybrid Parallel Computing Infrastructure.* Manuscript submitted for publication.

Electronics, Information Technology and Intellectualization – Song & Kwak (Eds)
© *2015 Taylor & Francis Group, London, ISBN 978-1-138-02741-1*

A Stable Quick Sort Algorithm (SQSA) for fragment-continuity-sequential data

Ruyun Wang, Di Cao, Zhentao Zang & Yinxian Yu
*College of Harbor, Coastal and Offshore Engineering, Hohai University, Nanjing, Jiangsu,
People's Republic of China*

ABSTRACT: In scientific computation, sometimes, it is necessary to sort large fragment-continuity-sequential data. Considering the characteristics of the data, a new sort algorithm is given by using the binary search for a homing position, two-way passed and directed homing. It is proved that the average efficiency of the new sort of algorithm for fragment-continuity-sequential data is relatively high.

1 INTRODUCTION

Because of the importance, in theory and in wider applications, sorting has been listed as one of the top ten issues which have the greatest influence on science and engineering (Dongarra J 2000).

According to the time complexity, sorting algorithms are classified into two classes: (i) $O(n^2)$ and (ii) $O(n \log n)$ (Babu DR 2011). The elementary sort, whose time complexity is $O(n^2)$, mainly include Bubble sort, Selection sort and Insertion sort (Biggar P & Gregg D 2005). Among them, the time complexity is $O(n^2)$ (Lee EA 1987), $O(n^2)$ (Murthy PK 2001) and $O(n^2)$ (Deng kai L & Shu M 1998) respectively. The improved sort algorithms, whose time complexity is about $O(n \log n)$, mainly include Shell's sort (Shell DL 1959), Quick sort (Hoare CAR 1962) and Heap sort (Williams JWJ 1964). Their time complexity is about $O(n \log_2 n)$ (Sedgewick R 1996), $O(n \ln n)$ (Hong-Wei H 2002) and $O(n \log n)$ (Biggar P & Gregg D 2005). However, when the initial data is partially ordered, Quick sort may reduce to Bubble sort and take $O(n^2)$ time.

As for the above improved sorting algorithms, though their efficiency is good, they are not stable (Huiqin Z 2009, Dechao W 2012). The elementary sorts are stable, but their efficiency is relatively low. In scientific computation, sometimes, it is necessary to sort the large fragment-continuity-sequential data (any neighbour data is increasing or decreasing, and all of the other data is greater or less than them). When using the above sorting algorithms to sort the data, it may occur that the time taken is long or the relative position of the same or equivalent elements changes.

Based on the above sorting algorithms, considering the characteristics of the fragment-continuity-sequential data, a new sorting algorithm is given by using binary search for homing position, two-way passed (passing the sorted data toward left or right side of the array) and directed homing.

2 BINARY SEARCH FOR HOMING POSITION

2.1 The idea of binary search

Same as the binary search (Zhi X 1989), the precondition of binary search for homing position is that the data is ordered (*increasingly or decreasingly*) and the idea of the algorithm is as follows:

Compare the element with the middle element of the sequence, if it is less than the middle element, using binary to lookup the homing position in the left side of middle element, otherwise, lookup in the right side until there is only one element in the left (or right) of the middle element. Then, the position between this element and the middle element is what we are looking for.

2.2 Analysis of time taken

Assuming the sequence length is n, it needs k times to find out the position. That is $2k^{-1} < n \leq 2^{k+1}$. Take the log to both sides of the inequality:

$$k - 1 < \log_2 n \leq k + 1 \tag{1}$$

In the formula (1), k as the times of finding position, is an integer, and the range of k is

$$\lfloor \log_2 n \rfloor - 1 \leq k \leq \lfloor \log_2 n \rfloor + 1 \tag{2}$$

Here $\lfloor \rfloor$ is the sign of roundness. When analysing the time taken, k can be $\lfloor \log_2 n \rfloor$.

2.3 Stability analysis

When using the binary search for homing position, the algorithm sets that, if the element is greater than or equal to the middle element of sequence, lookup the position in right side of the middle element. Then, it can ensure the relative position of the same or equivalent elements never change. That is, the binary search for homing position is stable.

3 SQSA SETS AND ANALYSIS

3.1 The idea of the algorithm

For the characteristics of the fragment-continuity-sequential data, the idea of the algorithm is as follows (show the ascending way only for illustration):

Put the small\large element of the first two data into the LS\RS (Left Sequence\Right Sequence) of SA (Storage Array). Give the logical variable an initial value which is used for deciding the homing direction. Because the positions and values of the endpoints of LS and RS exchange with their lengths changing, call them dynamic endpoint.

According to the four dynamic endpoints sort the data. To better adapt the characteristics of the fragment-continuity-sequential data, make the element be a dynamic endpoint after being put into the LS\RS. Then, if the elements behind the current element are ordered, these elements just compare with the dynamic endpoint to find out the homing positions, and avoid looking up blindly.

To ensure that the element is a dynamic endpoint after being put into the LS\RS, use the sort algorithm to sort the data for five cases.

(1) The element is less than the left dynamic endpoint of the LS. Move the LS to the right by one bit position. Put the element into the leftmost position of the LS.

(2) The element is less than the right-endpoint of LS. Use the binary search algorithm to find out the homing position, and move the elements of LS, which are greater than the element, to the RS. Give the logical variable false and put the element into the leftmost position of the RS.

(3) The element is greater than or equal to the right-endpoint of LS and less than the left-endpoint of RS. Put the element into the rightmost position of the LS if logical variable is true; otherwise put it into the leftmost position of the RS.

(4) The element is greater than or equal to the left-endpoint of LS. Use the binary search algorithm to find out the homing position, and move the elements of RS, which is less than or equal to the element, to the LS. Give the logical variable true and put the element into the rightmost position of the LS.

(5) The element is greater than or equals to the right dynamic endpoint of the RS. Move the RS to the left by one bit position. Put the element into the right-most position of the RS.

3.2 The algorithm

Assume that the data, whose size is n, is stored in the array a, b is an SA and $nsgn$ is a logical variable, whose initial value is true. The sort algorithm is as follows (show the ascending way only for illustration):

(1) If $a(1) \le a(2)$, given values: $b(1) = a(1)$, $b(n) = a(2)$, $nsgn = false$, otherwise, given values: $b(1) = a(2)$, $b(n) = a(1)$, $nsgn = true$, $n_0 = 1$, $n_1 = n$, $i = 3$. /*Now, the $b(1) \sim b(n_0)$ is the LS, the $b(n_1) \sim b(n)$ is the RS, $b(1)$, $b(n_0)$, $b(n_1)$ and $b(n)$ are dynamic endpoints.*/

(2) If $a(1) \le b(1)$, set $b(j+1) = b(j)$, $j = n_0 \ldots 1$, given value: $n_0 = n_0 + 1$, go to (14).

(3) If $a(i) < b(n_0)$, go to (6).

(4) If $a(i) < b(n_1)$, go to (11).

(5) If $a(i) < b(n)$, go to (12), otherwise, go to (13). /*(6) $\sim$ (9) is to show how to use the binary search algorithm for homing position.*/

(6) Given values: $m_l = 1$, $m_r = n_0$.

(7) Given value: $m_{mid} = (m_l + m_r)/2$.

(8) If $a(i) < b(m_{mid})$, given values: $m_r = m_{mid}$, otherwise, given values: $m_l = m_{mid}$.

(9) If $m_r - m_l > 1$, go to (7).

(10) Set: $b(n_1 - n_0 + j - 1)$, $j = n_0, \ldots, m_r$. Given values: $n_0 = m_l$, $n_1 = n_1 - n_0 + m_l - 1$, $b(n_1) = a(i)$, $nsgn = false$, go to (14).

(11) If $nsgn = true$, given values: $n_0 = n_0 + 1$, $b(n_0) = a(i)$, otherwise, given values: $n_1 = n_1 - 1$, $b(n_1) = a(i)$. Go to (14).

(12) Omitted. /*For the symmetry, the way of putting $a(i)$ into RS is very similar to (6)$\sim$(10)*/

(13) Set $b(j-1) = b(j)$, $j = n_1, \ldots, n$, $b(n) = a(i)$, $n_1 = n_1 - 1$.

(14) Given value: $i = i + 1$, if $i \le n$, go to (2).

(15) End. /*Now, $b(1) \sim b(n)$ is the ascending array.*/

3.3 Algorithm analysis

3.3.1 Analysis of time taken

For the data containing fragment-continuity sequences, the way to analyse the time taken of the sort algorithm is as follows:

Assuming there are m $(1 \le m \le n)$ sequences of data, and the average length of every sequence is about $\frac{n}{m}$. When i sequences have already been put into the SA, the probability of the $(i+1)$-th sequence to be put into SA in the any case is $\frac{1}{i+1}$.

First, distinguish which case the first element of every sequence belongs to. The probability of any sequence belongs to case α_i $(i = 1, 2, \ldots, 5)$ is as follow:

$$
\begin{cases}
\alpha_1 = \alpha_5 = \dfrac{1}{m-2}\left(\displaystyle\sum_{i=2}^{m-1}\dfrac{1}{i+1}\right) \\[2ex]
\alpha_2 = \alpha_4 = \dfrac{1}{2(m-2)}\left(\displaystyle\sum_{i=2}^{m-1}\dfrac{i-2}{i+1}\right) \\[2ex]
\alpha_3 = \dfrac{1}{m}\left(\displaystyle\sum_{i=1}^{m}\dfrac{1}{i}\right)
\end{cases} \tag{3}
$$

Then, for the m sequences, the average numbers of exchanges and comparisons in any case are as follows:

Case 1: Because of using the binary search for homing position, two-way passed and directed homing to sort, if the sequence is decreasing, all the elements of the sequence will be sorted in case 1. If the sequence is increasing, only the first element of the sequence will be sorted in case 1. In addition, the remaining elements of the sequence will be sorted in case 3. Considering the probability of the different sequence is 0.5, for the m sequences, the average numbers of exchanges and comparisons are as follows:

$$\begin{cases} C_1 = \alpha_1 \left[\sum_{j=1}^{m} \left(\frac{2n}{m} - 1 \right) \right] \\ M_1 = \alpha_1 \left\{ \sum_{j=1}^{m} \left[\frac{jn}{4m} \left(\frac{n}{m} + 1 \right) + \frac{n}{4m} - \frac{1}{2} \right] \right\} \end{cases} \quad (4)$$

Case 2: Only the first element of the sequence will use binary to find the homing position and be sorted in case 2 and the rest of the elements in the sequence will be sorted in case 3. Then, for the m sequences, the average numbers of exchanges and comparisons are as follows:

$$\begin{cases} C_2 = \alpha_2 \left\{ \sum_{j=1}^{m} \left[\log_2 \frac{(j-1)n}{2m} + \frac{3n}{m} - 1 \right] \right\} \\ M_2 = \alpha_2 \left[\sum_{j=1}^{m} \left(\frac{jn}{4m} + \frac{3n}{4m} \right) \right] \end{cases} \quad (5)$$

Case 3: Because of the way of the algorithm, all elements of the sequence will be sorted in case 3. Then, for the m sequences, the average numbers of exchanges and comparisons are as follows:

$$\begin{cases} C_3 = \alpha_3 \left(\sum_{j=1}^{m} \frac{3n}{m} \right) \\ M_3 = \alpha_3 \left(\sum_{j=1}^{m} \frac{n}{m} \right) \end{cases} \quad (6)$$

Case 4: Because of symmetry of the algorithm, compared with the comparisons of the case 2, the first element of every element extra to compare with $b(n_1)$ and $b(n)$. Then, for the m sequences, the average numbers of exchanges and comparisons are as follows:

$$\begin{cases} C_4 = \alpha_4 \left\{ \sum_{j=1}^{m} \left[\log_2 \frac{(j-1)n}{2m} + \frac{3n}{m} + 1 \right] \right\} \\ M_4 = \alpha_4 \left[\sum_{j=1}^{m} \left(\frac{jn}{4m} + \frac{3n}{4m} \right) \right] \end{cases} \quad (7)$$

Case 5: Because of the symmetry of the algorithm, compared with the comparisons of the case 1, the first element of every element extra compares with $b(n_0)$,

$b(n_1)$ and $b(n)$. Then, for the m sequences, the average numbers of exchanges and comparisons are as follows:

$$\begin{cases} C_5 = \alpha_5 \left[\sum_{j=1}^{m} \left(\frac{7n}{2m} + \frac{1}{2} \right) \right] \\ M_5 = \alpha_5 \left\{ \sum_{j=1}^{m} \left[\frac{jn}{4m} \left(\frac{n}{m} + 1 \right) + \frac{n}{4m} - \frac{1}{2} \right] \right\} \end{cases} \quad (8)$$

In conclusion, the average numbers of comparisons and exchanges of the algorithm are as follows:

$$\begin{cases} C = m \left(\sum_{k=1}^{5} C_k \right) \\ M = m \left(\sum_{k=1}^{5} M_k \right) \end{cases} \quad (9)$$

By the equations (3)~(8), the equations (9) can be replaced as follows:

$$\begin{cases} C = \frac{\alpha(17n - m)}{2} + (1 - 3\alpha) \left[m \log_2 \frac{n}{2m} + \log_2(m!) + 3n \right] \\ M = \frac{1}{8} \left[mn + 7n - \alpha \left(2n^2 + \frac{n^2}{m} - mn - 7n - 8m \right) \right] \end{cases} \quad (10)$$

Among them, α is equal to $\frac{1}{m} \left(\sum_{i=1}^{m-1} \frac{1}{i+1} \right)$.

When n is large and m is relatively large, $\alpha \approx \frac{\ln m}{m} \approx 0$, $\log_2(m!) \approx \log_2 \left[\sqrt{2\pi m} \left(\frac{m}{e} \right)^m \right]$. Then the main item of C is about $3n + m \log_2 m$, and the main item of M is about $\frac{mn}{8}$.

3.3.2 Stability analysis

Sorting stability is a process in which records, which have the same element, stay in the same relative order during the sort (Biggar P & Gregg D 2005).

When sorting, the algorithm uses the binary search for homing position, two-way passed and directed homing to sort the element in any case. During the process, the binary search for homing position is stable. When discussing the element with five cases, set the data ranges of all cases to be left closed-interval and right open-interval, ensure that the relative position of the same or equivalent elements never changes. That is, the algorithm is stable. We denote the above algorithm is SQSA for short.

4 TESTING AND DISCUSSING

We test the sorting efficiency of the SQSA for the fragment-continuity-sequential data by FORTRAN language in the system of the Win7 64-bit and 4G of the memory.

Use the random number generator to generate the integer, which is the size of the sequence, and then generate part ascending and descending sequences.

Table 1. The time taken of sorts for fragment-continuity-sequential-data (Unit: ms).

Size of data (n)	Bubble sort	Selection sort	Insertion sort	Shell's sort	Heap sort	SQSA
50000	4087.30	4229.12	851.80	3.51	11.78	0.48
100000	21104.90	24473.95	10404.65	8.05	26.10	0.65
500000	...	...	...	43.25	149.65	3.15
1000000	...	...	...	82.55	314.95	7.45

Combine these sequences into different double precision real fragment-continuity-sequential data. Test the elementary sorts, Shell's sort, Heap sort and SQSA. See Table 1, which is the result of the test (for each n, test 20 times and take an average value).

Because of the sequences of the data, which is a disadvantage to Quick Sort, we do not list its time taken in the Table 1.

As seen in Table 1: For the fragment-continuity-sequential data, the efficiency of the SQSA is much higher than that of elementary sorts, is about 10 times as high as that of Shell's sort and about 37 times as high as that of Heap sort.

5 CONCLUSIONS

Considering the characteristics of the fragment-continuity-sequential data, the SQSA is given by using binary search for homing position, two-way passed and directed homing to sort.

In ensuring the stability premise, SQSA adapts well to the characteristics of the fragment-continuity-sequential data and reduces the numbers of comparisons and exchanges. Compared with common sorts, it has a better efficiency for fragment-continuity-sequential data.

REFERENCES

Babu DR, Shankar RS, Kumar V P, et al (2011). Array-indexed sorting algorithm for natural numbers[C]// Communication Software and Networks (ICCSN), 2011 IEEE 3rd International Conference on. IEEE, 606–609.

Biggar P& Gregg D (2005). Sorting in the presence of branch prediction and caches [R]. Technical Report TCD-CS-2005-57 Department of Computer Science, University of Dublin, Trinity College, Dublin 2, Ireland.

Dechao W (2012). Analysis and Comparison of the Common Sorting Algorithms [J]. Modern Computer, (13):7–9.

Dengkai L, Shu M (1998). Computer Algorithm Design and Analysis [M]. China Railway Publishing House, 76–78.

Dongarra J, Sullivan F (2000). Guest editors' introduction: the top 10 algorithms [J]. Computing in Science & Engineering, 2(1): 22–23.

Hoare CAR (1962). Quicksort [J]. The Computer Journal, 5(1): 10–16.

Hongwei H, Jin XU (2002). A Study on Quicksort Algorithm [J]. Microelectronics computer, (6): 6–9.

Huiqin Z (2009). Analysis and Comparison of Commonly Used Sequencing Algorithms [J]. Computer Development & Applications, 22(6): 67–69.

Lee E A, Messerschmitt DG (1987). Synchronous data flow [J]. Proceedings of the IEEE, 75(9): 1235–1245.

Murthy P K, Bhattacharyya S (2001). Shared buffer implementations of signal processing systems using lifetime analysis techniques [J]. Computer-Aided Design of Integrated Circuits and Systems, IEEE Transactions on, 20(2): 177–198.

Sedgewick R (1996). Analysis of Shellsort and related algorithms [M]//Algorithms—ESA'96. Springer Berlin Heidelberg, 1–11.

Shell DL (1959). A high-speed sorting procedure [J]. Communications of the ACM, 2(7): 30–32.

Williams JWJ (1964). Algorithm-232-Heapsort [J]. Communications of the ACM, 7(6): 347–348.

Zhi X (1989). The Realization of Using Binary Algorithm to Improve Speed of Search [J]. Application Research of Computers, 1: 10–11.

Electronics, Information Technology and Intellectualization – Song & Kwak (Eds)
© 2015 Taylor & Francis Group, London, ISBN 978-1-138-02741-1

The design and implementation of network alarm and data fusion analysis systems based on cloud computing

Hongmin Li, Min Lu & Jianping Zhang
China Academy of Engineering Physics Institute of the Overall Project

Lin Huang
Computer College, SWUST, Mianyang, China

ABSTRACT: The current security experts focus on a difficult study which is the process-reproduce of the whole network attack by linking the independent pieces of infinite information from various kinds of security equipment. Based on the above requirements, this paper researches the analysis techniques of dealing with large scales of logs from different kinds of equipment, and constructs the hierarchical framework to fuse the pieces of information from the data layer to the feature level and the decision level. In the end, this paper implements alarm data analysis systems to verify these techniques.

Keywords: Cloud computing, data fusion, secret-related network

1   INTRODUCTION

With the rapid advance of information technology and the continuous development of information technology, the network has become an important platform for researching, production, and office working. As the network has some features, such as opening, interconnected and sharing etc., so the risk of being invaded is becoming serious. With hacker activities becoming more frequent, website backdoor, phishing, malicious programs, denial of service attacks have had a substantial growth, and APT attacks have become more frequently. The information of Secret-Related Network systems are facing serious challenges. In order to ensure the security of information systems, military networks deploy a large number of network security devices (such as firewalls, IDS), host monitoring systems, applications, and auditing for enhanced network security and network security audits. At the same time, these devices and systems produce a large number of log information which are different in structure and independent from each other and which cannot represent a full attack. These logs have only recorded some fragments of an attack, so it is important to know how to link these pieces of information to reproduce the entire network attack process which is currently the focus of the network security situation and is difficult to research. Based on these requirements, this paper designs and implements a cloud-based alarm data analysis system which is based on researching the multi-sources and massive data processing of log analysis techniques, constructing a hierarchical and data fusion framework.

2   KEY TECHNOLOGY RESEARCH

2.1   *Preprocessing techniques of massive logs*

Firstly the logs of firewalls, IDSs, host monitoring systems for multi-sources, need to do fusion analysis, log data for centralized collection and to be stored in the log collection server; then there is a to start data push service (optional wee small online traffic) and files of logs data should be pushed to Hadoop platform; finally the log files are written in HDFS cluster. For firewalls, intrusion detection systems and other equipment, the

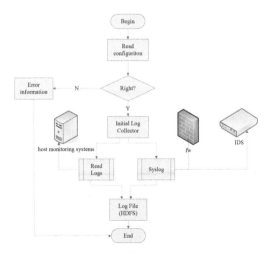

Figure 1.   The process of log data generation.

29

Table 1. All kinds of correspondence between log attributes and their meaning.

attributes	meaning	Remark
ID	Log ID	\
dev_type	Device Type	Firewall is expressed as fw, An Intrusion detection system is expressed as ids, and host monitoring system is expressed as hm, and then add the device number to consist device types, such as fw001, ids001, hm001 respectively.
event type	Event Type	Firewall configuration management class, access control class, connection class, etc., IDS itself defined event type, ftp, http, smtp, telnet, usb, print and other categories of host monitoring system.
priority	Event priority	Divided into eight: critical emergency (0), alarm alert (1), severely critical (2), error (3), warning (4), prompt notice (5), information (6), debug.
user	Login Username	\
src_ip	Source IP	\
src_port	Source port	\
dst_ip	Destination of IP	\
dst_port	Destination port	\
proto	Protocol type	\
op	Operation Command	\
time	Time of events recorded	\
result	Event results	Into success, failure.
msg	Event-related Information	\
inpkt	Number of received packets	\
outpkt	Send packets	\
sent	Send bytes	\
rcvd	Receive bytes	\
pc_name	Computer Name	\
pc_ip	Computer ip	\

log information is generated by the equipment and can be pushed through the way of the syslog protocol, log collection server use UDP or TCP port to listen to the log collection. For syslog host monitoring system which does not provide a way to push the log, you need to read the host surveillance system server log, and save it as a file and push it to the server. The process of log data generation is shown in Figure 1.

After the study of log data formats of firewalls, IDSs, and host monitoring systems, the logs are divided into four categories (i.e., management configuration exception classes, traffic anomaly classes, illegal operations, security attack classes) and normalized.

(1) Management Configuration Exception Classes, logs of Management Configuration, and exception classes are defined as logs extracted and analysed from management configuration related logs of firewalls, IDSs, and host monitoring systems, in order to find unusual configuration management operations. These logs can be normalized in the following formats after studying them:
LogManager (ID, dev_type, event_type priority, user, src_ip, op, time, result, msg)

(2) Traffic Anomaly Class
The size of each session sent and received, which is recorded in firewall logs, can be used to analyse the anomaly of traffic. Such logs will be normalized in the following formats:
LogFlow (ID, dev_type, event_type, priority, src_ip, src_port, dst_ip, dst_port, time, proto, inpkt, outpkt, sent, rcvd)

(3) Illegal Operations
Mainly through illegal logging host monitoring system to analyse the illegal operations, standardization of such logs like the following formats:
Log illegal (ID, dev_type, event type, user, pc, name, pc_ip, time, msg)

(4) Security Attacks Classes
A comprehensive analysis of firewall, IDS, host monitoring system logs can be used to find three categories of potential security attacks, which mainly related to the control class logs of firewall access control class logs, IDS logs and host monitoring, as follows standardized format:
LogSec (ID, dev_type, event_type, priority, src_ip, src_port, dst_ip, dst_port, time, proto)

Meaning of each attribute as indicated in Table 1.

In this paper, a HDFS file system is used to store raw logs of firewall, IDS, and host monitoring systems, and sets the log collection server immediately to transfer the log when detecting firewall, IDS, host monitoring system generates a log, in order to avoid a malicious hacker deleting the original log. Example of HDFS file block storage shown in Figure 2.

Configuration class fusion analysis process.

HDFS file blocks from the store can be seen in the example of Figure 2. Backup numbers of raw logs of the host Monitoring System (hm.log) is 3, are stored in DataNode 1, the DataNode 2, DataNode 4, and three nodes; backup number of firewall raw logs (fw.log) is 2, are stored in DataNode1, the DataNode 3 with two nodes; backup number of IDS raw logs (ids.log) is 2, are stored in DataNode 3, the DataNode 4 has two

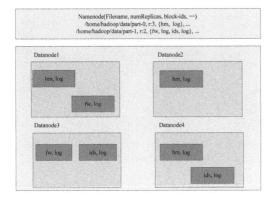

Figure 2. HDFS file block storage.

nodes. These information files nodes are recorded in NameNode, when one node fails, NameNode will read data from another node, in order to avoid data loss or corruption after a single point's failure.

2.2 Methods and process design of network alarm fusion analysis

Network Alarm fusion analysis uses the massive multi-source logs data of firewall, IDS, host monitoring system, with the help of the proposed rule strategy, combined with asset information, vulnerability information and other information associated with a comprehensive analysis of knowledge, through the integration of analysis to determine whether the real network attacks occur. Corresponding to the four types of logs designing four rules of standardization strategy, this article divides the true alert into four categories, namely: configuration management class alarms, traffic anomaly class alarms, illegal operations class alarms, and security attack class alarms. Like configuration management class alarms, the integration of analysis and process is designed as follows:

Using configuration management class rules to analyse standardized logs of firewall configuration class, you can effectively analyse and manage the IP which is illegally and logs of users, and to generate alarm information for users, to alert users the presence of illegal IP and users 'operation of firewall configuration'. Fusion analysis flow chart of Alarm Management Configuration class is shown in Figure 3:

① Stream of reading configuration management class from HDFS.
② Reading the log file stream line by line, according to the configuration management class rules match log file stream. Seeing if there exists a successful match with the rules, and the time of log occurs in normal working hours, then the log is normal behaviour. This is the end of this log analysis, determining the number of logs and if the log is more than 0, enter ②, otherwise enter ④; If the rule matches unsuccessfully, proceed to step ③.

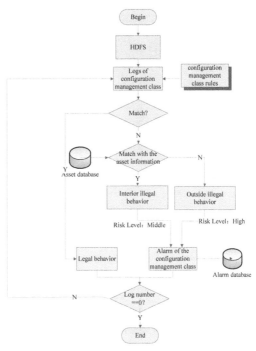

Figure 3. Alarm management.

③ Match with the asset library. If the match is successful, then the log from internal staff who manage the firewall configuration, belong to an internal management acts, marking its risk level as 'medium', and record in the configuration management alarm class database. This is the end of this entry log analysis, judging the number of logs, if the log number is more than 0, enter ②, and otherwise enter ④. If the match is unsuccessful, then the log management configuration is from the outside, belongs to the external ultra virus management behaviour, marking its risk level as 'high', record in configuration management alarm class database, then the end of this entry log, adjudging the number of logs, the log is more than 0, enter ②, otherwise enter ④.
④ Management Configuration class Alarm fusion analysis ends.

3 SYSTEM ARCHITECTURE DESIGN

3.1 Overall architecture and service levels

Hadoop-based network analysis system alarm uses the log collection server and pushes logs data of the network security devices to the Hadoop platform and then with the help of the Hadoop platform stores secures, and pretreats aggregate fusion analysis logs, and lastly finds the network's abnormal behaviour, aggressive behaviour, illegal behaviour and sends it to the alarm monitoring centre visual display. The system has the following characteristics:

Figure 4.　Overall system. Architecture diagram.

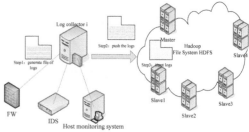

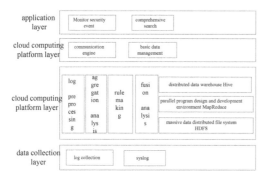

Figure 5.　Service-level system diagram.

1. With the increasing growth trend of network security device logs the size and the quantity produced, the system uses Hadoop platform to build a private cloud for data storage and raw log data can be safely and reliably stored.
2. The system uses Hadoop platforms to analysis and process logs, and to effectively improve the efficiency of the system of analysis and processing.
3. The system sets up an alarm monitoring centre, so that the user can draw real-time monitoring of security events from log analysis.

Figure 4 shows that the overall architecture diagram of alarm fusion analysis system based on Hadoop. The system is divided into three parts: the log collection terminal, log analysis, and the processing centre. The log collection system is responsible for collecting the logs of the firewalls, the intrusion detection systems, and the host monitoring systems and push logs to Hadoop platform. The log analysis processing centre is mainly responsible for the secure storage of raw logs, logs pretreatment, log aggregation, alarm fusion analysis, the generation of real attack alarms, and sends them to the alarm monitoring centre. The alarm monitoring centre is mainly for administrators to directly view the current exception, attacks, violations and other events.

(1) Data collection layer

Data collection layer is the data source of the whole system, because the log collection server unifies collection of all safety equipment and services focused on starting of data push, and finally writes the log file to the cluster in HDFS.

(2) Cloud computing platform layer

Hadoop, as a support, is the data processing centre of the whole system, the use of HDFSs to store

amounts of log data, ensuring reliable storage and parallel reading and writing; utilizing MapReduce parallel processing mechanism, the system provides a powerful data computing power; a cloud computing platform provides a series of application interface layers, in order to improve efficiency and stabilize data persistence support for the entire system.

(3) Adaptation Layer

Hadoop-based network alarm integration, and the analysis system adaptation layer, which is between the application layer and the server clusters, provides management and services for the entire system, and provides standardized interface and protocol for the application layer. The interactive communication engine and transmission of the entire system is responsible for data flow and control of flow. Basis data management is used for controlling the entire agent child node and object configuration.

(4) Application Layer

J2EE standard application layer using JSP, HTML, SSH (struts2 + spring + hibernate) technology, combined with powerful search function Hive which provides massive data, and provides a visual representation of the entire alarm system.

3.2　Design and implementation of log collection module

This paper uses a logs collection system for log data's collection. Log data collection program consists of three steps: generate log data file (firewall, IDS, host monitoring system); push the log data; log data is written to an HDFS. The specific process is shown in Figure 6.

Figure 6 depicts the flow of the log data's collection, firstly gathering logs from the firewall, the IDS, and the host monitoring system and storing it on the log collection server; then data (optional smaller online traffic morning) will be pushed to the log data file Hadoop platform and finally written to the log file in an HDFS cluster. Among them, the key code is as follows:

```
String target="hdfs://10.11.1.136:9000/home/hadoop/
data/orig_log/origfw1.txt";
FileinputStream fis=new FileinputStream (new File
("d:\\fw001.txt"));
```

Figure 6.　Log collection process.

Figure 7.   Illegal operations.

Configuration conf=new Configuration ();
File System fs=FileSystem.get (URI. Create (target), conf);
Output stream os=fs. Create (new Path (target));
IOUtils. Copy Bytes (fis, os, 4096, true);

Among Code, 10.11.1.136 is HDFS's address, origfw1.txt is original firewall log's location in Hadoop, and fw001.txt is a local file's location. Through these key parts of the code, logs can be pushed to the local file and stored in HDFS.

3.3   Design and implementation of alarm monitoring centre

After fusion analysis, the exception, illegal security incidents are extracted and stored in a MySQL database, foreground using J2EE specification JSP, HTML, SSH (struts2 + spring + hibernate) technology, combined with HIVE which provides powerful ability of data searching for the entire system alarm visual presentation.

Hive related data Example:

(1) Hive Import Map / Reduce operating data:
    LOAD DATA INPATH '/user/hduser/bin-output/part-r-00000' INTO TABLE tablename;
(2) Hive establish a connection with MySQL:
    connToHive=DriverManager.get   Connection ("jdbc:hive://10.11.1.136:10000/default", "hive", "MySql"); / / Get connection with the Hive
    connToMySQL=DriverManager.get Connection ("jdbc:mysql://10.11.1.136:3306/hive?use Unicode= true & character Encoding=UTF8", "root", "MySql"); / / Connect to MySQL
(3) Enter the HQL query:
    Human-computer interaction will analyse the results and show them to the administrator, which is good for the administrator direct control the current security status of the network such as regularities class. All the illegal information in host monitoring system will be screened and show a list of events in the category of illegal operations, the administrator can query multiple conditions filtration, filter out illegal operations concerned by the administrator. Illegal class event interface designed as shown in Figure 7.

4   CONCLUSIONS

The system's design is complete, and its validity and effectiveness in a classified network environment needs to be verified and revised to improve its design.

REFERENCES

DS theory classifier based on active alarm verification. HE Guang-yu, WEN Ying, Zhao, "computer engineering", 2009.02

Fan G., Jehu, Y. and Min, Y. "Design and implementation of a distributed ids alert aggregation model," in Computer Science Education, 2009. ICCSE'09. 4th International Conference on, 2009. 975–980

Han Jing Ling, Sun Min. Intrusion detection and alarm information fusion system construction and implementation of Computer Technology and evelopment, 2007 (6):159–162

Hofmann A. and Sick B. "Online intrusion alert aggregation with generative data stream modeling,"Dependable and Secure Computing, IEEE Transactions on, vol. 8, no. 2, 2011. 282–294

Intrusion detection alarm system design and implementation of the module validation. Left crystal, a new paragraph sea, in Shirley, "computer engineering", 2008.01

Jeong Jin Cheon and Tae-Young Choe. "Distributed Processing of Snort Alert Log using Hadoop", IJET, Vol. 5, No. 3.:2685–2690, 2013

Kruegel C, Robertson W. Alert Verification: Determining the Success of Intrusion Attempts [C]. Proceedings of the 1st Workshop on Detection of Intrusions and Malware & Vulnerability Assessment. Germany, 2004-07

National Computer Network Emergency Response Technical Team Coordination Center China Internet Network Security Report. CNCERT/CC.2012 years, Beijing, 2013: 13–17

Robert Richardson. 2010/2011 CSI Computer Crime and Security Survey. 2011

Valdes, A. and Skinner, K. Adaptive, Model-Based Monitoring for Cyber Attack Detection, RAID 2000 Conf. Oct. 2000. 80–92.

Wang Jing Xin, Wang Zhiying, DAI Kui-based IDS alarm validation study of multi-source security information, Computer Applications, 2007 Vol. 27 (8): 1910–1912

Wen, S., Xiang, Y. and Zhou, W. "A lightweight intrusion alert fusion system," in High Performance Computing and Communications (HPCC), 2010 12th IEEE International Conference on, 2010. 695–700

Wenjie Xu, etc. Application of Bayesian Network in Information Fusion Analysis of Four Diagnostic Methods of Traditional Chinese Medicine. 2010 IEEE International Conference on Bioinformatics and Biomedicine Workshops. 2010 IEEE. 694–697

Zang, T., Yun, X. and Zhang, Y. "A survey of alert fusion techniques for security incident," in Web-Age Information Management, 2008. WAIM'08. The Ninth International Conference on, 2008. 475–481

Zhang Ge, LEI Ying-jie, Xue Mei, Ann peace. Intuitive application of fuzzy comprehensive evaluation in a multi-source verification of alarms. Computer Systems 2011.07

Zhihong, T., Baoshan, Q., Jianwei, Y. and Hongli, Z. "Alert-clu: A Realtime alert aggregation and correlation system," in Cyber worlds, 2008 International Conference on, 2008. 778–781

Electronics, Information Technology and Intellectualization – Song & Kwak (Eds)
© 2015 Taylor & Francis Group, London, ISBN 978-1-138-02741-1

The development of an intelligentized information platform for tourism landscape management using GIS and GPS technology

K. Chang
National Chiayi University, Chiayi, Taiwan

H.I. Chen
Chinese Culture University, Taipei, Taiwan

ABSTRACT: Tourism destinations provide a wide spectrum of services such as accommodation, food services, recreation activities, and conservation. For those services, tourism landscapes often need to build complex systems of infrastructures and facilities such as water systems and sewage systems as well as handling data from various professions such as site surveyors, visitor survey data, ecological data, and construction projects. To make informed decisions, the site managers need to put those multiple factors together into consideration. However, the information of the service systems is often disconnected, incompatible, and difficult to use. Therefore, the intent of this study was to develop a mechanism that includes materials, tools, and processes for the establishment of a Web-based, visualized, and integrated information platform for tourism landscape management.

1 INTRODUCTION

Sustainable tourism settings, such as state parks, not only provide the public with recreation, but are also places that conserve treasured culture and nature resources. Sustainable tourism settings serve multiple purposes encompassing the demonstration of the force of nature, education, science, and exceptional tourism opportunities. The sizes of sustainable tourism settings vary. A middle size one may be approximately 8000 hectares in surface area.

To successfully serve such a wide spectrum of missions, the site manager often needs to make informed decisions among alternatives. Often, tourism destination management can be location sensitive. For managerial questions, such as, where should I place the new nature centre so that this development can be suitable for the local soil structure and not within an ecological sensitive area to raise the sustainability level of this development?; where are the campsites which have a full service sewage that I need to change in accordance with new environmental regulations? Those questions show the vital needs for an information platform. This tool development that is user-friendly and powerful for professional decision-making, deserves study and development.

A GIS is a computer system that manages data with a geographic component, including satellite data, aerial photographs, digital maps, and other sources in an automated system that supports data analysis, modelling, and display. Decisions in tourism landscape management are location sensitive, information sensitive, and require integrated readings of information to include the interactions among different information sets. Thus, this study uses the framework of a geographic information system and is assisted by a global positioning system to develop a Web-based integrated platform for tourism site management.

However, in current conventional practices, integrated information reading is limited by the incompatible data formats and carrying software among datasets. Management data can include data from larger scale of planning, smaller scale design details, facility locations, amenity features, ecological data, and visitor survey data. Each of those data sets often requires specialized software such as AutoCAD, ArcGIS, or ERDAS IMAGINE for reading or editing. Although this professional software is powerful in its functions, long term training, consistent practice, and costly purchase become the difficult prerequisites before site managers can conduct simple and sound information reading for their judgments.

Therefore, as an application of intelligentization, this study developed an information platform that was intended to explore a solution for professional information reading and retrieving in a user-friendly and informative platform for sustainable tourism site management.

2 STUDY METHOD

2.1 Data system design

Field investigations and interviews with site managers were conducted for data system design. A thirteen category data dictionary was designed for location and detailed feature data collection using GPS, including sewer systems (e.g. sewage treatment, sewer lift station), water system (e.g. water line, water valve), electric system (e.g. primary and secondary line and pylon), telecommunication cable system (e.g. telecom line, telephone), transportation system (e.g. road, parking spaces), roofed structure (e.g. building, roofed with no wall), land facility (e.g. campsite, playground, sign), golf course (e.g. cart path, yard mark), lake facility (e.g. boat ramp, dock), nature feature (e.g. stream, sensitive area), and administrative system (e.g. destination boundary, ownership), and management system (abandoned object, issue).

2.2 Data collection

The field data, image data, management data from the office, and other database e.g. governmental census were matched and corrected into one geodatabase.

In addition, based on the thirteen-category data dictionary, location and feature data were collected in two Trimble Geo 6000 GPS units. Prior to field data collection, the data dictionary was established in the desktop GPS management software, Pathfinder Office and then transported into database creation software-TerraSync in the GPS units. After GPS setting refinement, the researchers began to conduct data collection throughout twenty state parks to complete data collection. In addition, underground facilities, for example, water lines and sewer lines were collected based on the existing AutoCAD files in the site management office.

2.3 Database establishment

The data in the AutoCAD file format-dwg were post-processed through the techniques – georeferencing and correction algorithms – to fit the set projection before being entered into the geodatabase. The image data in the database include aerial photographs and Lidar (Light Detection and Ranging) imagery. The aerial photos were compressed into ECW (Enhanced Compression Wavelet) format. ECW format compression allows very large images with small RAM requirement in the computing environment. The compression ratios can be from 1:2–1:100. This compression can accelerate very large image performance through fine alterations while retaining their viewing quality. After the ECW compression, a Chche technique was conducted to improve the speed of the performance of multiple large images in a www environment. The Lidar data were processed and transformed into 1 metre contour lines in the geodatabases.

The field data were transformed from the format of GPS data-ssf to the geodatabase format-ship. Because the field data may contain noise data caused by the signal interruption from elevated obstacles, inspections and corrections were made for data quality.

2.4 Server application

ArcGIS Server was used as the server management tool for information publication and interactive computing on the internet. ArcGIS Server was used to integrate different data types (Nasser, 2014). The ArcSDE and SQL servers were set and provided the communication channels between geodatabase in ArcMap and ArcGIS server. The geodatabase requires a save in a shared directory and an importation into ArcSDE. In the interface of ArcGIS server manager, a map document file (mxd) is placed into a new service (Henrich & Luedecke 2007).

2.5 Interface design and platform

The interface design and platform construction used a connected set of tools, ArcGIS Viewer for Flex and ArcGIS API for Flex. The ArcGIS Viewer for Flex is programed as a framework while ArcGIS API for Flex requires advanced programing for customized operation and flexibility such as editing, geocoding, and routing (ESRI, 2013). In this study, the research team combined these two tools for developments. The functionality widgets were also programed or managed to retrieve specific information in the geodatabase. In addition, the fine resolution aerial imagery was processed using Cache technique to pre-catch the imagery in 24 scales from approximate 1:70 to 1:590000000.

Based on the comments from pilot user tests, the study team then setup suitable feature controls for each layer including label, type, and visibility to make judgments on look and feel features of the interface (van Elzakker & Wealands 2007).

2.6 Function installation

Tools and Functions: Figure 1 presents an overview of the viewer site with highlighted tool groups. These tool groups contain functions of information delivery for different aspects of park management and planning.

The following describe the interactive functions of the platform interface that includes:

Overview: 'Navigation bar', 'Base map slide bar', 'Area view', 'Interactive coordinates', and 'Function bar'.

Function bar: 'Quick Locate', 'Table of Contents', 'Bookmarks', 'Locate', 'Draw and Measure', 'Identify', 'Search Records', 'Chart-Age', and 'Print'.

Navigation bar: the Navigation bar is placed in the left side of the website, allowing zooming in, zooming out, pan, and change viewing scale.

Base map slide bar: the Base map slide bar is placed at the right upper corner of the website, allowing shifting base maps between street map, aerial photo, and topographic map. The slide option gives the control of transparency level when viewing.

Area view: an Area view tool is placed at the right lower corner of the website, showing the relative location of the current view. The users can choose to show or hide this tool by clicking the arrow sign at the corner of this tool.

Interactive coordinates: the tool for interactive coordinates is placed at the left bottom of the website. The users can obtain automatically generated coordinates in latitude and longitude with the cursor movement.

Function bar: the Function bar contains a series of tools for acquiring information on the Viewer site: (Ordering from left to right of the function bar on the Viewer):

'Quick Locate': this tool brings the specified site to the centre of the screen by entering the park name.

'Table of Contents': this tool provides the control of turning on or off layers, viewing layer legends, and managing layer visibility.

'Bookmarks': using this tool, the users can setup their own quick links to the extent of their desired or unfinished project site.

'Locate': this tool assists the user to find places by entering either address or coordinates of longitude and latitude.

'Draw and Measure': using this tool, the user can mark and/or measure length, perimeter, and area for the target object with the choice of various drawing options including line, freehand line, rectangle, circle, polygon, freehand polygon, and inset text. Also, this tool provides the users full control of colour choice, style, transparency level, font, and units.

'Identify': with the selection tools of point, line, and polygon, the user can obtain the information for the target objects such as elevation and facility attributes. The result tag within this tool allows the users to view complete records and photos of the object selected.

'Search Records': this tool provides the option to filter records by attributes. By choosing the target facility or infrastructure and formulating the user-defined criteria, the user can assess the records of objects from a database in order to learn about the details such as their locations, numbers, and their full records.

'Chart-Age': this sample tool shows the age composition of the interested areas and presents the information in a pie-chart for intuitive comprehension.

'Print': this tool provides the user options to export and print the view, a marked map, or search results on screen, to other formats, such as Pdf files, for communication with others or for the user-defined reference.

3 DISCUSSION AND CONCLUSION

Followed by the improvement of the ability to produce information, the level of data complexity and

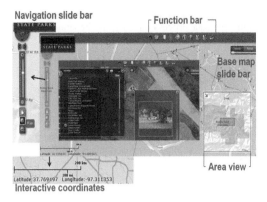

Figure 1. The interface overview of the information platform with highlighted tool groups.

abundance continuously increase with innovations and developments, such as the Lidar technology, remote sensing data, and weather data. However, from a management perspective, those essential and sophisticated information systems are often very difficult to use for their management tasks on a daily basis. The principal reasons can be insufficient access to the data and steep curing curve that requires using different professional software. This software was made by various companies and is difficult to communicate among information sets. Thus, the purpose of this study was to use GIS framework to establish a user-friendly information platform that can integrate various types of information in one web-based platform with friendly and fast access to facilitate the process of making informed decisions for tourism landscape management using intelligent information.

In the development of a Web-based GIS for fulfilling professionals' constant requests for spatial and operational information in order to sustain effective asset management, a development of a Web-based integrated information platform took place. Data for asset management, ecological data, political boundary data, site operational data, site infrastructure data of various types such as CAD drawings, GPS data, Excel, Access, Erdas Imaging data, raster images, and an ArcGIS database, were all processed into a single Web-based platform that can be deployed via the internet with interactive functionalities and visual effects.

In such a process, the study team found that the geographical platform framework can be a suitable base for establishing an information platform for tourism landscape management. The visual and location specific characteristics are the key factors to facilitate and promote the use for data. The research team also suggests that, when designing the interface, the look and feel factors play an important role for later use frequency (Berry 2005). In addition, although the visualized effect is one of keys for use, the very large image data can make the response speed become intolerable to the users. Thus, for developers who intend to establish such platform that needs to process very large geographic information and to make that information visualizable, the use of multiple techniques to accelerate the data process speed must be considered in development preparation.

REFERENCES

Berry, D. (2005, Feb 09). IBM: The user experience. Retrieved January, 5, 2009, from http://www.ibm.com/developerworks/web/library/w-berry/

ESRI, (2013, January). ArcGIS API for Flex. Retrieved June, 30, 2014, from https://developers.arcgis.com/flex/

ESRI, (2013, January). ArcGIS Viewer for Flex. Retrieved June 30, 2014, from http://resources.arcgis.com/en/communities/flex-viewer

Haklay, M., & Zafiri, A. (2008). Usability engineering for GIS: Learning from a screenshot. The Cartographic Journal, 45(2), 87–97.

Henrich, A., & Luedecke, V. (2007). Characteristics of geographic information needs. Paper presented at the Proceedings of the 4th ACM workshop on Geographical information, Lisbon, Portugal.

Nasser, H., (2014). Administering ArcGIS for Server Installing and configuring ArcGIS for server to publish, optimize, and secure GIS services. Birmingham, Packt Publishing Limited.

van Elzakker, C. P. J. M., & Wealands, K. (2007). Chapter 34: Use and users of multimedia cartography. In W. Cartwright, M. Peterson & G. Gartner (Eds.), Multimedia cartography. New York: Springer.

Electronics, Information Technology and Intellectualization – Song & Kwak (Eds)
© 2015 Taylor & Francis Group, London, ISBN 978-1-138-02741-1

The creation of an intelligent index system, taking Gongqing Digital Eco-City as an example

Xiaoqiong Li
Poyang Lake Eco-economy Research Center & Civil Engineering and Construction Institute of Jiujiang University, Jiujiang, Jiangxi Province, China

Wei Zhang
Jiujiang Branch of 707 Institutes of China Shipping Industry Corporation, Jiujiang, Jiangxi Province, China

Chunyuan Tao
Poyang Lake Eco-economy Research Center of Jiujiang University, Jiujiang, Jiangxi Province, China

Yingxiang Wang
Gongqing Management and Investment Ltd, Jiujiang, Jiangxi Province, China

ABSTRACT: The objective of creating the index system of Digital Eco-City is used for putting the Digital Eco-City ideals on a practicable level and guiding and evaluating the process and level of Digit-Eco-City's construction. It elaborates the rules, methods, and the framework of the index system. According to the convenience of implementing urban planning, Figure1 shows that the index system divides into two categories: the index of digital and ecological planning and the index of digital and ecological control. Index of digital and ecological planning includes 5 aspects, 3 index classes, and 67 detailed indices. Index of digital and ecological control includes 19 aspects, 2 index classes, and 162 detailed indices. It gives the answer that Gongqing Digital Eco-City has combined digital techniques, ecological techniques, and local features. So the index system of Gongqing Digital Eco-City has a clear index framework, detailed index numbers, and superior operability.

Keywords: Intelligent; Eco-City; Digital City

1 INTRODUCTION

From the perspective of city construction in human history, cities were green and ecological when they were only trade and political centres before they became industrial cities. However, three hundred years of industrialization has made green cities into grey cities. Although only 50% of world population lives in cities, 85% of the world's resources and energy are consumed there, and 85% of the world's wastes and carbon dioxides are discharged in cities [1]. Most dreams and masterpieces of humans are created in cities, but they become the most powerful weapons on the earth. The Digital Eco-City is based on energy conservation, environmental protection and sustainable development and new technologies that are created by the digital revolution are utilized to create a better life. The tight structure of a city can reduce the negative impacts of overcoming change in the macroeconomic conditions of society [2]. The combination of digital technology and ecological technology makes the Digit-Eco-City forefront in the development direction of an intelligent city, and is an inevitable choice in the course of human history.

How to construct and evaluate a Digital Eco-City has become one of the biggest problems.

2 OVERVIEW OF SINO-FINLAND DIGIECOCITY

The Digital Eco-City is located in the southwest of Gongqing Development Zone of Jiangxi province; the total land is 524.48 hectares and the total construction land covers 471.14 hectares with a population of 60000. At the beginning of city planning, digital infrastructure design and ecological design were integrated; the intelligent concept was adopted in every detail, which was the biggest difference. In the Digital Eco-City, digital services will become more and more popular in daily life. Digital services will reduce the size of buildings, for example, electronic banking and electronic health care reduce the reliance on banks and hospitals. Mobile technologies and remote office technologies also reduce the need for travel. Compared with flow of people and goods, electronic data have a smaller carbon footprint, therefore, intelligent design of digital services will also help to create an intelligent city. Digital Eco-City emphasizes more on the combination of ecological city and digital city in technology, this makes information technologies serve the construction of ecological city and city life in the future, and in this way, people can get convenience of digital city and a good environment [7].

3 THE CREATION OF AN INTELLIGENT INDEX SYSTEM FOR THE DIGIECOCITY

The goal of building an intelligent ecological index system for the Digital Eco-City is applying the concept of Digital Eco-City in specific operation, and using a comprehensive and reasonable index system to assess construction process and level of the ecological city [4]. Most of existing index systems in China are based on the index systems of the National Environmental Protection Department and the index system of China Cities Scientific Research Board, as well as research results in "The building of index system of liveable cities development in China and assessment results" by Eco-city demonstration evaluation group. Considering different geographical climate, city ecological base and levels of economic and social development, development and innovation in important areas of ecological city are encouraged [5].

3.1 The theory and method of the index system

According to the main content of the Digital Eco-City, conditions and features of Gongqing city and the construction of an intelligent ecological index system for the Digital Eco-City, are based on a complex ecosystem theory, "society-economy-nature" and digital information theory. Taking urban planning as a whole, the working system of urban planning includes two parts. They are making and are being implemented, each part having its own system. So the index system also includes planning and control categories. During the creating of an ecological index system, scientificity and operability need to be combined qualitatively and quantitatively. Accessibility and prospectiveness should also be combined. According to the above basic principles of building the intelligent ecological index system, the following methods, case reference methods, standard reference methods, index selection methods, clustering relationship methods, and conditions and characteristics analysis methods were mainly included [6].

3.2 The framework of the index system

According to the convenience of implementing urban planning, the index system was divided into two categories: the index of digital and ecological planning and the index of digital and ecological control. The index of digital and ecological planning included 5 aspects, 3 index classes, and 67 detailed indices. Figure 1 shows its framework. The index of digital and ecological control included 19 aspects, 2 index classes,

Figure 1. The framework of digital and ecological planning index system.

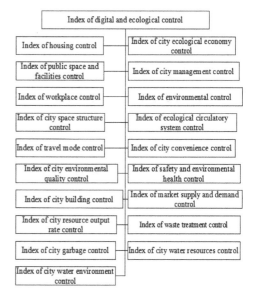

Figure 2. The framework of digital and ecological control index system.

and 162 detailed indices. The total detailed indices are 229. The framework is shown as Figure 2.

3.3 Key index for intelligent performance

DIGI concepts cover digitally integrated highly interactive information systems providing access to community services like health care, learning and trade, highly automated transport, logistics and building systems, and information systems to city management and service production [3]. The key index for intelligent performance is shown in table 1 and table 2.

4 CONCLUSIONS

Compared with other ecological and digital cities in China, the index system of Gongqing Digital

Table 1. Key index lists for intelligent performance.

Key performance indices	Definitions
Share of high speed IP address	IP address with download speed of more than 5 Mbps
3G/4G mobile phone users	The proportion of people using advanced mobile terminal
The percentage of the use of electronic government services	Percentage of population frequently using electronic government services
Schools with broadband access	The percentage of schools
Online business of enterprises	The percentage of enterprises
Electronic commerce	The percentage of turnover from electronic commerce
E-Commerce Applications	The proportion of people frequently using Internet business
The Internet and electronic bank	The proportion of adults using online banking
Telecommuting	The proportion of people who do not commute and work for over eight hours each day

Table 2. Target value of key index for intelligent performance.

Key performance indices	Target value
Share of high speed IP address	100% of families
3G/4G mobile phone users	(not including people using only voice terminal)
The percentage of the use of electronic government services	80% (not including laggards)
Schools with broadband access	100%
Online business of enterprises	90% of independent companies have websites (not including small and temporary enterprises); 60% of enterprises support electronic commerce (not including enterprises which do not depend on electronic commerce, such as salon service provider)
Electronic commerce	TBD uncertain (the figure changes every year, so it is not suitable to set a target)
E-Commerce Applications	80% (not including laggards)
The Internet and electronic bank	80% (not including laggards)
Telecommuting	20% of people who work outside Gongqing City

Eco-City enjoys great advantages which are suitable for resources characteristics in the region. The index system of Gongqing Digital Eco-City has four distinct characteristics. First: the integration of a digital technology index is the biggest difference between Gongqing Digital Eco-City and other domestic ecological cities, having the integration of digital infrastructure at the beginning of city planning, its purpose being to offer high quality life for local residents. Second: the index system has regional characteristics such as an index system containing integration degrees for the Communist Youth League, the culture of the Communist Youth League, a digital rate of the cultural resources of the Communist Youth League, an integration degree of Poyang Lake culture, and a digital rate of the cultural resources of Poyang Lake. Third: a clear classification of indices, based on planning and control management classification, which have great manoeuvrability. The number of indices is the largest in China and the index system is further refined on the basis of an ecological index systems from home and abroad, which offers technical support for assessment.

ACKNOWLEDGMENTS

This paper is the result of the Sino-Finland cooperation project 'Gongqing Digital Eco-City Urban Planning and Construction Technologies" which was approved by the Ministry of Science and Technology in December, 2009, the project number is 2010DFB90460.

REFERENCES

Chinese City Science Research Association, Report on China intelligent Ecological City's Development (2011), Beijing: China Architecture & Building Press, 2011.6, 162–168, 174–178, 181–186.

Digital Ecological City Limited Company (DigiEcoCity Ltd, abbreviated as "DEC"), Detailed Plan for Sino-Finland Gongqing Digital Ecological Industry Base, 2011.06, 9, 16–22, 35–38, 39–41.

Digital Ecological City Limited Company (DigiEcoCity Ltd, abbreviated as "DEC"), Conceptual Plan for Sino-Finland Gongqing Digital Ecological Industry Base, 2010.05, 3–5, 8–10.

Li Lin (editor), Guide on Construction of Digital City (1), Nanjing: Southeast University Press, 2010.3, 11–23, 44–47.

Li Lin (editor), Guide on Construction of Digital City (2), Nanjing: Southeast University press, 2010.10, 717–724, 1230–1231.

Wang Jiayao, Ning Lusheng, Zhang Zuxun (editors), Study on China Digital City Construction Scheme and Promotion Strategy, Beijing: Science Press, 2008.11, 3, 18–23, 45–46.

Zhang Quan, intelligent and Ecological City Planning, Beijing: China Architecture Industry Press, 2011.3, 130–142.

Electronics, Information Technology and Intellectualization – Song & Kwak (Eds)
© 2015 Taylor & Francis Group, London, ISBN 978-1-138-02741-1

Research on digital forensics and its relevant problems

X.G. Wang

Department of Computer Science, East China University of Political Science and Law, Shanghai, China

ABSTRACT: From the perspective of digital forensics, this article mainly discusses the forensic technology of electronic evidence, forensic standards of electronic evidence, and forensic procedures of electronic evidence. Furthermore, it analyses the existing problems of digital forensics and electronic evidence and relevant judicial problems in China. As well, it gives the relevant solution proposals at the end of this article.

Keywords: Digital Forensics; Electronic Evidence; Judicial Problems

1 INTRODUCTION

Computer Crime Forensics is also known as Digital Forensics. It refers to the confirming, protecting, extracting and archiving procedures of electronic evidence in computers and relevant peripherals. And the evidence obtained by digital forensics must be sufficiently reliable, convincing and acceptable to the court. Digital forensics is mostly used in criminal investigation, but is also applicable to civil disputes [1].

In China, there is no more than ten years of history in the judicial application of digital forensics and its electronic evidence in court. However, in countries with the relatively developed information technology of the Anglo-American law system (Common Law system) and continental law system (civil law system), the judicial application of such means has been used for around thirty years. At present, in foreign countries, especially in those having the Anglo-American law system, there have been many theoretical studies and practical experience on the research and application of digital forensics and its electronic evidence. Moreover, countries and organizations, such as UNCITRAL (United Nations Commission on International Trade Law), the United States and the United Kingdom, whose information technology and electronic legislation are of leading status, have successively issued legislation or model laws on electronic evidence. Because digital forensics and electronic evidence law theory research in China is still weak and the relevant legislation is inadequate, we are still in a plight over judicial practice concerning digital forensics. This article mainly presents discussions and research on digital forensics, electronic evidence and its present relevant legal issues. At the end of this article the author will give some proposals for its resolution.

2 DIGITAL FORENSICS TECHNOLOGIES

Digital Forensics or Computer Crime Forensics was formally put forward in the first international conference of the International Association Computer Specialists (ICACS), 1991 and in 2001 at the 13th International annual meeting of Forum of Incident Response and Security Teams (FIRST), it became the main topic for discussion.

2.1 Static digital forensics technology

Currently widely used digital forensics technology is a kind of static method. It extracts, analyses and obtains valid electronic evidence after the incident occurred. In comparison, static forensics technology is well developed, especially in the aspects of scene evidence extraction, analysis, appraisal, submission and compliance with legal procedure. There are many applicable methods and technology solutions such as disk image copy technology, data recovery technology, information searching and filtering technology, etc. which all have played an important role in the process of digital forensics.

Some static forensics tools developed by foreign security companies such as text search tools, drive image program, The Forensics Toolkit, The Coroner's Toolkit, ForensiX, NTI, EnCase and so on, also have been proved and accepted by forensics experts.

2.2 Dynamic digital forensics technology

After an incident has occurred, static forensics technology is a kind of static analysis of the target system. However, the continuous improvement of computer crime and the emergence of anti-digital forensics such as data erasure technology, data hiding technology and data encryption technology, etc., makes it difficult to obtain valid evidence by only applying static forensics technology.

The complete process of the intruders attacking the target system is divided into three stages: before the invasion, during the invasion and after the invasion. If the intruders' invasion evidence of the first two phases, before the invasion and the invasion, is maliciously damaged, then in the third stage (i.e., after the invasion) any forensics work will become very difficult. Therefore, we should obtain or transfer the criminal evidence as far as possible in the stages before invasion and during the invasion, in order to avoid evidence being damaged. This method is so called the dynamic digital forensics. So far, the theory study and tools software development of the dynamic forensics technology are still in the exploratory stage. Dynamic digital forensic technology compared with static technology is far from perfect.

2.3 Anti-digital forensics technology

Anti-digital forensic technology is to delete or hide evidence in order to make forensic investigation invalid. Nowadays anti-digital forensic technology mainly includes data erasure technology, data hiding technology and data encryption technology as well as anti-digital forensic technology aiming at networks, etc. [1–3].

2.4 The relevant technology of digital forensics

Computer crime is complicated. Considering the process of electronic evidence collection, the relevant technology of digital forensics is as follows [3, 4]:

(1) Electronic Evidence Monitoring Technology

Electronic evidence monitoring technology is to monitor electronic data in various system devices and storage medium, and analyse whether there is electronic data which can be used as evidence. In general, it involves event monitoring, crime monitoring, abnormity monitoring and log analysis, etc.

(2) Physical Evidence Collection Technology

When the electronic evidence forensics system monitors invasion, we shall immediately obtain physical evidence through electronic evidence monitoring technology, because it is the foundation of all evidence collection work. Commonly physical evidence collection technology includes secure acquisition technology of a computer system and data files, secure backup technology without damage to disk or other storage medium, recovery and reconstruction technology of deleted files, information mining technology of disk space, undistributed space and freedom space, information recovery technology of exchange files, cache files and temporary files, network flow data acquisition technology and so on.

(3) Electronic Evidence Collection and Preservation Technology

Electronic evidence collection technology refers to saving and preprocessing collected data according to authorized methods as well as using authorized software and hardware devices, and then completely transferring data from the target machine to forensics devices.

Preservation technology refers to the protection of electronic evidence and a complete set of forensic mechanisms. It requires secure transmission technology, compression technology without loss, data reduction and recovery technology, data encryption technology, digital abstract technology, digital signature technology, digital certification and so on.

(4) Electronic Evidence Processing and Appraisal Technology

Electronic evidence processing refers to some preprocessing work such as filtering, pattern matching, hiding, data mining, and so on, for collected electronic evidence.

Appraisal technology on the basis of preprocessing takes data analysis work by data statistics and data mining for the processed data. Its main aim is to give a clear and legal statement about attack time, attack target, attacker's identity, attack intention, attack means and the consequences, etc.

(5) Electronic Evidence Submitting Technology

Electronic evidence submitting technology refers to submitting electronic evidence and the corresponding documentation to the court in acceptable evidence form according to legal procedures. By comprehensively analysing the target computer system and carrying on the summary of tracking results, an analysis conclusion is given. The content of this conclusion should include the overall situation of the system, the information of file structures, file data and file author's situation, the records of any attempts to hide, delete, protect and encrypt information, and other relevant information found in the survey. It also includes the collecting time, location, device, collectors and witnesses. And at last it is submitted to judicial organizations in the form of evidence according to legal procedure.

3 DIGITAL FORENSICS STANDARDS AND DIGITAL FORENSICS PROCEDURE

3.1 Digital forensics standards

According to the easily destructive characteristics of electronic evidence, if we want to ensure that electric evidence is reliable, accurate, complete and complying with relevant laws and regulations, in either a static forensic process or a dynamic forensic process, some common principles must be abided. Violating these principles, whatever kind of forensic method it is, the evidence effect will be affected remarkably [2, 3, 5].

(1) Timeliness Principle

It requires collecting electronic data as soon as possible and making sure there is no damage so as to ensure the timeliness of electronic evidence.

(2) Evidence Preservation Principle

The integrity of the evidence chain must be insured, namely that when evidence is formally submitted to the court, we must be able to show that any changes from the initial state of any evidence, obtained for the state, has appeared in court.

(3) Legality Principle

It requires that the electronic evidence forensic process conforms to legal procedures and takes place in an open and fair manner, and that the resulting evidence is real, objective and reliable. Meanwhile, it also requires that all tool software used in the evidence collection process also complies with legal rules.

(4) Multiple Backup Principle

It requires that at least two copies of digital information containing electronic evidence should be made. The original digital information should be stored in a specialized legal place and safeguarded by legal special custody. Copies can be used in extraction and analysis by the electronic evidence forensics personnel.

(5) Environmental Safety Principle

Electronic evidence should be properly kept in preparation for restruction, test or display at any time. To be specific, this principle requires storage media or medium of electronic evidence should be kept away from any high magnetic field, high temperature, dust, humidity, corrosive chemical reagent, etc. With regard to the packaging of computer equipment and components, material like paper bags, which is not prone to generate static electricity, should be used as far as possible in order to prevent electrostatic degaussing. Environmental safety principle also requires the prevention of data being artificially damaged, which includes intentionally malicious behaviour or wrong operation behaviour.

(6) Process Management Principle

This requires investigators and custodians collectively to complete the transferring, keeping, unsealing and disassembling process of evidence medium. During this process, they must check the authenticity and the integrity of evidence at every step. Meanwhile they should also take photos and make detailed records as well as sign jointly. The whole process of inspecting and obtaining evidence must be supervised. Supervisors comprise computer experts, legal experts, etc.

3.2 *Digital forensics procedure*

Electronic evidence forensics has special requirements. In the process of digital forensics, considering the digital forensic standards, it should follow the following steps [3, 4]:

(1) Protecting evidence. It refers to identifying the suspect computer probably containing electronic evidence, meanwhile protecting and avoiding rebooting the system or running any applications on it.

(2) Obtaining evidence. It is to obtain potential evidence stored in a suspicious computer and backup the original driver.

(3) Transmitting evidence. If it is stored in a hard disk, we should avoid evidence parts being damaged and causing data cannot be read. Whereas it is transmitted through networks, we must ensure the safety of transmission.

(4) Backup evidence. Before the newly created backup drive is analysed, we must first generate a hash code so as to make the identified backup evidence identical with the original evidence.

(5) Analysing evidence. Its main aim is to find legal and effective digital evidence for future use in court.

(6) Summarizing evidence. In this step, we should print out the comprehensive analysis of the target computer system, including all the relevant files and documents found in the data. Then an analysis conclusion is given.

(7) Submitting evidence. Finally, it is to submit digital evidence in the court. This phase we should act according to the policies and regulations. The use of the evidence supervision chain during all steps will make it hard for the defence lawyer to claim that evidence in your keeping process has been tampered with.

4 EXISTING PROBLEMS AND SUGGESTIONS

4.1 *Existing problems*

(1) Digital forensics tools

Digital forensics need to use a lot of tool software, mainly involving the disk image copy, network detecting, data analysis and other tool software.

Now most forensics tools often used in China are replicas or Chinese versions of foreign tool software. However, the legality and validity of these tools in our country has not been evaluated and confirmed by law enforcement agencies. In this case, the scientificity, authenticity, validity and normativity of forensics cannot be guaranteed.

(2) Digital forensics personnel's quality

Technical personnel act as an important part of the entire electronic evidence forensics; their quality directly affects the quality of evidence forensics. Technical personnel's quality includes moral quality and technical quality. The moral quality determines whether the forensics is objective or not. Whether forensics personnel can have legal awareness and conform to the facts and obtain evidence in an impartial manner or not, it is very important for evidence to become effectively legal evidence in court. Technical quality refers to the skills of forensics personnel on technology related to computer, network and others. Therefore, forensics personnel need to have high level of technical skills and constantly update their knowledge structure as well as improve their level of knowledge. Otherwise they will not be able to meet the scientificity and authenticity requirements of forensic evidence.

(3) Digital forensics procedure

At present, the justice departments still roughly use the traditional forensic rules and procedures in electronic evidence forensics. Because there are huge differences in obtaining evidence between the electronic evidence and traditional evidence process, in order to ensure the electronic evidence is recognized by law, the electronic evidence collection and preservation, and analysis process should also comply with national legal norms and relevant technical standards and specifications. Therefore, taking account of the electronic evidence forensics procedure, a new scientific and legal working procedure should be established according to the characteristics of electronic evidence, so as to ensure the validity and legality of evidence in the process of electronic evidence fixing [6].

4.2 Relevant suggestions

From perspective seeking methods and measures to ensure and improve the electronic evidence effectivity, the author has the following suggestions, by taking into account the key problems affecting electronic evidence forensics in order to achieve scientific, authentic, effective, objectives and standard targets for electronic evidence forensics:

(1) Formulate the corresponding digital forensics tool standards

Currently in our country electronic evidence forensics tools mainly include foreign electronic evidence forensics tools or application software, developed by domestic large enterprises without specific standard of electronic evidence forensics tools as a guide. Furthermore, there is also no special organization to authenticate the forensics personnel qualification and research institutes of forensic tools. The quality of electronic evidence forensics tools directly relates to whether the case would be able to be solved and whether judicial activities are fair and impartial. Furthermore, if the reason is that the poor quality and functioning of electronic evidence forensics tools lead to a criminal not being accused or convicted, the dignity and justice of law will be damaged. Therefore, the author suggested that the corresponding tools standards should be established so as to ensure the quality of the forensic tools, which will make the application of the electronic evidence forensics tools more useful to the electronic evidence forensics work.

(2) Strengthen the evaluation, training and supervision of digital forensics personnel

The technical requirements of electronic evidence forensics work are much higher than those of the evidence collection of traditional criminal investigation, so the electronic evidence forensics personnel's professional quality should be regulated particularly and specifically. Therefore, a regular strict appraisal system should be established. If the forensics personnel do not pass the test, they should be given the training so as to achieve the level of technical requirements of forensics work. At the same time, given various requirements of all kinds of personnel, especially for some special technical work, the institute of forensics personnel should formulate a targeted plan of education and training so as to make sure each forensics personnel accumulate enough experience in discipline implementation and coping with unforeseen circumstances.

(3) Establish the legislation of electronic forensics

In the fifth chapter of the Criminal Procedure Law of the People's Republic of China (2012 amendment) [7], audio-visual material and electronic data has been included as the eighth type of evidence. It is an approval for the legal status of electronic evidence, which means that electronic evidence can be used as legal evidence whenever cases involve electronic data.

Although our judicial departments have successfully completed a great deal of electronic evidence forensics work, because of the lack of corresponding technical standards and operational specifications, there are still a lot of problems in forensic practice. For example, much forensic work cannot be carried out due to lack of necessary tools or methods and even cause permanent damage to the evidence. Because of the arbitrariness of operation in the electronic evidence forensics work, evidence chain quality cannot be guaranteed, which all cause considerable difficulties for subsequent forensic appraisal.

According to the framework put forward by Digital Forensics Research WorkShop (DFRWS), the electronic evidence forensics procedure regulations shall include evidence identification, evidence preservation, evidence verification, evidence analysis and submission.

In addition, because the research of electronic evidence forensics starts rather late in our country, there is no special law and regulations on electronic evidence forensics at the moment. Furthermore, some existing forensic specifications are made separately by some professional forensic institutes according to their own experience, so their legal validity is hard to be guaranteed. Therefore, it is necessary to establish the legislation of electronic evidence forensics to be complied with while carrying on electronic evidence forensics work. It makes the whole process of the electronic evidence from collection to submission in the court legally valid, which ensures not only the probative force of electronic evidence, but also the admissibility and certification of electronic evidence.

5 CONCLUSIONS

In discussing the relevant aspects of digital forensics involved in electronic evidence forensics technology, forensics standards and forensics procedures, this article mainly analyses relevant problems in the current practice of digital forensics and its relevant legislation in China. This article also proposes some corresponding suggestions for the resolution of the issues.

ACKNOWLEDGMENTS

This work is supported by National Social Science Foundation of China (No.11BFX125).

REFERENCES

Criminal Procedure Law of the People's Republic of China (2012 amendment), 2012.

Ding, L.P. and Wang, Y.J. Study on Relevant Law and Technology Issues about Computer Forensics, Journal of Software, vol.16 (2), pp260–275, 2005.

He, J.H. and Liu, P.X. Research on Electronic Evidence Act, Law Publishing House, Beijing, 2002.

Marcus, R. and Kate, K.S. The future of Computer forensics: A needs analysis survey, Computers and security, vol. 23 (l), pp12–16, 2004.

Parra, M. Computer forensics (2002), http://www.giac.org/practical/Moroni_Parra_GSEC.doc.

Patzakis, J.M. Computer forensics-from cottage industry to standard practice, Information System Control Journal, vol. 2 (2), pp. 5–7, 2001.

Wang, Y., Cannady, J. and Rosenbluth, J. Foundations of computer forensics: A technology for the fight against computer crime, Computer Law & Security Report, 21(2), pp. 119–127, 2005.

Electronics, Information Technology and Intellectualization – Song & Kwak (Eds)
© 2015 Taylor & Francis Group, London, ISBN 978-1-138-02741-1

Synchronization of fractional-order chaotic systems using adaptive linear feedback control

XianFeng Li & XiaoBo Rao
Department of Mathematics, Lanzhou Jiaotong University, Lanzhou, China

Hui Zhang
Department of Mechanics, Lanzhou Jiaotong University, Lanzhou, China

ABSTRACT: Based on the stability theory of fractional-order systems, a novel adaptive synchronization scheme is presented. The interaction terms are designed with linear feedback in variables, in which the convergence speed of feedback strength is regulated by a constant. In comparison with nonlinear feedback controls, the controller is simple but direct. Different levels of white noise are involved to test and verify the robustness of the scheme.

1 INTRODUCTION

Fractional-Order Differentials (FOD) have a long history over than 300 years, but these have been widely used in modelling realistic system for no more than 30 years. Fractional derivatives are able to model memory and hereditary effects observed in physics due to their non-local essence. From the end of the last century, it has been observed that nonlinear fractional-order derivatives can also exhibit chaotic motions, see (Cafagna & Grassi 2009, Yu, Li, Wang, & Yu 2009, Zhang, Zhou, Li, & Zhu 2009, Zhu, Zhou, & Zhang 2009, and references therein). Controlling and synchronizing the complex behaviour in the FOD systems using some forms of control mechanism have recently been the focus of much attention, see (Capone to, Dongola, Fortuna, & Petras 2010, Cai, Jing, & Zhang 2009, Luo & Wang 2013, and references therein). Although most of the control schemes have been verified to be effective in control theory, some are too complex to be applied in practice (Asheghan, Beheshtia, & Tavazoei 2011, Yu & Li 2008). In this paper, based on the stability theory of fractional-order system, we proposed a novel adaptive synchronization scheme for a class of FOD chaotic systems. The interaction terms are designed with linear feedback in variables. Differing from previous work, the feedback strength is not prescribed as a constant but adaptive to an updated law. The convergence speed of feedback strength is regulated by a constant. Compared with nonlinear feedback controls, the synchronization scheme is so simple that it can be realized in physics. The reliability of synchronization state is guaranteed with rigorous linear algebraic theorems and precisely numerical matrix computations. Numerical simulations of the fractional-order Lorenz system demonstrates the effectiveness of the proposed scheme.

2 PREREQUISITES AND MATHEMATICAL DESCRIPTIONS

The following theorems talk about the stability of FOD systems and the properties of computational matrices, which are prerequisites for our proposed scheme.

The stability of a FOD system is given in the first theorem.

Theorem 1 considering the following FOD system (Tavazoei & Haeri 2009)

$$d^\alpha x/dt^\alpha = Ax, \quad 0 < \alpha \leq 1 \tag{1}$$

where $x = (x_1, x_2, \ldots, x_n)^{\mathrm{T}}$ is the variable vector, A is the coefficient matrix. If $V(x)$ is a positive-definite function, but

$$d^\alpha V(x)/dt^\alpha = x^{\mathrm{T}} \cdot d^\alpha x/dt^\alpha = x^{\mathrm{T}} \cdot Ax \leq 0 \tag{2}$$

then, system (1) is asymptotically stable (Zhang & Yang 2010).

The following two basic algebraic theorems are introduced to guarantee the reliability of our proposed control scheme. Note that both Theorems 2 and 3 can be proofed by Jordan decomposition (Horn & Johnson 1990).

Theorem 2 Suppose $\lambda_i, i = 1, 2, \ldots, m$ are m different eigenvalues of a square matrix $A \in \mathcal{R}^{n \times n}$, is the summation of algebraic multiplicities of every λ_i, ρ is a constant. Then, $\lambda_i + \rho$, $i = 1, 2, \ldots, n$, are the corresponding different eigenvalues of matrix $A + \rho E$, where E is a n-order identity matrix.

Theorem 3 Let a square matrix $A \in \mathcal{R}^{n \times n}$ be strictly row diagonally dominant. Then the number of eigenvalues of A with positive (resp. negative) real part is equal to the number of positive (resp. negative) diagonal entries of A. Moreover, strictly diagonally dominant matrices are always non-singular.

Consider a class of fractional-order chaotic systems, which can be decomposed into two parts,

$$d^{\alpha}x/dt^{\alpha}=F(x)=Ax+f(x), \qquad (3)$$

where, A is composed of all of the parameters of $F(x)$. There are no any parameters in nonlinear vector $f(x)$.

System (3) is taken as the drive system. The response system is identical but is configured with a control input vector u,

$$d^{\alpha}y/dt^{\alpha}=F(y)=Ay+f(y)+u, \qquad (4)$$

where, $y=(y_1,y_2,\ldots,y_n)^{\mathrm{T}}$ is the state vector of the response system (2). Similarly, there are no any parameters in nonlinear vector $f(y)$.

The synchronization of the drive system (3) and the response system (4) refers to the error in variables is asymptotically stable with the action of interaction terms. Define e as the error vector in variable vectors y and x, i.e., $e = y - x$. Consequently, the error dynamical system can be described as

$$d^{\alpha}e/dt^{\alpha}=F(y)- F(x)=Ae+f(y)- f(x)+u. \qquad (5)$$

Due to the nonlinear terms, $f(y)$ and $f(x)$ are same in structure, the difference between them can be represented as

$$f(y)-f(x)=B_{x,y}(y-x)= B_{x,y}e, \qquad (6)$$

where $B_{x,y}$ is a $n \times n$ matrix, which relies on vectors y and x.

The control input vector u is designed with linear feedback,

$$u=k(t)e, \qquad (7)$$

and the feedback strength $k(t)$ is designed with following the updated law

$$d^{\alpha}k(t)/dt^{\alpha}=-\gamma e^{\mathrm{T}}e, \qquad (8)$$

where, γ will be prescribed as a positive constant to regulate the convergence speed. Different with the integer-order systems, the FOD function (8) may not converge to a constant but should be lower than a certain negative constant k^* as evolution time $t \to \infty$ once the synchronization state is achieved, $e \to 0$.

As a result, the error dynamical system (5) can be represented as

$$d^{\alpha}e/dt^{\alpha}=(A+B_{x,y}+k^*E)e+(k(t)-k^*)e. \qquad (9)$$

Remark 1: Suppose $\lambda_{\max}$ is the largest eigenvalue of matrix $A + B_{x,y} = (a_{ij} + b_{ij})$, $i,j = 1, 2, \ldots, n$, k^* is the certain negative constant such that $\lambda_{\max} + k^* \leq 0$. Note that the largest eigenvalue $\lambda_{\max}$ exists and is definitely bounded, if and only if both $f(x)$ and $f(y)$ are bounded due to $\lambda_{\max} \leq \max_i (\sum^n (a_{ij} + b_{ij}))$ (Zhang 2004).

We define $e_k = k(t) - k^*$, and choose $V(e,e_k) = 0.5(e^{\mathrm{T}}e + (e_k)^2/\gamma)$. Thus, we have,

$$
\begin{aligned}
d^{\alpha}V(e, e_k)/dt^{\alpha} &= \\
e^{\mathrm{T}} \cdot d^{\alpha}e/dt^{\alpha}+1/\gamma \cdot d^{\alpha}k(t)/dt^{\alpha} \\
&=e^{\mathrm{T}}(A+B_{x,y}+k^*E+e_ke)-e^{\mathrm{T}}e \\
&=e^{\mathrm{T}}(A+B_{x,y}+k^*E) \\
&\leq n(\lambda_{\max}+k^*)e^{\mathrm{T}}e \\
&=n(\lambda_{\max}+k^*)\|e\|_2 \leq 0.
\end{aligned} \qquad (10)
$$

According to Theorem 1, the error dynamical system (9) is asymptotically stable. In this case, the synchronization states of all variables can be achieved with the adaptive linear feedback control if the critical strength k^* is satisfied.

Remark 2: It is clear that the proposed synchronization scheme is suitable for not only a lot of FOD chaotic systems, such as Chen, Lü, Liu, and other Lorenz-like systems in their fractional-order counterparts, but integer-order systems since the differential order $0 < \alpha \leq 1$.

Remark 3: To guarantee the synchronization state taking place in all of variable synchronously, usually should be test many times. Moreover, the synchronization state (9) between the drive and the response systems is always monitored by the root-mean-square error (RMSE) in variables, where RMSE = Sqrt($e^{\mathrm{T}}e/n$).

3 NUMERICAL SIMULATIONS

The fractional-order chaotic Lorenz system is employed here to demonstrate the effectiveness of the proposed adaptive linear feedback scheme. The drive system is described as

$$
\begin{aligned}
d^{\alpha}x_1/dt^{\alpha}&=\sigma(x_2-x_1), \\
d^{\alpha}x_2/dt^{\alpha}&=-x_1x_3+rx_1-x_2, \\
d^{\alpha}x_3/dt^{\alpha}&=x_1x_2-bx_3+rx_1,
\end{aligned} \qquad (11)
$$

which generates a double-scroll chaotic attractor when parameters $\sigma = 10$, $r = 28$, and $b = 8/3$ with commensurate fractional-order $\alpha = 0.995$ (Zhou, Cheng, & Kang 2010).

Rewrite the drive system in the form of (3), in which, the variable vector $x = (x_1, x_2, x_3)^{\mathrm{T}}$, $a_{11} = -\sigma$, $a_{12} = \sigma$; $a_{21} = r$, $a_{22} = -1$; $a_{33} = -b$, the rest entries of matrix A are all 0. $f(x) = (0, -x_1x_3, x_1x_2)^{\mathrm{T}}$.

The response system is

$$
\begin{aligned}
d^{\alpha}y_1/dt^{\alpha}&=\sigma(y_2-y_1)+u_1+\delta N_1, \\
d^{\alpha}y_2/dt^{\alpha}&=-y_1y_3+ry_1-y_2+u_2+\delta N_2, \\
d^{\alpha}y_3/dt^{\alpha}&=y_1y_2-by_3+ry_1+u_3+\delta N_3,
\end{aligned} \qquad (12)
$$

50

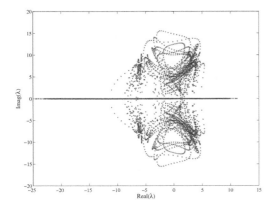

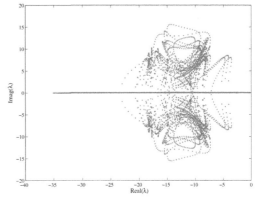

Figure 1. The distributions of all of numerical eigenvalues without k^*.

Figure 2. The distributions of all of numerical eigenvalues with k^*.

where, δ is a switch function. $\delta = 1$ when it is activated, otherwise $\delta = 0$. $N_i, i = 1, 2, 3$ are white noise on different levels.

We define the errors in variables $e_i = y_i - x_i$, $i = 1,2,3$ and select the adaptive linear controller $u = (u_1, u_2, u_3)$ as (7).

Similarly, we rewrite the response system in the form of (4), in which, the variable vector $y = (y_1, y_2, y_3)^T$, A is same as before, but $f(y) = (0, -y_1 y_3, y_1 y_2)^T$.

Hereby, the error dynamical system can be represented as

$$d^\alpha e/dt^\alpha = (A + B_{x,y} + k^* E)e + e_k e, \qquad (13)$$

where, $d^\alpha e/dt^\alpha = (d^\alpha e_1/dt^\alpha, d^\alpha e_3/dt^\alpha, d^\alpha e_3/dt^\alpha)^T$. The non-zero entries of $B_{x,y}$ are, $b_{21} = -y_3$, $b_{23} = -x_1$; $b_{31} = x_2$, $b_{32} = y_1$.

Suppose that λ_{max} is the numerical largest eigenvalues of matrix $A + B_{x,y}$, k^* is a constant such that $\lambda_{max} + k^* \leq 0$. In Figure 1, the distribution of all of numerical derived eigenvalues of matrix $A + B_{x,y}$ along with the attractors starting from different initial points is plotted on the complex plane. The largest eigenvalue of matrix is real, $\lambda_{max} = 11.1568$. Without loss of generality, k^* will be set slightly lower than λ_{max}, e.g., $k^* = 11.2$. The distribution of all of the numerical eigenvalues of matrix $A + B_{x,y} + k^* E$ is shown in Figure 2. As it shown, all of the eigenvalues have the negative real part along with the attractors even starting from very different initial points. In this case, the error dynamical system (14) should be stabilized once the feedback strength $k(t) \leq k^*$.

The evolutions of all of variables in drive and response systems are illustrated in Figure 3. In Figure 4, the semi-logy plotting of RMSE is shown in a long time interval. It can be seen that in a very short time interval, the magnitude of power is reduced to a very small negative number. They manifest that the trajectories of the response system will follow those of the drive system with the action of interaction terms (7) as the evolution time $t \to \infty$ (See Figure 3). From the numerical result shown in Figure 5, the evolutions

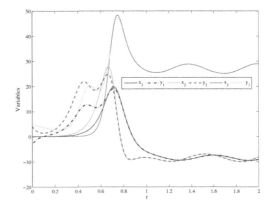

Figure 3. The evolutions of variables vs time t.

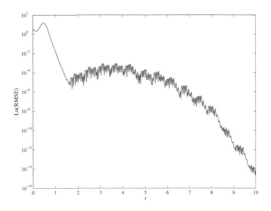

Figure 4. RMSE in variables vs time t.

of $k(t)$ regulated with $\gamma = 0.1$ are lower than k^* after a certain time $t^*(t^* \approx 0.5)$. We also tested the robustness of the proposed synchronization scheme. Figure 6 shows the bound of RMSE by interfering with different strengths of white noise.

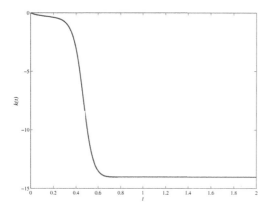

Figure 5. The evolutions of $k(t)$ vs time t.

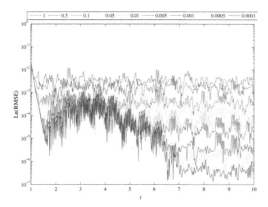

Figure 6. The robustness to different strengths of white noise.

4 CONCLUSIONS

In this paper, an effective synchronization scheme designed with adaptive linear feedback is proposed for a class of FOD chaotic systems. Rigorous theoretical and precisely numerical results have been presented to demonstrate their effectiveness. The robustness to different levels of white noise is tested and verified. We are sure that it is hopeful for realizing in practice due to their simple interactions.

ACKNOWLEDGMENTS

The research is supported by NNSFs of China (Grant No. 11161027, Grant No. 11262009), Key FSN of Gansu Province, China (Grant No. 1010RJZA067), and YSSF of Lanzhou Jiaotong University, China (Grant No. 2011026).

REFERENCES

Asheghan, M. M., M. T. H. Beheshtia, & M. S. Tavazoei (2011). Robust synchronization of perturbed chen's fractional-order chaotic systems. Commun. Nonlin. Sci. Numer. Simul. 16 (2), 1044–1051.

Cafagna, D. & G. Grassi (2009). Fractional-order chaos: A novel four-wing attractor in coupled Lorenz systems. Int. J. Bifur. Chaos 10 (19), 3329–3338.

Cai, N., Y.W. Jing, & S. Y. Zhang (2009). Generalized projective synchronization of different chaotic systems based on antisymmetric structure. Chaos Solitons Fractals 42 (2), 1190–1196.

Caponetto, R., G. Dongola, L. Fortuna, & I. Petras (2010). Fractional Order Systems: Modeling and Control Applications. Singapore: World Scientific Publishing Company.

Horn, R. A. & C. R. Johnson (1990). Matrix Analysis. Cambridge: Cambridge University Press.

Luo, C. & X. Y. Wang (2013). Chaos in the fractional-order complex Lorenz system and its synchronization. Nonlin. Dyn. 71(1–2), 241–257.

Tavazoei, M. S. & M. Haeri (2009). A note on the stability of fractional order systems. Math. Comput. Simul. 79 5), 1566–1576.

Yu, Y. G. & H. X. Li (2008). The synchronization of fractional-order Rössler hyperchaotic systems. Phys. A 387 (5–6), 1393–1403.

Yu, Y. G., H. X. Li, S. Wang, & J. Z. Yu (2009). Dynamic analysis of a fractional-order Lorenz chaotic system. Chaos Solitons Fractals 42 (2), 1181–1189.

Zhang, R. X. & S. P. Yang (2010). Adaptive synchronization of fractional-order chaotic systems. Chin. Phy. B 19 (2), 020510.

Zhang, W. W., S. B. Zhou, H. Li, & H. Zhu (2009). Chaos in a fractional-order Rössler system. Chaos Solitons Fractals 42 (3), 1684–1691.

Zhang, X. D. (2004). Matrix Analysis and Applications. Beijing: Tsinghua University Press.

Zhou, P., Y. M. Cheng, & F. Kang (2010). Synchronization between fractional-order chaotic systems and integer orders chaotic systems (fractional-order chaotic systems). Chin. Phy. B 19 (9), 090503.

Zhu, H., S. B. Zhou, & J. Zhang (2009). Chaos and synchronization of the fractional-order Chua's system. Chaos Solitons Fractals 39 (4), 1595–1603.

Electronics, Information Technology and Intellectualization – Song & Kwak (Eds)
© 2015 Taylor & Francis Group, London, ISBN 978-1-138-02741-1

A multi-utility framework for wireless sensor network design

Shengbin Liao
Engineering and Research Center of Information and Technology on Education, Huazhong Normal University, Wuhan, China

Cuitao Zhu
Department of Electronics and Information Engineering, Huazhong South-Center University for Nationalities, Wuhan, China

ABSTRACT: In this paper, we present a multi-utility framework and apply it to a wireless sensor network design. We often consider two performance metrics, i.e. the transmitting rate and the coverage area, and thus, we define the utility of the data rate and the utility of the coverage area. Due to the energy limits in wireless sensor networks, optimizing simultaneously the two types of utility leads to a multi-objective optimization problem. We use the algorithm of NSGA-II to solve the proposed multi-objective problem and the numerical results show that NSGA-II can approximate to the Pareto front of our model.

1 INTRODUCTION

A Wireless Sensor Network (WSN) usually consists of some sensor nodes which are located on an area for sensing and collecting data. When we plan and design a WSN, coverage area, rate data, energy consumption and network lifetime are considered. These metrics measure different performances and resource allocation. There are some inherent trade-offs because of conflicts between different design goals. Therefore, it is important to build a mathematic model to analyse any trade-off to optimize a WSN design.

Some multi-objective optimization frameworks were presented for addressing this problem [1]. The most common method is to optimize simultaneously these metrics and this leads to a multi-objective optimization framework. However, these metrics map different performances or quality of service. They usually have different dimensions and magnitudes, direct optimization for these metrics do not accurately reflect the trade-off between them.

In this context, some researchers proposed utility-based optimization which transforms different performance metrics into different kinds of utilities [2]. In this paper, we also adopt this idea to define the utilities of coverage area, rate data and energy consumption for WSN design. However, differentiating the classical utility-based optimization method, we will build a multi-utility framework by optimizing simultaneously different kinds of utilities instead of optimizing the weight sum of different kinds of utilities. The method optimizing the weight sum of different kinds of utilities belong to single objective optimization, however, our presented multi-utility framework is a multi-objective optimization problem.

Here are some of the works which adopt multi-objective frameworks for network design and performance analysis: G. Fernando et al considered IEEE 802.16 network design by using a multi-objective framework, they analyse the trade-off between the costs, the revenues, service coverage, overall interference, and throughput [3]. P. Chutima and W. Naruemon proposed a multi-objective model which combines three problems together, including the optimal access point placement, the frequency channel assignment and the power level assignment [4]. A.George et al. studied traffic engineering for future networks by solving a multi-objective optimization problem [5]. All of this work uses classical methods to solve multi-objective optimization problems. However, in this paper, we present a multi-utility problem which objective functions may be non-concave, thus, we will use the evolutionary algorithm NSGA-II to solve our problem. NSGA-II is a non-dominated sorting-based multi-objective evolutionary algorithm and has better performance compared to many other multi-objective optimization algorithms [6].

The rest of this paper is organized as follows: Section II presents a multi-utility framework for a wireless sensor network design. The proposed work using the NSGA-II for approximating Pareto front is presented in Section III. Section IV provides the numerical results. Finally, Section V concludes this paper.

2 MULTI-UTILITY FRAMEWORK

We consider a wireless sensor network with S sensor nodes and L logical wireless communication links, and assume every node i has m performance metrics $\{p_{i1}, \ldots, p_{im}\}$ such as data rate, energy consumption and delay, and so on. Associated with each performance metric p_{ij} is a particular utility function $u_i(p_{ij})$, which defines the node i of the satisfaction degree with respect to the corresponding metric p_{i1}. Thus the total utilities of users associated the metric $p_j = (p_{ij}, \ldots, p_{Sj})$ is defined as $U_j = \sum_{i=1}^{S} u_i(p_{ij})$. Further, from the network service provider's point of view, its goal is to simultaneously maximize the multi-utilities under some network resource constraints. Hence, based on the conceptions of multi-objective optimization, we propose a multi-utility framework as follows:

$$\max U(p_1, \ldots, p_m) = (U_1(p_1), \ldots, U_m(p_m))^T$$
$$s.t. \quad (p_1, \ldots, p_m) \in \omega \tag{1}$$

where ω denotes the decision space, T denotes the vector transpose.

Model (1) is a multi-objective optimization problem, the objective functions of which denote different kinds of utility functions. Usually, there is no an optimal solution for model (1) because its objectives may be conflicting. Thus, for model (1), we hope to obtain optimal solutions of trade-off, i.e. Pareto optimal solution [7]. However, there may have many trade-off schemes, i.e. there may have many Pareto optimal solutions. We call the set of all Pareto optimal solutions *Pareto front*; therefore, for model (1), our aim is to design some algorithms to approximate its Pareto front.

Next, we will specialize in the model (1) to design and optimize a wireless sensor networks design. When we design a WSN, we need consider two issues, one is coverage area which can be expressed by transport distance, the other one is energy consumption; the latter is related to data rate and transport distance. Therefore, in this section, we will associate coverage area and data rate with two different utility functions.

The utility function associated with data rate is the classical network utility function which can be formulated by logarithmic function [8]. For the definition of utility function of coverage area, we assume that each sensor node has the same maximum transmission radius and each sensor node transmits its data to a neighbouring node. For applications in WSNs, we hope that each sensor can cover and sense a large area, so, we adopt the logarithm of coverage area as its utility function. It has the form as follows:

$$U_{ca} = \sum_{i=1}^{N} \log(\pi r_i^2) \tag{2}$$

where r_i denotes the transmission radius of sensor node i. Then, we can instantiate the model (1) in wireless sensor network with two kinds of utility functions as following:

$$\max (U_{dr}, U_{ca})^T = (\sum_{i=1}^{N} \log(x_i), \sum_{i=1}^{N} \log(\pi r_i^2))^T$$
$$s.t. \quad x = (x_1, \ldots, x_N)^T \in \phi, r = (r_1, \ldots, r_N)^T \in \varphi \tag{3}$$

where x is the data rate vector of sensor nodes and r is the transmission radius vector of sensor nodes, U_{dr} and U_{ca} denote the utility of data rate and coverage area respectively. T is the transpose of a vector. ϕ and φ denote the decision spaces of x and r, respectively.

We assume that the initial energy of each sensor node is E_0, and the lifetime of the considered WSN is T_0. That means that we will simultaneously optimize the two objectives in mode (3) with the constraints energy consumption and the network lifetime. For the energy consumption, we use the following power consumption model for the data transmission of nodes:

$$p_s = \sum_{l \in L_{out}(s)} \sum_{i \in S(l)} p_{sl} x_i$$

where $L_{out}(s)$ denotes the outgoing link set from node s, and $S(l)$ denotes the set of nodes using the link l in its route. p_{sl} is the power depletion transmitting the unit data over link l for node s, which can be expressed as $p_{sl} = \alpha + \beta d_{sl}^n$, here α and β are constants, d_{sl} is the transmitting distance.

Based on the above assumption, we can specialize in the model (3) as

$$\max (U_{dr}, U_{ca})^T = (\sum_{i=1}^{N} \log(x_i), \sum_{i=1}^{N} \log(\pi r_i^2))^T$$
$$s.t. \quad x = (x_1, \ldots, x_N)^T \geq 0, r = (r_1, \ldots, r_N)^T \geq 0$$
$$E_0 \leq T_0 p_s, \forall s \in S \tag{4}$$

In the model (4), we do not consider the receiving energy consumption for the simplicity of notations. Moreover, the transmitting distance will be the coverage radius.

3 APPROXIMATE PARETO FRONT

As illustrated in the above section, model (4) is a multi-objective optimization problem with contradictory objectives. Next we will approximate the Pareto front of model (4) by Non-Dominated Sorting in Genetic Algorithms II (NSGA-II) [9]. NSGA-II is a non-dominated sorting-based multi-objective evolutionary algorithm, which is proved to be better than many other multi-objective optimization algorithms [9][10]. NSGA-II is a population-based algorithm, the output of which is a Pareto optimal set of solutions. The goal of using NSAG-II to solve model (4) is to obtain an evenly distributed set of solutions, which approximates the Pareto front of model (4) as following:

The algorithm of NSGA-II works as follows [9] [10]:

Step 1 Initialization: set initial parent population P_0 of size N, and P_0 is sorted based on the non-domination. Moreover, the value of the fitness of each solution in P_0 is set to its non-domination value. Then, selection, recombination, and mutation operators are used to create the child population Q_0 of size N.

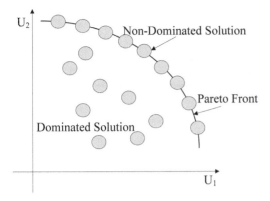

Figure 1. Evenly distributed set of Paretooptimal solutions.

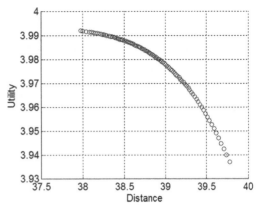

Figure 3. Trade-off between utility and distance.

Figure 2. Network topology.

Step 2 Update: For $t \geq 1$, do

Step 2.1 $R_t = P_t \cup Q_t$

Step 2.2 R_t is sorted based on the non-domination, the output is $F = (F_1, F_2 \ldots)$, where $F_1, F_2 \ldots$ the non-dominated fronts of is R_t.

Step 2.3 Compute crowding distance in F_i

Step 2.4 $P_{(t+1)} = P_{(t+1)} \cup F_i$

Step 2.5 Apply selection operator to $P_{(t+1)}$, and select parents from $P_{(t+1)}$, generate the child population Q_t using the crossover and mutation operators. $t = t + 1$;

Step 3 Stopping criteria: if stopping criteria is satisfied, then the algorithm stops and outputs P_t. Otherwise, go to Step 2.

For information about the non-domination sort, the computation of crowding distance, and selection operator in the above algorithm, refer to [9] [10].

4 NUMERICAL RESULTS

In this section, we present the numerical experiments with a linearly deployed sensor network as following:

We assume that the length of every link is equal, and $\alpha = 50$ nj/bit, $\beta = 0.0013$ pj/b/m4, and the pass loss constant $n = 4$. We also assume that the initial energy $E0 = 1000$ J and $T0 = 1200$ s. The following Figure 3 gives the trade-off information between network utility and transmitting distance. From Figure 3, we can see that as the transmitting distance increases, the total network utility will decrease. That means that there is inherent trade-off between data rate and transmitting radius. As plotted in Figure 2, when we design a linear wireless network, the energy of each sensor node is fixed. If we want the coverage area to belarge, i.e. the transmitting distance is large, the total utility will be small, and this will correspond to those points on bottom-right. If we want the total utility to be large, the

coverage area will be small, and this will correspond to those points on above-left.

5 CONCLUSION

A multi-utility framework for wireless sensor design was presented in this paper. When we design a wireless sensor network, there are many performance metrics to be considered. Due to trade-off between some of performance metrics, there is no best solution for a design model. In this paper, we proposed a scheme for metrics of data rate and coverage area for a linear network. We used the algorithm of NSGA-II to solve our multi-utility model, and it can approximate to thePareto front. Moreover, numerical experiments show Pareto optimal solutions obtain by NSGA-II can evenly distribute on the Pareto front.

ACKNOWLEDGEMENT

This work was supported in part by the National Natural Science Foundation of China through the grants No. 61072051 and No. 61202470, the key science and technology project of Wuhan under grant No. 2014010202010108 and the Doctoral Program of Higher Education in China under grant No. 20110141120046.

REFERENCES

Shibo He, Jiming Chen, Weiqiang Xu, Youxian Sun, PreethaThulasiraman and Xuemin (Sherman) Shen, A Framework for Stochastic Multi-Objective Optimization in Wireless Sensor Networks, EURASIP Journal of Wireless Communication and Networking, Volume 2010 (2010), Article ID 430615, 10 pages doi:10.1155/2010/430615

Chunlin Li, Layuan Li, Utility-based QoSoptimisation strategy for multi-criteria scheduling on the grid, Journal of parallel and distributed computing, 67 (2007) 142–153.

Fernando, G. Alpar, J. Jie, Z. A multi-objective optimization framework for IEEE 802.16e network design and performance analysis, IEEE Journal on selected areas in communications, 27(2) (2009) 202–216.

Chutima, P. Naruemon, W. Network design and optimization for quality of services in wireless local area networks using a multi-objective approach, WSEAS Transactions on communications, 8(7) (2008) 827–836.

George, A. Kostas, T. Panagiotis, V. Dimitrios, K. Panagiotis, D. Multi-objective traffic engineering for future networks, IEEE communications letters, 16(1) (2012) 101–103.

Tan, K. Khor, E., Lee T., Multiobjective Evolutionary Algorithms and Applications, ser. Advanced Information and Knowledge Processing. Berlin, Germany: Springer-Verlag, 2005.

Miettinet, K. Nonlinear Multi-objective Optimization. Norwell, MA: Kluwer, 1999.

Mo J. Walrand, J. Fair end-to-end window-based congestion control, IEEE/ACM Transactions on networking, 8(5) (2000) 556–567.

Deb, K. Pratap, A. Agarwal, S., Meyarivan, T. A fast elitist multi-objective genetic algorithm: NSGA-II, IEEE Transaction on evolutionary computation 2 (2002) 182–197.

Deb K, Agarwal S, Pratap A, Meyarivan T. A fast elitist non-dominated sorting genetic algorithm for multi-objective optimization: NSGA-II. In Proceedings of Parallel Problem Solving from Nature VI Conference, Paris, September 2000; 849–858.

Research and practice of developing English majors' competence in e-commerce

Xuanzhu Liu

School of Foreign Language, Shenyang University, Shenyang, Liaoning, China

ABSTRACT: Business English education focuses on the training of skill talents; its goal is to cultivate the comprehensive ability of students in e-commerce (electronic commerce). It is necessary to break traditional ideas, to find accurate positioning, to strengthen comprehensive training, to improve their professional ability and practical skills, and enhance competitive employment in order to construct a scientific training system of the competence of business English majors in e-commerce.

1 INTRODUCTION

In the background of the high-speed development of the industry, talents of e-commerce, credit systems, and distribution are becoming the three major bottlenecks affecting the development of China's e-commerce industry. Our country e-commerce enterprises employed more than two million people directly until 2012 which figure does not include more than fourteen million employed indirectly.

Along with the policy heating and continuous improvement of the environment in e-commerce, half of the thirty million small and medium-sized enterprises in China will become involved in the development of e-commerce, and the demand for e-commerce talent will become tighter. According to the analysis of social demand for talent cultivation industry and professional ability, e-commerce industry (Group) position is very extensive, mainly including: e-commerce commissioners, network operations managers, website planning, network promotion, website development, online marketing specialists, network editors, online customer services, online advertising sales representatives, research members BD, marketing, sales engineers, administrative secretaries, extensive post groups to post a lot of talent demand. Data shows that the Ministry of Education has approved 339 universities and 650 vocational settings for electronic businesses professionals since the year 2000. Every year there are more than 80 thousand graduates each year. E-commerce talent can only meet the industries demand for talent of half a month.

A broad stage is provided for business English majors as development of China's increasingly frequent business activities in the world. Economic globalization offers great demands for talents in business English. All kinds of professionals attach great importance to the training of skills, the cultivation

of students' ability in e-commerce in prominent positions, active research, and in practice.

Developing e-commerce capability becomes one of the professional goals of business English, just like the focus on training students' comprehensive quality and ability. In recent years, scholars have conducted a study and exploration of comprehensive e-commerce capability of business English major, and formed an initial formation for the training strategy of the business English professionals. They base e-commerce on the ability to break through traditional teaching ideas and grasp the basic skills in international business activities.

2 ANALYSIS OF THE MAIN FEATURES OF TALENTS IN E-COMMERCE IN CHINA

2.1 *The rapid expansion of the industry and the huge talent gap*

With the rapid expansion of the industry in the field of e-commerce, it has a huge gap in talent. In certain individual domains it is difficult to recruit enough talent for e-commerce. The vast majority of enterprises experience some recruitment pressure for talents in e-commerce. Half of the pressure mostly comes from the rapid growth of the enterprise. According to the development of the industry, more than half of the companies need to mount a large-scale recruitment each year.

2.2 *Scarcity and mobility of e-commerce talents*

A great demand gap is caused by the flow of talent; many enterprises are facing a large lack of personnel. E-commerce is a new industry, which inevitably faces the problems of the scarcity of skilled talent, enterprises intense competition for talent, and talent mobility. At the same time, professional training systems in universities and enterprise practices are seriously out of line, which results in a large number of students graduating from college out being illequipped for the development of enterprises.

2.3 *The business enterprise human resource high cost*

With the rapid growth of e-commerce industry, enterprise human resources are facing pressures. The pressure of salary growth has become one of the key problems of the HR sector that plagues enterprises. With the further development of the business, enterprise will gradually mature, and pay more attention to the construction of an enterprise culture and its system.

2.4 *The increases of enterprise human resources management difficulty*

At present, e-commerce enterprises are mainly made up of people who were born after 1980 or 1990. Their work enthusiasm and creativity is promoting enterprise, continuous innovation, and development, but the group has a distinct personality and values beyond reality, so this presents a huge challenge to enterprise management. Because most of the business enterprises are in an initial stage, enterprise systems, enterprise culture, and other aspects are not perfect; at the same time, the rate of staff turnover is higher than in other industries due to working pressure, fast-paced change, and overtime working hours

3 THE FUTURE OF E-COMMERCE TALENTS IN CHINA

Nowadays, the scale of economic growth is greater than the value growth factors in China's e-commerce enterprise. The growth has been slow and unstable. E-commerce field in the future will only strengthen the 'values' to achieve real growth. It looks for e-commerce and traditional commerce service differentiation, with high-end, value-added services reflecting industry's existence value.

Application: Try to connect B2B and production supply organically, and strive to become the modern productive service industry in the leading industry. Deepen the e-commerce application, promote the integration of e-commerce platform and internal business and management information system, realize the enterprise collaborative ability from 'the procurement, production, circulation, storage, sales of' supply chain integration, promote enterprise online collaborative R & D, design and manufacturing, and enhance in product, finished product, the stock of products of organic conversion cohesion in the fields of key industrial raw materials, equipment manufacturing, consumer goods, electronics and information, national defence science and technology. At the same time,

there is a need to strengthen the cooperative ability of the downstream industrial chain enterprise and promote product distribution after sale service level.

Innovation: The industry should try to promote the integration and promotion of the value-added services of prominent B2C platform information, and realize strategic transformation to achieve a 'C to B' business model. B2C platform is to complete the business between the enterprise and the user 'in the data form of face to face' change in consumption patterns, accumulated massive user base information and transaction data, as a bridge between enterprises and consumers bond. B2C e-commerce platform should give full play to its information aggregation mining ability, realize enterprise scale homogenization of production capacity and the consumer demand for personalized organic docking, and guide the business enterprise to the 'flexible on-demand production, personalized and accurate positioning, social core focus' in a gradual transition.

Loyalty: E-Commerce needs to try to play the role of a third party credit platform service ability in C2C, and reshape the long-term development foundation and healthy competition in the market. C2C e-commerce platform is based on a third party credit service system that the credit management system is for the establishment of 'product quality, service businesses, buyers of consumer credit' as the core, and to make the transaction information transparent and symmetric. It is driven by the short-term interests of the PPC to change ideas and forms an 'honesty ranking' as the cornerstone of the search service function. There is a need to combine credit management and the C2C market threshold of access, to prize honesty, to punish illegal awards and to create a healthy, benign, competitive market environment.

In addition, with the innovation of information technology and business model innovation being demand driven, the future of e-commerce and social networking, mobile Internet, location services and other emerging platforms will lever each other to interpret the 'SoLoMo' cross platform integration of external development trends of expansion.

In the future, the development of e-commerce in China will face a bottleneck stage in long-term development, but we should believe that new business models of e-commerce will mature, with value-added service values throughout multiple levels of industry, markets, and enterprise active subject E-commerce in China will shine in the broader arena.

4 E-COMMERCE ABILITY TRAINING STRATEGY FOR ENGLISH MAJORS

We have to improve e-commerce ability for English majors from knowledge structure of the international business occupation post in order to meet what the job or occupation requires and strengthen occupational skill training, and enable students to have occupation post adaptability.

4.1 To determine the business English major students' training objectives

Business English has to cultivate a good occupation quality and occupation morale, and grasp the basic knowledge and theory in the field of international business. It needs to have a certain business operational skills and professional ability, to be able to use English as a working language, and to participate in applied talents in international business and international business activities. The basic knowledge skills of business English talents are communicational skills and business skills.

4.2 To follow the principle of business ability training in e-commerce for English major students

The first is the principle of pertinence: to determine the overall framework of the teaching system and specific training programs according to the characteristics of business English talents at technician level, and a knowledge of the ability structure of students' future posts. The second is the principle of practicality: to start from the actual, practical perspective base of the teaching system, security mechanism and evaluation system according to the needs of society and students industry needs. The third is the principle of integration: theory of teaching and the practice of teaching, must be relevant to language proficiency and business skills, and promote each other. The last is the principle of combining the professional skills and comprehensive ability. Teaching system should include the students' future job skills, include comprehensive ability and professional ability, such as information collection and processing ability, and include new knowledge acquisition ability, problem analysing and solving skills, and teamwork and social activity ability.

4.3 Construct business ability system of e-commerce for English major students

Training objectives of business English majors is two talents ('skills' and 'application type'), namely training or training qualified people for the post with essential ability. Based on this demand, reasonable construction of the business English major students 'e-commerce' training system is imperative.

Firstly: Software security. Set the practical business English course and connect the 'English' and 'business' as two disciplines together. Set the curriculum, training, skills and the knowledge required to interface between curriculum consolidation and development, professional skills and the ecommerce capability. For example, the design should include occupation understanding, business communication such as the 'professional cognition training', business correspondence, import and export documents, e-commerce and other contents of the 'e-commerce practice', etc. It is important to emphasize speaking English and practical business training, and to embody professional business English comprehensive application ability; to explore students' ability in e-commerce training and the implementation of an approach that is a constructive and creative.

Secondly: Hardware security. Modern teaching equipment (including speech room, multifunctional language laboratory, business training room, the e-commerce chamber) opens up a new realm of classroom teaching, with the characteristics of directness, images, and a large amount of information, which greatly stimulates the learning interest of the students. Through presenting and listing business activities, understand the basic procedures of business operation; learn to deal with the related problems.

Thirdly: Community support. Through the various communities including English corner, English salon, entrepreneurs, lectures or seminars on business knowledge contest, the English speech contest, business translation and writing competitions, encourage students to participate in practice. From such activities, students can find their own value, discover their potential ability, find learning motivation, and improve their comprehensive ability.

5 CONCLUSION

Business English talents require moral, intellectual and physical development with a solid foundation in English and strong language communication ability, a mastery of international business professional knowledge, and a familiarity with the laws and regulations of international commerce. Students should have a scientific world outlook, outlook on life, good occupation moral behavioural norms, certain innovation spirit, practical ability and the initial start-up capacity. They should also have a basic business knowledge and actual work ability in an international environment to operate all kinds of business activities. Net exists everywhere in the world, so the capacity of e-commerce training will undoubtedly become the first priority among the priorities of business English majors.

REFERENCES

Almarza, G.G. (1996). Student foreign language teacher's knowledge growth. A. In D. Freeman and J.C. Richards (eds.). Teacher Learning in Language Teaching. C. New York: Cambridge University Press. 50–78.
Bao Liu. (2011). The actuality and expectation of electronic business payment in China. Journal of Chongqing RTV University, 72–74.
Fengrong Zhai. (2009). The present situation of China's e-commerce and its future development.Journal of Yantai University (Philosophy and Social Science Edition). 62–67.
Heider, E.R. & D.C. Oliver. (1999). The structure of colour space in naming and memory of two languages. J. Foreign Language Teaching and Research. 62–67.

Electronics, Information Technology and Intellectualization – Song & Kwak (Eds)
© 2015 Taylor & Francis Group, London, ISBN 978-1-138-02741-1

Discussion and analysis about the integration of information technology in Community Correction work

X.X. Zhang & R. Gang
University of Science and Technology Liaoning, Liaoning, China

ABSTRACT: In order to improve means of Community Correction work, exploring modern information technology and applying it to Community Correction work is the main measure to enhance Community Correction work's effectiveness. From the perspective of information technology and the connotation of Community Correction exploration, this paper expounds the necessity of integrating information technology with Community Correction, argues the measures to integrate information technology with Community Correction work, points out urgent problems to be solved about integrating information technology and Community Correction work and also puts forward some suggestions about applying information technology to Community Correction work.

Keywords: Community Correction work, information technology, integration

1 INTRODUCTION

In order to improve the means of Community Correction work, after doing some survey and research on this program, our team discovered that the lack of information technology applied to this program is one of the basic problems in our Community Correction work at present. Therefore, our team sorts out and analyses this problem by combining it with practice, hoping to bring help and reference to our country's Community Correction work's long-term development.

2 CONNOTATION OF INFORMATION TECHNOLOGY AND COMMUNITY CORRECTION WORK

2.1 Definition of information technology

Information technology is used in the process of obtaining information, processing, handling and disseminating it. Nowadays, with the help of computer technology and communication technology, people can reach a wider world. People use computer technology to realize obtaining, processing, handling and dissemination working content and working purpose, which is the most simple and effective approach. Hence, adopting modern information technology devices in work is the demand of information age reform. (He, Rongjie in Education technology, Information technology and curriculum integration.)

2.2 Connotation of Community Correction

In our country, the universally accepted definition of Community Correction was put forward by the Department of Justice research group on the Community Correction system. The definition describes that Community Correction is a kind of punishment in contrast to prison correction. Community Correction refers to a non-imprisonment penalty that allows the eligible offender to be paroled, to serve outside prison, under surveillance, but deprived of political rights. The offender serves a suspended sentence, the period of which is decided by the community with the help of the law-enforcing departments and the relevant social organizations, civil groups and volunteers in social organizations, in order to correct the offenders' criminal mind and behaviour, so that they can return to society smoothly. Community Correction is based on non-custodial penalties and takes free punishment as its centre. It is the outcome of the conception of human social punishment and non-imprisonment punishment. Community Correction is not only the trend of different counties' non-imprisonment punishment in the world but also an important measure for us to build a harmonious society. (Li, Hanyuan & Zheng, Lei. Analysis of problems in the execution of Community Correction.)

3 THE NECESSITY OF APPLYING INFORMATION TECHNOLOGY TO COMMUNITY CORRECTION WORK

3.1 The need of the rapid development of society

With the information technology's rapid development, it provides a material basis for people taking advantage of modern information technology means from space and information, which has greatly promoted the development of society. It is urgent to enhance the publicity quality and publicity results of Community Correction work by using information technology.

3.2 *The need of propaganda about Community Correction system*

Nowadays, there is not enough promotion of the relevant knowledge about Community Correction. Only 42.6 per cent of the population having heard of the existence of the Community Correction system, but even they have known little about the relevant knowledge of it and our country current policy of combining punishment with leniency. What is worse is that 87 per cent of community workers do not fully realize the importance of community correlation. Therefore, we must reinforce using information technology means to enhance the concept, nature, meaning and propaganda about the relevant knowledge of Community Correction.

3.3 *The need of cooperating between departments in Community Correction*

Community Correction work involves many departments and institutions. In order to ensure that the data about Community Correction offenders' transference, reception, supervision, assessment, reward and punishment, lifting the correction, and their commitment into custody etc. is consistent and shared, and in order to make Community Correction work on the offenders more efficiently, Community Correction work must be managed through the Internet, so that the following different functional departments, public security bureaus, procurators, courts, and judicial departments can work organically together.

3.4 *The need of supervising the regulatory target*

It's necessary to set up an information network for Community Correction centre to collect investigating information files about Community Correction crimes, sentences, correction periods, whether to extended sentences or not, whether to have intermediate people's court rulings or not, whether to have the Prison Administration approval for renewal procedures or not, whether to have regular check-ins and relieve correction information, so that the phenomenon of oversight, lack of control, omissions, and carelessness to the correction offenders can be avoided. (Guo, Jian'an & Zheng, Xiaoze. on Community Correction.)

4 THE MEASURES TO INTEGRATE INFORMATION TECHNOLOGY AND COMMUNITY CORRECTION WORK

Integrating information technology and Community Correction work should combine information technology, information resources, measures and human resources with Community Correction work, in order to complete transforming community offenders and integrate them smoothly into society.

4.1 *Establishing electronic files*

In an effort to fully record the personal information of the correction targets in detail and reasonably formulate corrective measures to rectify the targets, it is necessary to provide electronic standardization for correction offenders' files to record the information about correction offenders in detail and to set up reasonable measures. Consequently, this method can regularly display the basic information and contents of supervision in detail. Besides, establishing electronic files provides a reference for realizing the Community Correction's effective management through evaluation indicators on the offenders' admission of guilt, correction performance, family background, and personal experience and character.

4.2 *Establishing GPS positioning system management*

Administrative staff can locate correction offenders at any time by a GPS position system and choose to replay correction offenders' trails in a certain period, so that administrative staff can regularize offenders' effective monitoring areas and find abnormal conditions to prevent the offenders from escaping being supervised. If the correction offenders are out of the monitored area, the system can give a reminder to both administrative staff and offenders in order to prevent the offenders from committing crimes once more.

For example, real-time tracking and option for playback, using GPS positioning systems to track and position correction offenders, saving this information on the track record, then selecting it to playback will fulfil all the needs to realize the goal of supervision. (Ren, Hanqing. & Xiong, Yuelin. Information technology education applications of Community Correction work.)

4.3 *Establishing data analysis module*

Data analysis module can analyse correction offender's basic information, which can maximize the development of data resource's function and help to make good use of the data. This can also provide some basis for administrators working on correction measures and giving some assistance, and it also provides support to learn about the correction of offenders' changing situation over time with standard data.

4.4 *Establishing information sharing platform*

An information sharing platform provides Community Correction work's participating departments and their personnel a working and interactivity information communication platform. Under the leadership of the politics and law committee and the justice bureau, court, procurator ate, public security bureau, civil affairs bureau, labour bureau, social security bureau etc. constitutions can participate together. All the constitutions can then use the information sharing platform to set their own management organization information platforms and by checking other institutions' management organizational information platforms, they can then realize collaborative development, management and efficient links with different institution's correction management work, in cases where oversight, is out of control, and poor supervision leads to careless correction.

5 URGENT PROBLEMS TO BE RESOLVED ABOUT INTEGRATING INFORMATION TECHNOLOGY AND COMMUNITY CORRECTION WORK

5.1 *Problems in information construction*

Informational Community Correction is mainly based on the construction of an information management platform, and its information management system, as the important corporate system of the judicial administrative organ, does not have contact with 'the judicial administrative work foundation information management platform' or 'the released people's information management system' and so forth at present. This causes a waste of resource and may affect the effectiveness of the Community Correction's work. (Wu, Yingai. in making efforts to promote Community Correction work development.)

5.2 *The problem of using mobile phone location technique*

The Ministry of Justice has required that Community Correction institutions should use GPS positioning systems to supervise at least 70 per cent of offenders. This requirement has achieved good results. In the beginning, most Community Correction institutions built information management systems by putting chips in mobile phones to supervise the position of the correction offender, but the GPS chip for cellular phone might have lost positional function if the phone was out of a battery, the power off, or was shut down or the mobile number had changed. Besides, GPS positioning systems only rely on the phones and system devices and it would not rely on who uses it, which would cause the possibility of it being out of control or missing control.

5.3 *The problems of funds investment and personnel's quality*

Most regions do not include the Community Correction work's investment in the budget, so that the devices for Community Correction work are not enough. Some offices of Justice even do not have the necessary office equipment or enough experience, or studying and training places. Community Correction staff do not have a high level of ability to use and maintain information management platforms. These problems certainly would affect the quality of Community Correction work.

6 SOME SUGGESTIONS ABOUT ESTABLISHING COMMUNITY CORRECTION WORK

With the continuous development of economic society, continuous innovation of science and technology and community administration's gradual improvement, Community Correction's information construction will be a long-term and continuous task.

6.1 *Reasonable planning and continually complete information process*

Setting Community Correction information process is a continuous, dynamic, and self-improvement process, which needs to be developed and put into use. During continuously improving information system processes, information technology can be actually operated as a good tool and means, based on full research and guided by local needs, reasonable planning and implementation step by step.

6.2 *Effective using information technology and enhancing superintendence strength*

Aiming at problems that current positioning technology is not able to benefit and is not compulsory in practical utilization, we suggest introducing advanced foreign technology or creating new position terminals, such as electronic wristbands and implanted chip. Sometimes correction offenders need to report personal information by making a phone call or in writing, these methods are easily counterfeited, so we suggest adopting vocal print, fingerprint, video recognition technology to enforce recognition and supervision. (Guo, Weihe. in Three problems about innovation of Community Correction work.)

6.3 *Strengthen co-operation between departments*

Informational Community Correction work's development needs to solve problems about information data's source and sharing. This requires support and cooperation from different departments and all social sectors to integrate all resources, to bring Community Correction work's superiority into full play and realize the ultimate goal which is personality correction and a reintegration into society. Information sharing not only benefits local Community Correction institution but also lays a foundation for the remote trusteeship of correction offenders on a national scale in the future.

6.4 *Increasing funds to improve information construction and enhancing staff member training*

It is necessary to increase input into the manpower and material resources of Community Correction work. For example, computers for departments, finger print scanners, and some necessary office equipment are needed. In addition, the centres must arrange to have professional people to be responsible for developing and using informational Community Correction work. Each informational Community Correction setting outlay must be enrolled into budgets for specific operation expenses, and the expense must be paid based on implementation scheduling in time. At the same time, it is better to reinforce training for Community Correction working staff. The Community Correction working staff needs to improve their skills in operating the computer and increase their ability to use the system, so that they promote a comprehensive capability.

7 CONCLUSIONS

In conclusion, in integrating information technology with Community Correction work it is important to fully use modern information technology functions, to promote establishing information management platforms in cities, counties (districts) and judicial offices; to improve the functions of platforms, to widen the application fields of information technology, and to make full use of information technology in track inspection, in warning alerts, SMS alerts, network education, etc. Thus we can implement standardization and networks in Community Correction and improve the effectiveness of Community Correction using modern information technology to develop Community Correction work which will not only saves human resource but also satisfy the need of improving Community Correction workers. Therefore, continuously exploring and improving Community Correction work by modern information technology is a strategy measures for improving correction quality, developing Community Correction work and promoting social harmony. (Pan, Jun. in Nowadays improvement and prospect of Community Correction system in China).

ACKNOWLEDGEMENTS

This work was supported by a grant from the Innovation and Entrepreneurship Training Plan of Liaoning Province.

REFERENCES

Guo Jian'an & Zheng Xiaoze. 2004. On Community Correction. *Beijing: LowPress.*

Guo, Weihe. 2011. Three problems about innovation of Community Correction work. *Social Affairs*: 6.

He, Rongjie. 2010. Education technology, Information technology and curriculum integration. *Modern Educational technology* 20: (41–42).

Li Hanyuan & Zheng Lei. 2014. Analysis of problems in the execution of Community Correction. *Journal of Xinyu University* 1: (96–97).

Pan, Jun. 2012. Nowadays improvement and prospect of Community Correction system in China. *East China University of Political Science and Law.*

Ren, Hanqing. & Xiong, Yuelin. 2012. Information technology education applications of Community Correction work. *Journal of Fujian Radio & TV University* (2):17.

Wu, Yingai. 2012. Making efforts to promote Community Correction work development. *Chinese Justice:* 11.

Electronics, Information Technology and Intellectualization – Song & Kwak (Eds)
© 2015 Taylor & Francis Group, London, ISBN 978-1-138-02741-1

A kind of visualization spatial clustering algorithm

Guangquan Fan
Management Science and Engineering College, Hebei University of Economics and Business Shijiazhuang, China

Liping Ma
Computer Center, Hebei University of Economics and Business, Shijiazhuang, China

ABSTRACT: The paper presents a new kind of Visualization Spatial Clustering Algorithm based on a spatial neighbour relationship. This algorithm utilizes Delaunay triangulation to find the Delaunay neighbour relation, and then through interactivities with the user, finished the clustering process visually. Through the experiments on two dimension datasets, satisfactory effect has been gained. It is a fast, visual clustering algorithm without parameters to be specified by the user and can discover arbitrary shape clusters.

1 INTRODUCTION

With the application and development of GPS, GIS and remote sensing technology, a large amount of data related to space is growing fast. However, though database technology can realize the spatial data input, editing, statistical analysis and query processing, the valuable pattern and model hidden in these large databases, cannot be found. Spatial Data Mining refers to the process of extracting implicit, non-trivial, previously unknown, and potentially useful space rules (spatial association rules, spatial characteristic rules, spatial discriminate rules, spatial evolution rules, spatial topological rules etc.) from the massive spatial database (data warehouse). It is an important branch of data mining. The spatial clustering analysis is an important method of spatial data mining. It not only can be used as an independent tool to explore the distribution of data, but also as a data pre-processing of other data mining methods.

Spatial neighbour relations are an important spatial relationship in GIS and there are a lot of applications in natural surface interpolation, spatial neighbour queries, and in the extracting of topographic characteristic lines. Delaunay triangulation is a triangulated irregular network (TIN – Triangulated Irregular Network), usually used in the modelling of a digital surface. A continuous triangle surface can be generated through the irregular distribution of data points to approximate the terrain surface. Delaunay triangulation is the dual graph of Voronoi graph in the closest point meaning. It is a kind of triangulation of the planar point set in which the circum-circle of every triangle does not contain any other point of the point set.

The Delaunay triangle network can directly describe the neighbour relation of spatial objects and its data structure is relatively simple. For complex spatial objects, we can also generate its Delaunay triangle network (constraint) easily. Therefore, it is a good method to describe spatial neighbour relationship and reasoning.

Now there are many mature algorithms to get the Delaunay triangulation of a planar point set. This paper presents and realizes a new kind of visual spatial clustering algorithm based on spatial neighbour relationships.

2 THE BASIC PRINCIPLE OF THE ALGORITHM

2.1 Relative definitions

The most important Delaunay neighbour relationship is first-order Delaunay neighbour relationship. In order to describe the degree of first-order Delaunay neighbour relationship, we firstly give the definition of first-order Delaunay neighbour distance and first-order Delaunay neighbour degree.

Definition 1 (first-order Delaunay neighbour distance): Let A and B be the points in the discrete spatial point set P and they are first-order Delaunay neighbour, first-order Delaunay neighbour distance between A (x_A,y_A) and B (x_B,y_B) is Euclidean distance between A and B.

$$d_1(A, B) = \sqrt{(x_B - x_A)^2 + (y_B - y_A)^2}$$

Definition 2 (first-order Delaunay neighbour degree): Let A and B be the points in the discrete spatial point set P and they are first-order Delaunay neighbour, the first-order Delaunay neighbour degree between A and B is:

$$\delta_1(A, B) = \frac{\text{MaxLength} - d_1(A, B)}{\text{MaxLength}}$$

there into:

$$\text{MaxLength} = \text{MAX}(d_1(X, Y) \mid X, Y \in P)$$

According to the above formula we see, first-order Delaunay neighbour degree $\delta \in [0, 1]$. Because the first-order Delaunay degree has eliminated the dimension, so it can objectively describe first-order Delaunay neighbour degree.

2.2 *Basic thought of the algorithm*

Each edge of the Delaunay triangle network is a first-order Delaunay neighbour relation. The length of the edge is inversely proportional to the first-order Delaunay neighbour degree.

The algorithm first used Delaunay triangulation algorithm of planar point set to get every triangle, and then get the edge set in which every edge is not repeated in the Delaunay triangle network, thus the spatial neighbour relation is obtained. On the screen we first display all the vertices in planar point set P and the Delaunay edge that first-order neighbour degree is bigger than any arbitrary threshold θ. In the graph, if the two points are connected through line segment, they belong to the same class. So that when the users drag the slider to adjust the threshold θ, those edges will be filtered, of which the first-order Delaunay neighbour degrees are less than the threshold θ. The edges with strong neighbour relationship remain. So the user can see the clustering result visually and can adjust the threshold θ dynamically. Thus, the visualization of the clustering effect is realized.

3 DESIGN AND IMPLEMENTATION OF THE ALGORITHM

3.1 *Design of the algorithm*

According to the above basic thought, the Visualization clustering algorithm is designed as follows:

Algorithm name: Visualization clustering algorithm based on spatial neighbour relationship
Input: planar point set P
Output: planar point set p' with clustering identification for every point

Algorithm description:

Step 1: Calculate Delaunay triangulation of planar point set P, get the set S of Delaunay triangles
Step 2: Extract the edges of each Delaunay triangle, and remove repeated edges to obtain first-order Delaunay neighbour relation R (that is the set of all Delaunay triangulation edges)

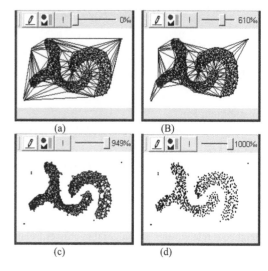

(a)　　　　　　　　(B)

(c)　　　　　　　　(d)

Figure 1. Visual clustering process.

Step 3: Calculate the length of each edge of Delaunay neighbour relationship R and neighbour degree δ of every pair of adjacent points;
Step 4: Set up an initial first-order Delaunay neighbour degree threshold θ (users can drag the slider to adjust it easily)
Step 5: Adjust the first-order Delaunay neighbour degree threshold θ;
Step 6: Display the points of planar point set P and the edges of which the first-order Delaunay neighbour degrees are greater than θ in Delaunay neighbour relationship R, and observe whether the result is satisfactory. If it is satisfactory then go to step 7, otherwise go to step 5;
Step 7: Identify of clustering, that is, identify the points connected together as the same class, and output the clustering result visually (draw the points in the same class with the same colour).

The algorithm is an iterative process, and needs the participation of the user. The whole clustering process is completed by the constant interaction of the user and the computer.

3.2 *Implementation of the algorithm*

The algorithm is realized under Jbuilder 9, VC++ 6 and Matlab 7.0.1.

Figure 1 shows some screenshots during the performance of the algorithm. When the threshold θ of first-order Delaunay degree is adjusted, the different visual modality will be displayed.

Through the visual clustering process in Figure 1 we can see that, when the threshold θ of first-order Delaunay degree is equal to 0, all of the first-order Delaunay neighbour relation can be displayed and that all points of the point set are connected to each other by a neighbouring relationship (the lines in the Figure 1); with the increase of the value of θ, the weak neighbour relationship will be removed gradually, only leaving

66

Figure 2. Visualization of clustering result.

the strong neighbour relationship. When the threshold θ is equal to 950 per cent (in the example) and the clustering effect is satisfactory, then we can proceed to the next step of the operation. That is to say, we will identify categories to points of the planar discrete point set.

3.3 *Identify clustering*

The process of identify clustering is to select the neighbour relations which are of the first-order Delaunay neighbour degrees, which are greater than or equal to neighbour degree threshold δ, and obtain the strong spatial neighbour relation R′. Then we can scan each binary relation (line segment) (A, B). The points A and B, which are associated with the binary relation, are classified as being of the same class.

There are four situations to be considered:

– If A and B have no class label, a new class label will be created and be assigned to point A and B.
– If one of the two points has class label, then the other point will get the same class label.
– If point A and B have the same class label, in this case, we do not need any process.
– If points A and B both have class labels, but they do not belong to the same class, then we need to merge the two classes. That is to say, the class labels of the points which have the same class labels as point A (B) are assigned to the class label of point B (A).

In this way, in order to finish clustering identification, we only need to scan all the 2-tuples in the neighbour relationships one time. Finally, we need to display the result of identify clustering visually.

According to the number of clustering, the algorithm can dynamically generate a colour array in which the differences of the colours are the biggest and assign automatically a different colour to every clustering. Figure 2 is the visualization display of one clustering result.

4 ANALYSIS OF EFFICIENCY AND CHARACTERISTICS OF THE ALGORITHM

4.1 *Efficiency of the algorithm*

The execution of the algorithm depends on the solution to the Delaunay triangulation. Now, the research on

the theory and algorithm of Delaunay triangulation is mature, methods are also more.

The efficiency of the Delaunay triangulation algorithm is relatively fast. After the execution of the Delaunay triangulation algorithm, we can get a set of triangles. To obtain the first-order Delaunay neighbour relation, we need only scan all triangles, and add all edges of the triangles into a database table, then utilize SQL to filter out duplicate edges. The time complexity is linear, and the efficiency is relatively high.

While determining the neighbour degree threshold in the interaction with the user, only the Delaunay strong neighbour relationships are selected and displayed on the computer screen. This procedure is also relatively fast, just as fast as displaying the vector map.

After determining the threshold of neighbour degree, the user needs to identify clustering in the last step of the algorithm. This procedure also needs only to scan each 2-tuples of the strong neighbour relationships selected. The time complexity is also linear.

In conclusion, except for the Delaunay triangulation, the time complexity of each step is always linear. The implementation efficiency is relatively high and is more suitable for the clustering of massive spatial data sets.

4.2 *Characteristics of the algorithm*

Through the above analysis, we can see that the clustering algorithm has the following advantages:

– Due to the use of the mature Delaunay triangulation algorithm, the algorithm runs faster and is suitable for processing large spatial databases.
– The user need not specify the threshold of clustering parameters. The threshold is determined during the interaction with the user, and the user's burden is reduced.
– It is a visual clustering algorithm. We identify the data points in one piece visually as a class, and achieve the effect of WYSIWYG.
– The algorithm can discover outliers automatically. When the algorithm finishes execution, the data points, of which the class IDs have not been modified, are the outliers. They are automatically classified as a class and have the initial class label 0.
– It can discover clusters of arbitrary shape. The algorithm is essentially a kind of clustering algorithm based on distance and the relative adjacent data points in space are clustered into one group.

5 CONCLUSIONS

This paper presents and realizes a visual clustering algorithm based on spatial neighbour relationships, and finishes the experiment on two-dimensional discrete point sets, and obtains satisfactory results. This shows the effectiveness of the algorithm. Further research is to extend the algorithm to the high dimension space.

REFERENCES

Dai, XiaoYan & Guo, ZhongYang & Li, QingFen & Wu, JianPing 2003. An Overview of Spatial Clustering Analysis and Its Application. Shanghai Geology (4):41–46.

Du, XiaoChu & Guo, QingSheng 2004. Spatial Neighbourhood Relation Reasoning Based on the Delaunay Triangulation. Science of Surveying and Mapping 28(6):65–67.

Han, J., Kamber, M. 2001, Data Mining: Comcepts and Techniques. San Francisco: Academic Press.

Lee, D.T. & Schachter, B. J. 1980. Two algorithms for constructing a Delaunay triangulation. International Journal of Computer and Information Sciences 9 (3):219–242.

Wang, JiaYe & Wang, WenPing & Tu, ChangHe & Yang, ChengLei 2011. Computational Geometry and Its Applications.Beijing: Science Press.

Xie, ZengGuang 2012. Divide-and-conquer algorithm for constructing Delaunay triangulation of planar points. Computer Engineering and Design 33(7):2652–2658.

Zhou, PeiDe 2011. Computational geometry – Algorithm Design and Analysis (Fourth Edition). Beijing: Tsinghua University Press.

Zhou, XiaoYun & Liu, ShenQuan 1996. A Robust Algorithm for Constrained Delaunay Triangulation. Chinese Journal of Computers 19(8):615–624.

Zhou, XueMei & Li, YingFei 2013. Based Bowyer-Watson triangulation generation algorithm research. Computer Engineering and Applications 49 (6):198–201.

Electronics, Information Technology and Intellectualization – Song & Kwak (Eds)
© 2015 Taylor & Francis Group, London, ISBN 978-1-138-02741-1

A dynamic packet assignment algorithm based on ECC

Yue-qian Zhang & Lei Yang
College of Electrical Engineering, Zhejiang University, Hangzhou, China

ABSTRACT: In order to decrease the Bit Error Rate (BER) and promote data-transmission efficiency, this paper has focused on the method for dynamically assigning the data packet size. The error probability of point to point communication was firstly analysed, then the adaptive packet assignment model based on the Error Correcting Code (ECC) was built and a maximum searching algorithm was used to search an appropriate packet size. A dynamic packet assignment algorithm was subsequently proposed. The hardware structure based on a STM32 single chip computer was designed to examine the proposed algorithm. Experimental results indicate that the transmission efficiency of the proposed algorithm is approximately 70 per cent and the transmission efficiency of the common algorithm is approximately 30 per cent, which verifies that this algorithm is feasible for improving the data-transmission efficiency.

1 INTRODUCTION

With the advancement of information communication technology, wireless network is gradually applied in more and more electronic equipment. Thus wireless information channels become crowded, mutual interference among different channels become severe and the Bit Error Rate (BER) inevitably increases[1]. The common method adopts Error Correcting Codes (ECC) to improve the reliability of information transmission[2]. This method adds some certain codes on the transmitting side according to the constraint relationship among information words and corrects the errors at the receive side due to the same constraint relationship. However, the data is divided and picketed in the fixed length so that the bit error rate cannot efficiently decrease. Consequently, it is necessary research a new algorithm which can dynamically adjust the length of data packets and the quantity of ECC[3]. In this algorithm, the length of data packets or other parameters are synchronously changed with the communication environment in real time in order to decrease the data resend times and increase the data-transmission efficiency.

In recent years, many scholars have devoted themselves to wireless communication protocol and proposed several new algorithms. In data link layers, the current algorithms are focused on lowering the bit error rate[4−6]. In document [4], aiming at the characteristics of a high-speed modulation channel, Qin Yan et al adopted interweave technology on the basis of the classic hamming code to transform the possible continuous error codes into one single stochastic error, simplify the complexity of error control algorithm, and improve the coding efficiency and the error correcting performance. In document [5], Deng Chunjian et al proposed an error control coding method to improve

the reliability of serial data transmission, used a parallel algorithm structure, deduced the generating coding matrix and efficiently reduced the bit error rate. In document [6], Gou Sheng-nan proposed an adaptive error control mechanism for wireless network data transmission in order to solve the problem of the limited wireless bandwidth. This mechanism adaptively selected the most appropriate error control technology by the structure of link layer frames and the communication length. Because communication interference is not completely eliminated, errors always exist in the communication process. However, these error data can be used to assess the quality of transmission and adjust the size of the data packet in order to increase the ratio of available data. Based on this viewpoint, a new fast algorithm is proposed to approach the maximum of data-transmission efficiency by keeping a small quantity of errors in the communication.

Aiming at the problem that an addition of the error correcting codes is not efficient in decreasing the bit error rate, the error probability of point to point communication is firstly analysed, then the adaptive packet separation model, based on ECC, is built and a maximum searching algorithm is proposed to dynamically adjust the size and structure of packet. Experimental results show that this algorithm efficiently decreases the data resend times and increase the data-transmission efficiency.

2 MATHEMATICAL MODEL ANALYSIS

2.1 *The error probability analysis of point to point communication*

An incident whether one bit is right or not is independent, thus this incident satisfies the random probability distribution function. There is one data of n bits and the probability that the bit i is wrong is $p(i)$.

The probability that all bits are right equals Equation 1:

$$P(n,0) = \prod_{i=1}^{n}[1-p(i)] \tag{1}$$

The probability that there is only one bit is wrong equals Equation 2:

$$P(n,1) = \sum_{i=1}^{n}\frac{P(n,0)}{[1-p(i)]} \cdot p(i) = \{\prod_{i=1}^{n}[1-p(i)]\} \cdot \{\sum_{i=1}^{n}[\frac{p(i)}{1-p(i)}]\} \tag{2}$$

The probability that no more than one bit is wrong equals Equation 3:

$$P(n,0)+P(n,1) = P(n,0) \cdot \{1+\sum_{i=1}^{n}\frac{p(i)}{1-p(i)}\} = \{\prod_{i=1}^{n}[1-p(i)]\} \cdot \{1+\sum_{i=1}^{n}\frac{p(i)}{1-p(i)}\} \tag{3}$$

Let the effective length of one data is l_{data} bits and this data is equally divided into n pieces (if l_{data} is not divisible by n, the remainder is made up of one packet. So the number of the total packets is $n+1$). Every piece of data adds m bits of redundant codes and e bits of error correcting codes.

The length of error correcting codes has two cases which are decided by Equation 4 and Equation 5.

$$e = (2+\log_2 u)(u=2^k, k=3,4,5,\cdots) \tag{4}$$

$$e = (3+\log_2 u)(u=2^k+c, k=3,4,5,\cdots, c\in(0,2^k)) \tag{5}$$

where u are the bits of one piece of data and they satisfy $u < len$. Len is the effective data length of one packet.

Thus it can be known that the total data length of all packets is $L = n(m+e)+a$, the data length of one packet is $l_{pakage} = m+e+fix(l_{data}/n)$ and the data length of the last packet is $l' = m+e'+(l_{data} \bmod (l_{data}/n))$. fix is a function of rounding down to the nearest whole unit.

When the data is transmitted by a wireless network, one bit of this data may go wrong and the probability is supposed as p. So the probability that less than two bits of error correcting codes go wrong equals Equation 6.

$$P_D = (1-p)^{(u-1)}(1-p+up) \tag{6}$$

The probability that all redundant codes always keep right equals Equation 7.

$$P_M = (1-p)^m \tag{7}$$

The probability that all error correcting codes always keep right equals Equation 8.

$$P_E = (1-p)^e \tag{8}$$

We suppose that a piece of data is divided into n packets. Then packet $i(i<n)$ includes several integral data units and the number is $unit = fix(fix(a/n)/u)$

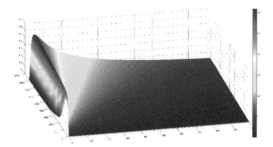

Figure 1. The Matlab simulation of energy efficiency expectation and data length.

and the data length of the last unit in packet i is $unit = fix(fix(a/n)/u)$. Packet n includes several integral data units and the number is $unit' = fix((a \bmod(a/n))/u)$ and the data length of the last unit in packet n is $u' = (a \bmod(a/n))\bmod u$.

During the actual communication, several packets are transmitted and received to send a piece of data. We suppose that P equals the probability that a piece of data does not need to resend. Some of its packets need to resend and the number of these packets is X. So it can be known that the probability that one packet is resend right equals Equation 9.

$$P = (1-p)^{l_{pakage}-unit-1}(1-p+up)^{unit}(1-p+u'p) \tag{9}$$

According to the derivation above, it can be known that when a piece of data is packed and transmitted, the mathematical expectation of the actual data length equals Equation 10.

$$l_{expect} = \frac{l}{P}[1-(1-P)^x]-xl(1-P)^x \tag{10}$$

If the resend times are unlimited then Equation 11 can be concluded by Equation 10.

$$\lim_{x\to\infty} l_{expect} = \frac{l}{P} \tag{11}$$

We suppose that when a bit of data is transmitted, energy E_0 is needed. Thus when a piece of data is sent, the energy expectation equals Equation 12 and the energy efficiency expectation equals Equation 13.

$$E_{Send} = \frac{E_0 l}{P} \tag{12}$$

$$\eta = \frac{E_{data}}{E_{Send}} = \frac{l_{data}}{nE_1+E_2} \tag{13}$$

2.2 The maximum search algorithm

Single chip computers (SCM) or others microprocessors have limited memory and calculation speed, so a simple algorithm is used to search the maximum instead of exhaustive search.

For this kind of function curve, the most simple search algorithm is $2N + 1$ ($N \geq 2$) points method. In this method, the computer gets $2N + 1$ points which equally separate an interval [a, b] (these points include interval endpoints and the value of the last point rounds down to the whole value when this interval length is not divisible by $2N + 1$, calculates the corresponding function values of these points and searches the maximum point by comparing their function values. The maximum point is fixed as one central point, its two consecutive points are fixed as the new interval endpoints and $2N + 1$ new division points are gotten. In this new interval, only $2N - 2$ points need to calculate their corresponding function values. Because in the proposed protocol, the values in all axes are positive integers. Thus when the maximum search ends, the length of the last search interval is not more than 2.

In the maximum search algorithm, the search lower bound is not less than the search lower bound when the data cache is full and the search upper bound is not more than the search upper bound when every bite is independently packed in case a packet without any data is transmitted. Thus the values of interval endpoints a and b equal Equation 14.

$$a = \frac{l_{data}}{l_{storage\_max}}, b = l_{data} \qquad (14)$$

where $l_{storage\_max}$ is the data length when the data cache is full.

For any integer $N \geq 2$, the total number of intervals is $2N$. Then in one maximum search calculation, the number of the maximum point is 1, the number of new search interval endpoints is 2 and only $2N-2$ points need to calculate their corresponding function values in next search calculation. Because the new search interval, after an iteration, shrinks into $1/N$ of the previous interval, the convergence rate is $1/N$. Through x iterations, the length of the search interval equals equation $(1/N)^x \cdot (b-a)$.

In the iteration process, a constraint condition which equals $(1/N)^x \cdot (b-a) \leq 2$ is always satisfied.

Thus the total iteration times equals Equation 15 and the times of using energy efficiency function to finish the maximum search equals $x(2N-2)+3$.

$$x \geq \log_N (\frac{b-a}{2}) \qquad (15)$$

We suppose that the abscissas of all division points compose the series. Then at the initial state, $X(n) = a + (b-a)(n-1)/(N+1), n = 1, 2, \ldots, 2N+1$. Because the values of $n = 1$ and $n = 2N+1$ are known, $2N-1$ points in the series need to calculate. Considering the condition that 3 points do not need to calculate their corresponding function values in every iteration, the optimized time complexity equals $O[\log_N((b-a)/2)]$. Thus, it can be known that when the number of points N is certain, the total time for maximum search logarithmically grows with the length of search interval.

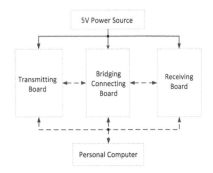

Figure 2. The hardware structure of a dynamic packet assignment algorithm.

In actual tests, the accuracy of the proposed maximum search algorithm is 0.91, the error points are not more than 4 points. The maximum function value error is not more than 5.471687e–04. The total iteration times is x = 9 and the times of using energy efficiency function are 21. When N severally equals 5, 8, 9, 10, 11, the searched maximum point has no errors. The iteration times × severally equals 4, 3, 3, 3. The times of using energy efficiency function severally equals 35, 45, 51, 57. Thus when the searched maximum point need no errors, we select N = 5 in order to make the search time and the memory overhead minimal.

In the above test results, the searched maximum point may be false. It is because the energy efficiency function is nonlinear. This function curve does not evenly increase or decrease, so when the search interval is divided at the same sample size, these division points may not include the actual maximum point in the function curve. However, through the following experiment, it can be known that the above search errors hardly have any impact on the maximum search results. Therefore, in order to reduce the calculation, we do not need to correct the above search errors in the proposed fast algorithm.

3 THE HARDWARE STRUCTURE OF DYNAMIC PACKET ASSIGNMENT ALGORITHM

Figure 2 is the hardware structure of dynamic packet assignment algorithm. As shown in Figure 2, the hardware structure consists of three STM32 development boards. NRF24L01 chips are installed in the left and right development boards. The middle STM32 development board is used as a bridging connection board to interconnect the left and right boards and the universal asynchronous transmitter (UART) is applied to communicate the bridging connection board with a personal computer.

Figure 2 is the hardware structure of a dynamic packet assignment algorithm. As shown in Figure 2, the hardware structure consists of three STM32 development boards. NRF24L01 chips are installed in the left and right development boards. The middle STM32

Table 1. Experimental results of two packet assignment algorithms.

The proposed algorithm			The common algorithm		
Payload /Byte	Total Tx /Byte	Efficiency (%)	Payload /Byte	Total Tx /Byte	Efficiency (%)
32667	51342	63.63	32667	102798	31.78
81361	118154	68.86	81361	263904	30.83
125105	173551	72.09	125105	407710	30.68
144705	202838	71.34	144705	451204	32.07
193386	264289	73.17	193386	626589	30.86
246933	349349	70.68	246933	829088	29.78
329246	481986	68.31	329246	1075597	30.61
434404	623156	69.71	434404	1437497	30.22
4144422	5888209	70.39	4144422	13575297	30.53

development board is used as a bridging connection board to interconnect the left and right boards and the universal asynchronous transmitter (UART) is applied to communicate the bridging connection board with a personal computer.

Considering that BER of wireless communication module is low and its ECC is difficult to capture, the bridging connection board is used to simulate the wireless communication environment and BER is set by the personal computer. The left and right development boards are connected with the bridging connection board by UART. Thus, the communication efficiency is obtained by reading the transmitting and receiving data efficiency recorded in the left and right board.

4 EXPERIMENTAL RESULTS

In order to examine the proposed dynamic packet assignment algorithm, an experiment was designed to compare the proposed algorithm with the common algorithm that fixes the packet size. In the experiment, the BER always equalled 10^{-9}. By changing the payload length, the total transmission data length and transmission efficiency were obtained.

In table 1, The *Payload* is the payload length, *Total Tx* is the total transmission data length and *Efficiency* is the transmission efficiency. It can be seen that the transmission efficiency of the proposed algorithm is approximately 70 per cent and that the transmission efficiency of the common algorithm is approximately 30 per cent. The proposed algorithm has an obvious advantage over the common algorithm in transmission efficiency. In other words, the proposed algorithm has a faster transmission speed. If BER equals 10^{-3} instead of 10^{-9}, the efficiency of the proposed algorithm is 48 per cent while the efficiency of the common one decrease to 3.4 per cent. Thus, the proposed algorithm has the more obvious effect on efficiency when there are severe errors in the data transmission process.

5 CONCLUSIONS

In this paper, the dynamic packet assignment algorithm can flexibly adjust the packet size and correct a bit of error code according to the communication environment. By keeping the number of error data and controlling the packet size, the ratio of available data increases and transmission efficiency improves a lot. Besides, the proposed algorithm has an obvious advantage on the common algorithm in transmission efficiency and transmission speed, especially when the signal interference is passive. In other words, the proposed algorithm can be applied in the industrial communication environment and effectively avoids the impacts of industrial noise. Thus, the dynamic packet assignment algorithm can be applied in point to point communication and obviously improves communication quality.

REFERENCES

Deng, C. An, Y. Lv, Y. 2012. Error control of high speed serial display data transmission. *Optics and Precision Engineering* 20(3): 632–641.

Gou, S. 2011. Performance study of an Error Control Mechanism for Wireless Sensor Networks. *Computer Simulation* 28(5): 167–170.

Liu, K. & Xu, H. 2012. Research on Reliability of Industrial Wireless Communication. *Computer Simulation* 29(12): 123–126.

Qin, Y. Wu, R. Su T. 2012. Research on High-Speed Data Communication and Error Control. *Fire Control Radar Technology* 41(3): 54–59.

Teng, X. Zhang, L. Jiang, Z. 2000. Analysis and simulation of Analysis and simulation of bit error performance for GMSK in frequency-selective fading channel. *Chinese Journal of Radio Science* 15(1): 1–6.

Wang, H. Li, F. Wu, M. 2011. Study on wireless sensor network adaptive error controlling technology. *Hign Technology Letters* 21(5): 465–470.

Electronics, Information Technology and Intellectualization – Song & Kwak (Eds)
© 2015 Taylor & Francis Group, London, ISBN 978-1-138-02741-1

A novel foreign object debris classification method for runway security

B. Niu, H.B. Gu & Z.X. Gao

College of Civil Aviation, Nanjing University of Aeronautics and Astronautics, Nanjing, China
Engineering Technology Research Center of Flight Simulation and Advanced Training,
Nanjing University of Aeronautics and Astronautics, Nanjing, China

ABSTRACT: Foreign Object Debris (FOD) on the runway is harmful to aircraft. Currently, most of the FOD monitoring is still carried out by human activity, which is inefficient and unreliable and also takes up valuable time while the runway is not in use. This paper proposed a foreign object debris classification methodology based on ordinarily Gabor multi-support vector machine (named OGMSVM for short). The Gabor wavelets, whose kernels are similar to the 2D receptive field profiles of the mammalian cortical simple cells, exhibit desirable characteristics of spatial locality and orientation selectivity. When the Gabor features were obtained, we constructed binary SVMs to OGMSVM by applying ordinal pair wise partitioning strategy. The results of the proposed method were compared with other conversional MSVMs and they showed the novel approach outperformed other multi-class classifications. The experiments' results showed that foreign object debris can be classified accurately.

Keywords: Foreign object debris, Gabor wavelet, Support Vector Machine (SVM), Multi-class classification, Ordinal pair wise partitioning

1 INTRODUCTION

Foreign Object Debris (FOD) presence on airport runways, taxiways, aprons and ramps poses a significant threat to the safety of air travel. FOD has the potential to damage aircraft during critical phases of flight, which can lead to catastrophic loss of life and the aircraft, or increased maintenance and operating costs (Michael, 2009). But it typically does not receive the same level of attention as an actual accident. Since the Air France Concord tragedy in July 25, 2000, was caused by a 43 cm (17 inch) metal strip that had fallen from an aircraft which took off five minutes earlier, FOD has become of increasing international interest. Many countries have sought to develop a technology that would have prevented the Concord disaster.

The International Civil Aviation Organization (ICAO) upgraded two clauses pertaining to FOD on the runway from Recommended Practice to Standards in July 2005. European Organization for the Safety of Air Navigation (Euro control) had initiated the Airport Runway Debris Detection as an innovation action and released a preliminary assessment of FOD Detection technologies in 2006. Moreover the Federal Aviation Administration (FAA) has been conducting research of the four leading systems from Qinetiq (a millimetre wave radar system), Stratech (a high resolution intelligent vision system), Xsight (a combination radar and camera system), and Trex Enterprise (a mobile millimetre wave radar and infrared camera system) to evaluate the available technologies for debris detection during 2007 and 2008 (Niu et al., 2013). The runway is a difficult environment for installing other equipment. Different FOD has different threats to the aircraft, so classification of FOD is important. With no EMI/EMC (electromagnetic interface/ compatibility) issues and lower-cost, we think the best performance can be obtained from mobile camera sensors and we propose a novel methodology using Gabor wavelets and multi-class SVM to classify foreign object debris.

In this study, a novel approach for Multi-class Support Vector Machine (MSVM) is proposed. We term the approach Ordinal Gabor Multi-class Support Vector Machine (or OGMSVM for short). Although combining several binary SVM classifiers, the approach we proposed is different from the traditional methods. First, Gabor wavelet is introduced to construct Gabor feature. Second, the Ordinal Pairwise Partitioning (OPP) approach is used to extend the binary GSVMs (Kwon et al., 1997). The OGMSVM uses fewer classifiers but may be more accurate than traditional methods. To validate the effectiveness of the proposed approach, we apply the approach to FOD classification. The results of the proposed approach and the traditional MSVM methodology were compared. We also compare the results of the approach with other traditional classification techniques.

The rest of this paper is organized as follows: Section 2 introduces a Gabor wavelet and the methodology of extracting a Gabor feature from FOD. In section 3, the approach for ordinal Gabor multi-class classification is proposed. Section 4 describes the data and experiments for validating the approach. Also, the empirical results are summarized and discussed. The final section presents the conclusions and future research direction on this study.

2 GABOR FEATURE EXTRACTION

2.1 Gabor wavelets

Gabor wavelets are often used in recognition applications, such as face recognition, fingerprint recognition, character recognition, etc because of their similarity to a human vision system (Gabor, 1946, Granlund, 1978). This feature based method aims to extract frequency and orientation information from an image. A family of complex Gabor wavelets can be defined as follows:

$$\Psi_{u,v}(k,z) = \frac{\|k_{u,v}\|^2}{\sigma^2} \exp\left(-\frac{\|k_{u,v}\|^2 \|z\|^2}{2\sigma^2}\right).$$

$$\left[\exp(ik_{u,v}z) - \exp\left(-\frac{\sigma^2}{2}\right)\right] \tag{1}$$

where the parameter u and v define the orientation of the Gabor kernels and the scale of the Gabor kernels, $z = (x,y)$, $\|\cdot\|$ denotes the norm operator, and the wave vector $k_{u,v}$ is defined as follows:

$$k_{u,v} = k_v e^{i\varphi_u} \tag{2}$$

where $k_v = k_{max}/f^v$ and $\varphi_u = u\pi/U$, k_{max} is the maximum frequency, and f is the spacing factor between kernels in different central frequencies (Bissi et al. 2013, Zhang & Liu 2013, Liu & Wechsler 2002).

By scaling and rotation via the wave vector $k_{u,v}$, all the Gabor kernels in Eq. (1) can be generated from the mother wavelet, so they are all self-similar. Each kernel is a product of a Gaussian envelope and a complex plane wave, while the first term in the square brackets in Eq. (1) determines the oscillatory part of the kernel and the second term compensates for the DC value. The parameter σ determines the ratio of the Gaussian window width to wavelength.

$v \in \{0, 1, 2, \ldots, V-1\}$ is scale label, and $u \in \{0, 1, 2, \ldots, U-1\}$ is orientation label. In most applications, Gabor wavelets with five different scales: $V = 5$ and eight orientations: $U = 8$ are used. With the following parameters: $k_{max} = \pi/2, f = \sqrt{2}, \sigma = 2\pi$, the kernels exhibit desirable characteristics of spatial frequency, spatial locality, and orientation selectivity. The Gabor kernels are shown in Figure 1.

2.2 Representation of Gabor feature

The Gabor wavelet transformation of an image is the convolution of the image with a family of Gabor kernels as defined by Eq. (1). Let $I(x,y)$ be the gray level

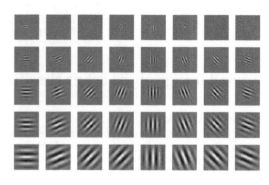

Figure 1. Gabor kernels (Real part with five scales and eight orientations).

distribution of an image, and define the convolution output of image I and a Gabor kernel $\Psi_{u,v}$ as follows:

$$G(z) = I(z) * \Psi(k,z) \tag{3}$$

where $z = (x,y)$, and * denotes the convolution operator. $G(z)$ is the convolution result. As described $V = 5$, $U = 8$, 40 Gabor filters are made, the set $S = \{G_{u,v}(k,z) : u \in \{0, \ldots, 7\}, v \in \{0, \ldots, 4\}\}$ forms the Gabor wavelet representation of the image $I(z)$. Where $G_{u,v}(k,z) = |I(z) * G(k,z)|$, all the Gabor features can be described as $G(I) = G = (G_{0,0}G_{0,1} \ldots G_{7,4})$. When the convolution results over each pixel of the image are concatenated to form an augmented feature vector, the size of the vector could be very large. Taking an image of size 128×128 for example, the convolution results will give $128 \times 128 \times 5 \times 8 = 655360$ features. Due to the large number of convolution operations, the computation and memory cost of feature extraction is also necessarily high. So each $G_{u,v}(k,z)$ is down-sampled by a factor r to reduce the space dimension. For a 128×128 image, the vector dimension is 10240 when the down-sampling factor $r = 64$.

3 A NOVEL APPROACH OF OGSVM FOR FOD CLASSIFICATION

In this section, we propose a novel approach of multi-class SVM technique named Ordinal Gabor Multi-class Support Vector Machine (OGMSVM), which is a hybrid algorithm that applies the Gabor wavelet and Ordinal Pairwise Partitioning (OPP) technique to multi-class SVM.

OPP is an approach for improving the performance of Artificial Neural Network (ANN) models of ordinal multi-class classifications, which was proposed by Kwon et al. (1997). When several binary ANN classifiers were combined to deal with multi-class classification problems, it performed better than a single ANN classifier. Thus, a new method named Ordinal Pair wise Partitioning (OPP) was proposed to consider the order of classes when combining several binary ANN classifiers. SVMs are originally binary classifiers; hence they are well-suited to the use of a new type of multi-class SVM model (Kim & Ahn, 2012).

In this study, the Ordinal Pair wise Partitioning (OPP) approach is used as a tool for upgrading the Gabor multi-class SVM. Depending on the methods applied for portioning and fusion, four types of OGMSVM are introduced. For the portioning method, there are the One-Against-The-Next and One-Against-Followers approaches. The One-Against-The-Next method is as same as One-Against-One, but more efficient. In One-Against-One method, all the classifiers for each pair of classes should be developed. However, in One-Against-The-Next, the binary classifiers are constructed only for the pairs $\{(i, i + 1): i = 1, 2, \ldots, k - 1\}$, where k is the total number of classes. Consequently, One-Against-The-Next constructs only $k - 1$ classifiers, when there are k classes.

In contrast to One-Against-The-Next, One-Against-Followers is similar to – but slightly more efficient than – One-Against-All. In case of One-Against-Followers, the classifiers are constructed for the pairs $\{(i, j): i = 1, 2, \ldots, k - 1, j = U_{m=i+1}^{k} m\}$, where k is the total number of classes. As a result, where there are k classes, One-Against-Followers also construct only $k - 1$ classifiers, although One-Against-All constructs k classifiers.

Regarding methods of fusion, there are forward and backward methods, named in accordance with the 'direction of reasoning'. The forward method fuses the classifiers in the forward direction, which determines the highest level of classes first, and the lowest level last. By contrast, the backward method combines the classifiers in reverse direction, which determines the lowest level of class first, and the highest level last.

The process of OGMSVM consists of two phases: (1) preparation and (2) interpretation. In the preparation phase, OGMSVM constructs the individual classifiers using the Gabor features. To elaborate, OGMSVM first divides the whole training dataset into $k - 1$ groups, in accordance with the partitioning method. Then, OGMSVM trains $k - 1$ GSVM models with each of the above data-subsets. For example, when the One-Against-The-Next approach is used for four-level classification problems, the first phase of OGMSVM produces three classification models: model 1 for the pair of classes $(1, 2)$; model 2 for the pair of classes $(2, 3)$; model 3 for the pair of classes $(3, 4)$.

In the interpretation phase, OGMSVM determines the class for the input data using the classifiers that are built in the first phase. To do this, it fuses the classifiers either in the forward or backward direction. In the case of the above example, the forward method begins with model 1. If a test datum is put into class 1 by model 1, then it is deemed "class 1". Otherwise, the test datum is passes on to model 2. If it is put into class 2 by model 2, then it is deemed "class 2". Otherwise, model 3 applies. In model 3, the test datum is finally classified as either "class 3" or "class 4". Using

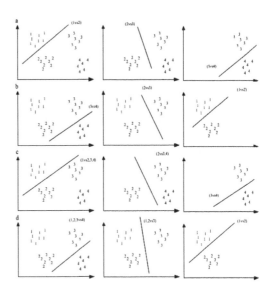

Figure 2. Difference between four types of OGMSVM: graphical presentation. (a) One-Against-The-Next + Forward approach, (b) One-Against-The-Next + Backward approach, (c) One-Against-Followers + Forward approach and (d) One-Against-Followers + Backward approach.

Figure 3. Some of typical FODs.

the same reasoning, the backward method starts with model 3. That is, if a test datum is put into class 4, then we regard it as belonging to "class 4". Otherwise, the test datum is passed on to the next model. The remaining procedure is the same as that of the forward method but in reverse order.

Figure 2 graphically shows the mechanism of each type of OGMSVM.

4 EXPERIMENTAL DESIGN AND RESULTS

In our experiments, FOD is divided into four classes, metal, plastic, paper and stone. Each kind has different Gabor features, 100 samples are collected for each kind. The video images are taken under different illumination conditions. Figure 3 shows some of typical FODs.

The normalized FOD images with a resolution of 64×64, 40 Gabor filters are used to extract features, the Gabor feature vector of each FOD image will be $64 \times 64 \times 5 \times 8 = 163840$ dimensions. Down-sampling Gabor feature vector by factor $r = 64$, the vector dimension is reduced to 2560. In addition, we normalized all the data in the range $[new\_min_x, new\_max_x]$ by computing Eq. (4):

$$x' = new\_min_x + \frac{x - min_x}{max_x - min_x} \cdot$$
$$(new\_max_x - new\_min_x)$$

(4)

Normalization is often applied to enhance the performance of the prediction model because it ensures that the larger value input features do not overwhelm smaller input features. In this study, all features of our datasets range from 0 to 1. To overcome the scarcity of samples, we adopted the fivefold cross validation. All the images were divided into five subsets with an equal number of images from each class. One of the five subsets was selected as the test set and the other four subsets were put together to form a training set. Every image appeared once in a test set, and appeared four times in a training set. The average accuracy across all five trails was computed.

To thoroughly validate the superiority of our model's performance, we applied our proposed model (OGMSVM) as well as six MSVM techniques, namely: (1) One-Against-All; (2) One-Against-One; (3) DAGSVM; (4) ECOC; (5) the method of Weston and Watkins; (6) the method of Crammer and Singer. The Gaussian radial basis function was used as a kernel function in SVM. We adopted libSVM 3.16 provided by Chang and Lin for SVM classification (Chang & Lin 2001). BSVM 2.06 software was used to execute multi-class classification as per the methods of Weston and Watkins and of Crammer and Singer (Hsu & Lin 2006).

The parameters C and σ^2 play an important roles in the performance of the Gaussian radial basis kernel classifier. In order to find the optimal parameters, we used grid search method proposed by Lin et al. and this strategy can improve the generalization of classifiers.

To compare the performance of each algorithm, we adopted the hit-ratio as the performance measure. Simply put, the hit-ratio means the ratio of the corrected case over all cases. The hit-ratio is defined in

$$CR = \frac{1}{n}\sum_{k=1}^{n} CA_i; \quad CA_i = 1 \; if \; PO_i = AO_i, \; 0 \; otherwise \text{(5)}$$

In Eq. (5), CR is the classification accuracy rate of the test data, CA_i is the classification accuracy of the ith case of the test data denoted by either 1 or 0 (correct means 1 while incorrect means 0). PO_i is the predicted outcome for the ith case, and AO_i is the actual outcome for the ith case.

From table 1, we can conclude that the forward in One-Against-Followers plays the best performance, the average accuracy is 94.50%. The method of One-Against-Followers performs better than One-Against-The-Next, the backward strategy is 87.50% > 85.00%, the forward strategy is 94.50% > 93.75%. In the same partitioning method, the forward strategy outperformed the backward strategy. For One-Against-The-Next the accuracy is 93.75% > 85.00%, and for

Table 1. Classification accuracy using OGMSVM (%).

	One-Against-The-Next		One-Against-Followers	
Dataset	Forward	Backward	Forward	Backward
1	91.25	87.50	92.50	86.25
2	95.00	83.75	95.00	88.75
3	96.25	85.00	96.25	87.50
4	92.50	86.25	93.75	90.00
5	93.75	82.50	95.00	85.00
Average	93.75	85.00	94.50	87.50

Table 2. Classification accuracy using normal GMSVM (%).

Dataset	One-Against-One	One-Against-All	DAG-SVM	ECOC	Weston and Watkins	Crammer and Singer
1	88.75	81.25	95.00	81.25	83.75	86.25
2	95.00	80.00	90.00	80.00	85.00	82.50
3	91.25	83.75	92.50	83.75	86.25	81.25
4	93.75	86.25	96.25	87.50	87.50	87.50
5	91.25	82.50	93.75	86.25	88.75	83.75
Average	92.00	82.75	93.50	83.75	86.25	84.25

One-Against-Followers the accuracy is 94.50% > 87.50%.

Table 2 presents the performance of other GMSVMs. Among the traditional Gabor MSVMs, DAGSVM performs the best, the accuracy is 93.50%, One-Against-One came second with the accuracy 92.00%, while One-Against-All method performs the worst, and the accuracy is only 82.75%. The penultimate method is ECOC with the accuracy 83.75%. The methods of Weston and Watkins, Crammer and Singer classify all the data at once, Weston and Watkins strategy outgoes Crammer and Singer method, with the accuracy 86.25% > 84.25%, but underperformed the DAGSVM and One-Against-One approaches.

In this experiment, compared with the approaches of constructing several binary classifiers, we used only three classifiers, which were the fewest. The other method used seven (ECOC), six (One-Against-One), six (DAGSVM), four (One-Against-All) classifiers respectively.

In order to validate that Gabor feature played an important part in the classifiers, we used Principal Component Analysis (PCA) and Linear Discriminant Analysis (LDA) to extract FOD features, the same classifiers were used. Table 3 to Table 6 showed the classification performance by using PCA and LDA as feature extracting methods respectively. The accuracy is an average of five validations.

Comparing from Table 1 to Table 6, we can conclude that using Gabor wavelet to extract FOD feature, then constructing GMSVM, performed better than PCA and LDA methods.

Table 3. Classification accuracy using PCA+OMSVM (%).

One-Against-The-Next		One-Against-Followers	
Forward	Backward	Forward	Backward
83.00	80.75	84.25	81.25

Table 4. Classification accuracy using PCA + MSVM (%).

One-Against-One	One-Against-All	DAG-SVM	ECOC	Weston and Watkins	Crammer and Singer
82.00	79.75	82.50	80.75	81.25	81.00

Table 5. Classification accuracy using LDA + OMSVM (%).

One-Against-The-Next		One-Against-Followers	
Forward	Backward	Forward	Backward
87.00	82.00	87.75	83.00

Table 6. Classification accuracy using LDA + MSVM (%).

One-Against-One	One-Against-All	DAG-SVM	ECOC	Weston and Watkins	Crammer and Singer
85.50	80.75	86.50	82.00	84.25	82.50

5 CONCLUSIONS

In this study, we proposed a novel algorithm called OGMSVM for FOD recognition. This method uses Gabor wavelet and order-information in ordinal multi-classification. As the experiments showed, OGMSVM outperformed other SVMs by introducing the Gabor feature and an ordinal pairwise partitioning approach. In our experiments, using a Gabor wavelet to extract FOD features is better than using PCA and LDA approach. Also, One-Against-Followers with forward strategy was more effective. And the proposed method requires the fewest classifiers among the multi-SVMs, which are designed by combing several binary classifiers.

However, further study should be continued: it is important for designing a better dimensionality reduction method to reduce the dimension of Gabor feature. The relationship of sample size used for training and the classifiers' performance should be developed. The computation time of the proposed method and other classifiers should be conducted in our future experiments.

The author would like to acknowledge the assistance and support of Lukou International Airport in Nanjing. The project is supported by Funding of Jiangsu Innovation Program for Graduate Education and the Fundamental Research Funds for the Central Universities (No.CXZZ12_0164). Moreover, this project is also supported by the Fundamental Research Funds for the Central Universities (NS2012060).

REFERENCES

Bissi, L., Baruffa, G. Placidi, P. et al. 2013. Automated defect detection in uniform and structured fabrics using Gabor filters and PCA. *Journal of Visual Communication and Image Representation*, 24(7):838–845.

Chang C.C. & Lin, C.J. 2001. LIBSVM: a library for support vector machines. Software available at http://www.csie.ntu.edu.tw/~cjlin/libsvm.

Gabor, D. 1946. Theory of communications. *Journal of Institution of Electrical Engineers*, 93(26): 429–157.

Granlund, G. H. 1978. In search of a general picture processing operator. *Computer Graphics and Image processing*, 8(2): 155–173.

Hsu, C.W. & Lin, C.J. 2006. BSVM: a SVM library for solution of large classification and regression problems. Software available at http://www.csie.ntu.edu.tw/~cjlin/bsvm.

Kwon, Y.S., Han, I., Lee, K.C. 1997. Ordinal pairwise partitioning (OPP) approach to neural networks training in bond rating. *Intelligent System in Accounting Finance and Management*, 6:23–40.

Kim, K.J., Ahn, H. 2012. A corporate credit rating model using multi-class support vector machines with an ordinal pairwise partitioning approach. *Computers and Operations Research*, 39:1800–1811.

Liu, C. & Wechsler, H. 2002. Gabor Feature based Classification using the Enhanced Fisher Linear Discriminant Model for Face Recognition. *IEEE Transactions on Image Processing*, 11(4): 467–476.

Michael J. 2009. Airport Foreign Object Debris (FOD) Detection Equipment. *Advisory Circular*, Federal Aviation Administration.

Niu, B. Gu, H.B., Sun, J., et al., 2013. Research of FOD recognition based on Gabor wavelets and SVM classification. *Journal of Information & Computational Science*, 6: 1633–1640.

Zhang, Y. & Liu, C.C. 2013. Gabor feature-based face recognition on product gamma manifold via region weighting. *Neurocomputing*, 117(6):1–11.

Electronics, Information Technology and Intellectualization – Song & Kwak (Eds)
© 2015 Taylor & Francis Group, London, ISBN 978-1-138-02741-1

Feature selection in medical text classification based on Differential Evolution Algorithm

Huaying Zhou, Qirui Zhang, Hexian Wang & Dan Zhang
College of Medical Information Engineering, Guangdong Pharmaceutical University, Guangzhou, China

ABSTRACT: Artificial intelligence algorithms have been widely applied in text classification and effective feature selection is essential to make the algorithm more efficient and accurate. This paper puts forward a new feature selection algorithm based on Differential Evolution (DE) Algorithm for medical text classification. It believes that the objective function based on Jeffries-Matusita distance and evolution operation can be sure to become rapidly the global optimal solution, which will speed up searching for the most suitable feature subsets among feature collections. The experimental results show that the classification accuracy in medical documents was improved effectively and feature space is much reduced. Compared with Genetic Algorithm (GA), the proposed method can select the best feature subsets for classification with limited evolutionary generations.

1 INTRODUCTION

Text Classification (TC) is the processing and automatically classifying predefined categories in order to free text documents and this is the core content of the text information search and mining. With the rapid development of the biomedical field and computer information technology, a large amount of biomedical knowledge is widely available online. In order to effectively manage and use all kinds of medical text information, text classification based on medical text content is particularly important. However the high-dimensional nature of the text data may make the calculation very complicated. Therefore, the feature dimension reduction is an integral part of the process of text classification.

The most important problem in text classification is the high dimensionality of the feature space. The initial feature set is made up of tens of thousands of medical terms (words or phrases) that come from the medical texts. So it is highly desirable to reduce the feature set without affecting the classification accuracy. Feature selection is trying to pick a subset of features that are relevant to the target concept.

Ideally, feature selection methods search through the subsets of features, and try to find the best one among the competing 2^N candidate subsets according to some evaluation function. However this procedure is exhaustive as it tries to find only the best one. It may be too costly and practically prohibitive, even for a medium-sized feature set size (N).

The focus in this paper is to propose a new method of feature selection for medical text classification according to the characteristics of medical documents. The rest of this article is organized as follows: Section II describes the differential evolution algorithm. Section III describes Feature Selection Using DE. Section IV are Summary and Conclusions.

2 DIFFERENTIAL EVOLUTION ALGORITHM (DE)

Many feature selection methods have been tackled in a variety of ways, such as Document Frequency(DF), Information Gain (IG), Mutual Information (MI), X^2 statistics, Expected Cross Entropy (ECE), Immune Cloning Selection Algorithms (ICSA) and Genetic Algorithms (GA). Following the neural network and genetic algorithms, differential evolution algorithm is a new bionics topic. It includes mutation, crossover, and selection operation.

Storn and Price propose a powerful computational implementation of the differential evolution that originates from the genetic algorithm. The differential evolution algorithm doesn't need to have an encoding and decoding operation as a genetic algorithm does. It establishes the idea that the difference vector, the crossover principle, and the selection operator can be sure to gain the property of rapid convergence to the global optimum. The DE algorithm includes three important steps, as follows.

2.1 Mutation operation

In the differential evolution algorithm, the process of mutation needs to use a plurality of individual linear combinations from their parents. Variation vector a_i is defined as follow.

$$a'_i = a_{r_1} + F \cdot (a_{r_2} - a_{r_3}), \quad i = 1,2,...,NP \tag{1}$$

$\{a_{r_1}, a_{r_2}, a_{r_3}\}$ are three different individuals selected from their parents randomly, and $r_1 \neq r_2 \neq r_3$ i. F (called scaling factor) is a constant factor and its value is between [0, 2].

2.2 Crossover operation

The crossover operation is designed by recombining the mutation feature vector (called a_i) and the target feature vector (called a_i) in order to improve the diversity of the new individual. The new crossover vector a''_{ij} is defined as follow.

$$a_{i,j}'' = \begin{cases} a_{i,j}', & randb \leq CR, or, j = rand_j \\ a_{i,j}, & randb > CR, or, j \neq rand_j \end{cases}$$
$$i = 1,...,NP, j = 1,...,NP \tag{2}$$

randb is randomized from [0,1]. *CR* is a constant factor between [0,1].

2.3 Selection operation

The selection operation is the choice of mode of a kind of 'greed'. Only when the fitness value of a''_i (the new vector) is better than that of the a'_i (the target vector), a_i will it be accepted. If the optimization problem is conversed to a function, such as $\max f(a)$, so the selection operation is defined as follow.

$$a_i^{t+1} = \begin{cases} a_i'', & f(a_i'') > f(a_i') \\ a_i', & others \end{cases} \tag{3}$$

In this paper, we propose a new feature selection method, called the feature selection, based on a differential evolution algorithm (DE). This establishes the idea that we select a subset of M features from a set of N features, where $M < N$, such that the value of a criterion function is optimized over all subsets of size M. Typically, an evaluation function tries to measure the discriminating ability of a feature or a subset to distinguish the different class labels. So it can be described as that selecting d features from D features to construct the feature subset which its similarity is the largest one. The largest similarity is served as the biggest distance among the training examples that come from different text categories. The similarity formula J_M based on the average Jeffries-Matusita (J-M) distance is defined as follow:

$$J_M = \sum_{i=1}^{c} \sum_{j=i+1}^{c} \sqrt{2\left(1 - \exp\left(-B_{ij}\right)\right)} \tag{4}$$

B_{ij} is the Bhattacharyya distance between the class i and class j. It is defined as follow:

$$B_{ij} = 1/8\left(M_i - M_j\right)^T \left[\frac{\left(V_i + V_j\right)}{2}\right]^{-1} \left(M_i - M_j\right) + 1/2\ln\frac{\left|\left(V_i + V_j\right)/2\right|}{\sqrt{|V_i||V_j|}} \tag{5}$$

where M_i denotes the average vector of the documents which come from class i, M_j denotes the average vector of the documents coming from class j, V_i denotes the matrix covariance of class i, and V_j denotes the matrix covariance of class j.

So the main task of the feature selection is to select a part of feature terms, by which we can gain the largest J_M.

The procedure of feature selection based on DE for medical text classification is designed as follows:

The first step: randomly select N_p (population size) feature sets to make up of the original vector group $A(a_0)$. Each of the vector is represented a kind of feature combination and all of them are between 0~1. Each vector is described by a d-dimensional vector $(a_{w1}, a_{w2}, \ldots, a_{wd})$, where d represents the original length of the feature vector. $a_{wi} = 1$ represents that the corresponding index term is selected; $a_{wi} = 0$ *represents* that the corresponding index terms is not selected.

The second step: express each vector to the corresponding index terms combination and we can gain the new training documents, then calculate the affinity $\{J_M(A((a_0))\}$ by (4).

The third step: judge whether the termination condition is satisfied. If it is not satisfied, go to the follow steps.

The fourth step: mutate the a_k by Eq. (1), a'_k can be get.

The fifth step: cross the a'_k by Eq. (2), a''_k can be get.

The sixth step: express each vector to the corresponding index terms combination and we can gain the new training documents, then calculate the affinity $\{J_M(A(a''_k))\}$ by Eq. (4).

The seventh step: select the vector a''_k by Eq. (3). If one vector a''_k has the largest J_M, (a''_k) $= \max\{J_M(a_{ij})|j = 1, 2, \ldots L, q_i\}$ (q_i representing the vector dimensional), that is to say $J_M(a_i) < J_M(a''_k)$, $a_i \in A(k)$, in the new group $A''(k)$. Meanwhile, select some vectors who have the larger affinity to form the next generation population $A(k + 1)$ according to the pre-set percentage.

The eighth step: according to the new generation population $A(k + 1)$, calculate the affinity $\{J_M(A(k + 1))\}$ by Eq. (4).

The ninth step: $k = k + 1$, then go to the third step.

3 FEATURE SELECTION USING DE

3.1 Data sets

In order to test the performance of the feature selection method based on differential evolution algorithm (DE) in experiments, we set up a wide range of medical data consisting of 4803 documents from CNKI, CQVIP and WanFangData in 2000~2012. From the dataset, we select five categories for training and testing. The category structure is shown in Table 1. In addition, the number of each category is also shown in Table 1. Its category distribution is more uniform, so it is a balanced data set.

Table 1. The category distribution.

Item	S_C	C_C	NS_C	RS_C	DS_C
Category Label	1	2	3	4	5
Examples	857	1330	1008	949	659

where: S_C means category of stomatology, C_C means a cardiovascular category, NS_C means a category of the nervous system, RS_C means a category of the respiratory system, and DS_C means a category of the digestive system

3.2 Classification algorithm

In our experiments, we use the KNN (K-nearest neighbour classification algorithm) method to classify the dataset. The KNN has been widely used in text classification.

The principle of the KNN algorithm is very simple: to give a test text, the system selects k nearest neighbours among the training texts, and uses the category rules of the k neighbours to measure the test texts. The decision rule in KNN can be written as:

$$y(x,c_i) = \sum_{d_j \in kNN} sim(x,d_j) y(d_j,c_i) - b_i \qquad (6)$$

where $y(d_j,c_i)$ means text d_j belonging to category c_i ($y=1$ means YES, and $y=0$ means NO); $sim(x,d_j)$ represents the similarity between the test text x and the training text d_j; b_i means the category-specific threshold of the binary decisions. The cosine value of two vectors is used to measure the similarity between of the two documents.

3.3 Performance evaluation

The paper uses recall (r), precision (p) and F1 measure to evaluate the classification performance. In our experiments, micro-averaging is used, i.e., these scores are computed globally over all the n × m binary decisions where n is the number of total test documents, and m is the number of categories.

3.4 Differential Evolution (DE) algorithm initialization

In this paper, we randomly set the dataset into a training set and a testing set according to the proportion of 2:1 and the subset of a feature is constructed based on DE. The population size is set to 20, the length of individual coding is set to 80, and the maximum generation of evolution is set to 60.

3.5 Experimental results and analysis

In this study, we used the Naïve Bayes (NB) classifier and the K-nearest neighbour's (KNN) classifier to test the feature selection method of differential evolution algorithm (DE), and compared it with the feature selection method of genetic algorithm (GA). The experimental results are shown in Table 2 and Figure 1.

Table 2. The classification performance.

Item	S_C	C_C	NS_C	RS_C	DS_C
DE_F1	76.40%	73.80%	77.50%	70.20%	79.50%
GA_F1	73.30%	69.60%	75.90%	68.30%	74.80%

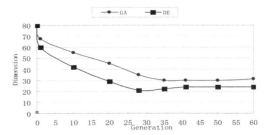

Figure 1. Comparison of classification performance on different featured selection methods

As shown in Table 2, based on the DE, the average F_1 measure came up to 75.48% and the largest of F_1 79.50% was obtained on the category of the digestive system. While basing the performance on the GA, the average F_1 measure was only 72.38% and the largest F_1 75.90%. The experiments show that the feature selection method based on DE can be successfully applied to medical text classification, which indicates that the feature selection method is feasible.

It can be seen from Figure 1, compared with the genetic algorithm (GA), that the DE can get the better performance of classification with fewer features and make the classifier more efficient with fewer evolutionary generations.

In these experiments, we find that the problem is how to define the standard of the feature weight and this is very important. The more accurate the weight, the better performance the classifier can get. So we can get the conclusion that the standard of the feature weight is close to medical terminology. Therefore, in further work, we will study the problem how to achieve the standardization of feature weight for medical text classification.

4 SUMMARY AND CONCLUSIONS

In this paper, we propose a new method of feature selection, based on differential evolution algorithm (DE), for medical text classification. In this method, the mutation operation can expand the searching space effectively; the crossover operation can maintain the diversity from different feature terms. The selection operation can reduce effectively the size of the population through local optimization. The experiment's performance shows that the method of feature selection based on DE has the best classification performance. Next, we will study the problem of how to define the standard of the feature weight in the medical field.

REFERENCES

Ali Wagdy Mohamed, Hegazy Zaher Sabey, "Constrained optimization based on modified differential evolution algorithm" Information Sciences 2012(194). pp. 171–208.

Chakraborty U. Advance in Differential Evolution. New York: Springer, 2008.

Elissa, K. "Title of paper if known," unpublished. Kira, K. and Rendell, L.A., "The feature selection problem: Traditional methods and a new algorithm". In: Proceedings of Ninth National Conference on Artificial Intelligence, pp. 129–134, 1992.

Hua-ying Zhou, Qi-rui Zhang, Man Luo, He-xian Wang. "Feature Selection in Medical Text Classification Based on Immune Algorithm". China.2010 the 3rd International Conference on Advanced Computer Theory and Engineering (ICACTE 2010). August, 2010. pp. 212–216, Chengdu.

Liuling Dai, Heyan Huang, and Zhaoxiong Chen." A comparative study on feature selection in Chinese text categorization". Journal of Chinese Information Processing, vol. 18, January 2004, pp. 26–32.

Matthieu Weber, Ferrante Neri, Ville Tirronen, "A study on scale factor in distributed differential evolution." Information Sciences 2011(181), pp. 2488–2511.

Qirui Zhang, Jinghua Tan, HuayingZhou, Weiye Tao, Kejing He. "Machine learning methods for medical text categorization". Proceedings of 2009 Pacific-Asia Conference on Circuits, Communications and System, 2009, pp. 162–165.

Qirui Zhang, Jinghua Tan, Man Luo, Hexian Wang. "A Medical Text Classification System Based on Immune Algorithm". 2008 International Conference on Future BioMedical Information Engineering, 2008, pp. 345–348.

Qirui Zhang, Jinghua Tan, Weiye Tao, Hexian Wang, "A Web Page Classifier Based on Immune Algorithm". Journal of Computational Information Systems, 2009, pp. 179–185.

Qirui Zhang, Man Luo, Hexian Wang, Jinghua Tan. "A Hyper-lipidemia Information Analysis System Based on Immune Algorithm". Proceedings of 2010 International Conference on Computer Application and System Modelling, 2010, pp. 421–424.

Raymer M L, Punch W F, Goodman E D, et al. "Dimensionality reduction using genetic algorithms" [J]. IEEE Transactions on Evolutionary Computation, 2000, 4(2), pp. 164–171.

Storn R, Price Differential evolution – a simple and efficient adaptive scheme for global optimization over continuous spaces. Berkley: International Computer Science Institute, 1995. http://www.cs.umsl.edu/~uday/DE-CEC2006.

Yan Xu, Jintao Li, Bin Wang and Chunming Sun, "A category resolve power-based feature selection method," Journal of Software, vol. 19, January 2008, pp. 82–89.

Yiming Yang, O. P. Jan (1997). "A comparative study on feature selection in text categorization". Proceeding of ICML-97, 14th International Conference on Machine Learning. pp. 412–420.

Yong Wang, Zixing Cai, Qingfu Zhang, "Enhancing the search ability of differential evolution through orthogonal crossover.". Information Sciences 2012(185), pp. 153–177.

Electronics, Information Technology and Intellectualization – Song & Kwak (Eds)
© *2015 Taylor & Francis Group, London, ISBN 978-1-138-02741-1*

Improving autocorrelation performance of 2-Dimensional Coupled Logistic (2DCL) sequence using phase space methods

Bin Chen
School of Software and Communication Engineering, JiangXi University of Finance and Economics, NanChang, China

ABSTRACT: The well-known Logistic Chaotic sequence has been widely used. The 2-Dimensional Coupled Logistic (2DCL) sequence is more complex, and could be used more widely. However, the 2DCL sequence autocorrelation performance is poor, so in this paper, by using recently presented phase space methods and APAS theorem, the 2DCL sequence autocorrelation performance is improved. Meanwhile, the sequence's other performances did not deteriorate.

1 INTRODUCTION

The Logistic chaotic sequence is widely used as a kind of pseudorandom sequence[1–25], which is not too complex, and therefore can be easily attacked. A 2-Dimentional coupled Logistic (2DCL) sequence has been presented at some conferences[26–28], and the 2DCL sequence is more complex, and could be used more widely. However, the 2DCL sequence autocorrelation function has some problems and its applications are limited.

The autocorrelation function is very important for signal detection and identification in communication and radar[1–25]. Recently, some rules between autocorrelation and phase space trajectory have been revealed in the Autocorrelation Phase-space Axis Symmetric (APAS) theorem[1,25] by using phase space methods. We can therefore use the APAS theorem to improve the 2DCL sequence autocorrelation performance.

2 THE AUTOCORRELATION PROBLEM OF A 2DCL SEQUENCE

The dynamic equation of a 2DCL map[26–28] is given by

$$\begin{cases} x_1(n+1) = 4(1-u)x_1(n)[1-x_1(n)]+ux_1(n)x_2(n) \\ x_2(n+1) = 4(1-u)x_2(n)[1-x_2(n)]+ux_2(n)x_1(n) \\ \\ x(n+1) = x_1(n+1)-0.5 \end{cases} \quad , \quad (1)$$

where $x_1(n), x_2(n)$ are the state variables of the two coupled Logistic maps respectively. u is the coupled factor, and $(1-u)$ are also the parameters of the two coupled Logistic maps. The whole system output variable is $x(n)$ whose value area is $x(n) \in [-0.5, 0.5]$.

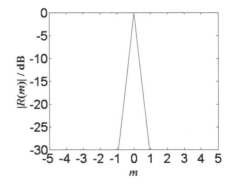

Figure 1. Normalized autocorrelation of Logistic sequences.

Given suitable initial values of $x_1(0)$, $x_2(0)$ and suitable coupled factor u, as n increases, a chaotic sequence $x(n)$ can be obtained.

In communication and radar, the autocorrelation function shape of a good performance signal should be like a sharp needle: that is, the autocorrelation function has a sharp and prominent main peak, and has no obvious side lobes, as shown in Figure 1. This means that the signal can be easily detected and identified[1–14].

The autocorrelation function of a Logistic sequence is depicted in Figure 1. Its main peak is sharp and prominent, and the function has no obvious side lobes, so the performance is good. On the other hand, the 2DCL autocorrelation function has a blunt main peak and some obvious side lobes, so its performance is poor, as shown in Figure 2. The 2DCL sequence then needs to be reformed in order to improve its autocorrelation performance.

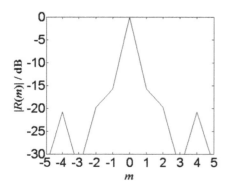

Figure 2.　Normalized autocorrelation of 2DCL map.

3　APAS THEOREM

The dynamic equation of a discrete chaotic system is given by

$$x(n+1) = \varphi[x(n)].\tag{2}$$

The autocorrelation function $r(m)$ of the chaotic sequence $\{x(n)\}$ with length N is defined as

$$r(m) = \sum_{n=1}^{N-m} x(n)x(n+m).\tag{3}$$

If $m = 0$, the autocorrelation function $r(m)$ reaches its maximum $r(0)$

$$r(0) = \sum_{n=1}^{N} x(n)^2\quad.\tag{4}$$

Then, the normalized autocorrelation function $R(m)$ is defined as

$$R(m) = r(m)/r(0).\tag{5}$$

We list the following Autocorrelation Phase-space Axis Symmetric (APAS) theorem[1,25]. The APAS theorem was presented recently, and was proved and verified in Ref. [1, 25].

Theorem 1: Given a stationary ergodic discrete real dynamic system:

$$x(n+1) = \varphi[x(n)]\quad,\tag{6}$$

where $\varphi(x)$ is a single valued function, the value area of $x(n)$ is $[-a, a]$, and 'a' is a positive real number, the mean of $\{x(n)\}$ is zero, the values of $x(n)$ are statistically balanced: that is $f[x(n)] = f[-x(n)]$, where $f[x(n)]$ is the probability density function of $x(n)$.

If $\varphi(x)$ satisfies any one of the following conditions:

$$x(n+1) = \varphi[x(n)] = \varphi[-x(n)]\quad,\tag{7}$$

$$f[x(n), x(n+1)] = f[-x(n), x(n+1)],\tag{8}$$

we reach the following conclusion.

If $N \to \infty$, and delay $m \neq 0$, for any other delay m, the normalized autocorrelation function of $\{x(n)\}$ is almost zero. That is,

$$\lim_{N\to\infty} R(m) = \lim_{N\to\infty} \frac{1}{r(0)} \sum_{n=1}^{N-1} x(n)x(n+m) = 0$$

$$m \in N,\quad m \neq 0\tag{9}$$

Theorem 2: Assume a stationary ergodic discrete real dynamic system:

$$x(n+1) = \varphi[x(n)]\quad,$$
$$\tag{10}$$

where the inverse function $\varphi^{-1}(x)$ of $\varphi(x)$ is single valued, the value area of $x(n)$ is $[-a, a]$, and 'a' is positive real, the mean of $\{x(n)\}$ is zero, the values of $x(n)$ are statistically balanced: that is $f[x(n)] = f[-x(n)]$, where $f[x(n)]$ is the probability density function of $x(n)$.

If $\varphi(x)$ satisfies any one of the following conditions:

$$x(n) = \varphi^{-1}[x(n+1)] = \varphi^{-1}[-x(n+1)]\quad,\tag{11}$$

$$f[x(n), x(n+1)] = f[x(n), -x(n+1)]\quad,\tag{12}$$

we reach the following conclusion.

If $N \to \infty$, and delay $m \neq 0$, for any other delay m, the normalized autocorrelation function of $\{x(n)\}$ is almost zero. That is,

$$\lim_{N\to\infty} R(m) = \lim_{N\to\infty} \frac{1}{r(0)} \sum_{n=1}^{N-1} x(n)x(n+m) = 0$$

$$m \in N,\quad m \neq 0\tag{13}$$

Combining Theorem 1 and Theorem 2, the Autocorrelation Phase-space Axis Symmetric (APAS) theorem is obtained. The APAS theorem was proved and verified in Ref. [1, 25].

In order to assess a chaotic sequence's autocorrelation performance, we can observe the shape of the delay 1 phase space trajectory of the chaotic sequence. If the trajectory shape is x- or y-axis symmetrical, then according to the APAS theorem, its autocorrelation performance is good, while the opposite is not true.

The delay 1 phase trajectory and the value distribution of the Logistic sequence is shown in Figures 3(a) and (b) respectively; its trajectory is +, − value balanced and y-axis symmetrical, which satisfies the conditions of the APAS theorem. It demonstrates that its autocorrelation performance is good, as shown in Figure 1, and this accords with the APAS theorem. For other chaotic sequences that satisfy the conditions of the APAS theorem, such as a Tent sequence, we get the same result.

Studying a 2DCL map[26−28], although its trajectory seems to be y-axis symmetrical, as shown in Figure 4(a), its value distribution is not +, − value balanced, as shown in Figure 4 (b). Therefore, its delay

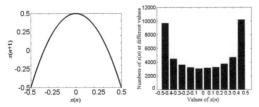

(a) Delay 1 phase space trajectory. (b) Value distribution.

Figure 3. Delay 1 phase space trajectory and value distribution of Logistic sequence.

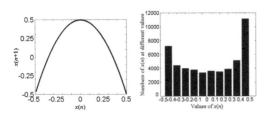

(a) Delay 1 phase space trajectory. (b) Value distribution.

Figure 4. Delay 1 phase space trajectory and value distribution of 2DCL sequence.

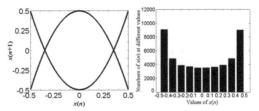

(a) Delay 1 phase space trajectory. (b) Value distribution.

(c) Normalized autocorrelation of S-2DCL sequence.

Figure 5. Delay 1 phase space trajectory, value distribution and autocorrelation of S-2DCL sequence.

1 phase trajectory is neither x-axis symmetrical nor y-axis symmetrical, so it does not satisfy the conditions of the APAS theorem; its autocorrelation performance is therefore poor, as shown in Figure 2, and this accords with the APAS theorem, too.

4 IMPROVE THE AUTOCORRELATION PERFORMANCE OF 2DCL SEQUENCE

In order to improve the 2DCL sequence autocorrelation performance, based on the APAS theorem, we modified the phase trajectory structure of 2DCL sequence and changed its delay 1 trajectory into an x- and y-axis symmetrical trajectory. The changed sequence is called Symmetrical 2DCL(S-2DCL) sequence, and its trajectory and value distribution are shown in Figure 5(a) and (b) respectively.

The dynamic structure equation of S-2DCL map is defined as:

$$\begin{cases} x_1(n+1) = 4(1-u)x_1(n)[1-x_1(n)] + ux_1(n)x_2(n) \\ x_2(n+1) = 4(1-u)x_2(n)[1-x_2(n)] + ux_2(n)x_1(n) \\ \\ z(n+1) = c - d|z(n)| \\ \\ \begin{cases} x(n+1) = x_1(n+1) - 0.5 & \text{if } z(n+1) \ge 0 \\ x(n+1) = -[x_1(n+1) - 0.5] & \text{if } z(n+1) < 0 \end{cases} \end{cases} \tag{14}$$

where $x_1(n), x_2(n), z(n)$ are the state variables, and u, c, d are the parameters. The whole system output variable is $x(n)$ whose value area is $x(n) \in [-0.5, 0.5]$.

Choosing suitable initial values of $x_1(0)$, $x_2(0)$, $z(0)$, and suitable parameter values for u, c, d, such as, $u = 0.005, c = 0.5, d = 1.999, x(n)$ evolves in a chaotic state, and satisfies $x(n) \in [-0.5, 0.5]$.

For the S-2DCL sequence, as shown in Figure 5(a), (b), its value distribution is $+, -$ balanced, and its delay 1 phase trajectory structure is x-axis and y-axis symmetrical; it therefore satisfies the APAS theorem conditions. The sequence autocorrelation performance is therefore good: that is, the autocorrelation function has sharp and prominent main peak, and has no obvious side lobes, as shown in Figure 5(c).

Other characteristics of the S-2DCL sequence, such as chaotic property, ergodicity, spectrum, cross-correlation, Lyapunov exponent, etc., have also been studied and they have not deteriorated either.

Thus, the S-2DCL sequence has good autocorrelation performance and other characteristics, plus it is more complex than the well-known Logistic sequence. The sequence could therefore be used more widely.

5 CONCLUSIONS

The autocorrelation performance of 2DCL sequence is poor and its applications are limited.

In this paper, in order to improve the autocorrelation performance of a 2DCL sequence based on APAS theorem, the 2DCL map has been changed into an S-2DCL map. The S-2DCL sequence has excellent autocorrelation performance, whereas its other performance characteristics did not deteriorate. This means that this sequence could be used more widely.

ACKNOWLEDGMENT

This work was supported by the National Natural Science Foundation of China (Grant No. 61362024) and the Natural Science Foundation of Jiangxi Provincial, China (Grant No. 20114BAB201021).

REFERENCES

Bucolo M., Caponetto R., Fortuna L., Frasca M., Rizzo A., 2002. Does Chaos Work Better Than Noise? *Circuits and Systems Magazine, IEEE*, 2(3): 4–19.

Chen B., 2011. Assessment and improvement of autocorrelation performance of chaotic sequences using a phase space method. *Science China Information Sciences*, 54 (12): 2647–2659.

Chen B., Zhou Z. O., Liu G. H., Zhang Y., Tang J., 2008. Application of Chaos Series as Noise Source in Noise Radar. *Modern Radar*, 30(5): 24–28.

Chon K. H., Kanters J. K., Iyengar N., etc., 1997. Detection of Chaotic Determinism in Stochastic Short Time Series. *IEEE/EMBS Proc.-19th International Conference*, 10: 275–277.

Chen B., Tang J., Cai P., etc., 2010. Research on autocorrelation of chaotic sequence by phase space method. *2010 IEEE International Conference on Wireless Communications, Networking and Information Security*, Beijing, China, 6(2): 6–12.

Dawood M., and Narayanan R. M., 2000. Ambiguity function of an ultrawideband random noise radar. *Proc. IEEE Antennas and Propagation Soc. Int. Symp.*, Salt Lake City, UT, USA, 4: 2142–2145.

Flores B. C., Solis E. A., and Thomas G., 2003. Assessment of chaos-based FM signals for range-doppler imaging. *IEE Proc.-Radar Sonar Navig.*, 150(4): 313–322.

Flores B. C., Solis E. A., and Thomas G., 2002. Chaotic signals for wideband radar imaging. *Proc. SPIE-Int. Soc. Opt. Eng.*, 4727: 100–111.

Gambi E., Chiaraluce F., Spinsante S., 2008. Chaos-Based Radars for Automotive Applications: Theoretical Issues and Numerical Simulation. *IEEE Transactions on Vehicular Technology*, 57(6): 3858–3863.

Gu H., Liu G., Zhu X., Su W., and Li X., 1997. A new kind of noise-radar random binary phase coded CW radar. *Proc. IEEE National Radar Conf. Syracuse*, NY, USA: 202–206.

Jose A. R., Eduardo R., Juan C. E., Hector P., 2008. Correlation analysis of chaotic trajectories from Chua's system. *Chaos, Solitons and Fractals*, 36: 1157–1169.

Kohda T., Tsuneda A., 1997. Statistics of chaotic binary sequences. *IEEE Transactions on information theory*, 43(1): 104–112.

Kohda T., 2002. Information sources using chaotic dynamics. *Proceedings of the IEEE*, 90(5): 641–661.

Liu Z., Zhu X. H, Hu W., Jiang F., 2007. Principles of chaotic signal radar. *International journal of bifurcation and chaos in applied sciences and engineering*, 17(5): 1735–1739.

Myers J., and Flores B. C., 1999. Radar imaging via random FM correlations. *Proc. SPIE-Int. Soc. Opt. Eng.*, 3721: 130–139.

May R. M., 1976. Simple mathematical models with very complicated dynamics. *Nature*, 261: 459–467.

Rovatti R., Mazzini G., Setti G., Giovanardi A., 2002. Statistical modeling and design of discrete-time chaotic processes: advanced finite-dimensional tools and applications. *Proceedings of the IEEE*, 90(5): 820–841.

Sune R. J., 2004. Noise radar using random phase and frequency modulation. *IEEE Trans. Geosci. Remote Sens.*, 42(11): 2370–2384.

Sakaguchi H., Tomita K., 1987. Bifurcation of the coupled Logistic map. *Progress of Theoret. Phys.*, 78(3): 305–309.

Setti G., Mazzini G., Rovatti R., Callegari S., 2002. Statistical modeling of discrete-time chaotic processes: basic finite-dimensional tools and applications. *Proceedings of the IEEE*, 90(5): 662–690.

Syuji M., Miki U. K., Kei E., Mika I. etc., 2008. New developments in large deviation statistics and time correlation calculations in chaotic dynamics and stochastic processes. *IEICE Technical Report*, 3: 37–42.

Vijayaraghavan V., Henry L., 2005. A novel chaos-based high-resolution imaging technique and its application to through-the-wall imaging. *IEEE Signal Processing Letters*, 12(7): 528–531.

Wu X., Liu W., Zhao L., and Fu J., 2001. Chaotic phase code for radar pulse compression. *Proc. IEEE National Radar Conf.*, Atlanta, GA, USA: 279–283.

Xu Y., Narayanan R. M., Xu X., and Curtis J., O. 2001. Polarimetric processing of coherent random noise radar data for buried object detection. *IEEE Trans. Geosci. Remote Sens.*, 39(4): 467–478.

Yang Y. T., Wang M. Z., Zhang Z. M., 2007. Research on Digital Sound Spreading-Frequency Modulation Based on Hyperchaos Sequence. *Journal of UESTC*, 9: 699–702.

Zhong Q. C., Zhu Q. X., Zhang P. L., 2009. Multiple Chaotic Maps Encryption System. *Journal of UESTC*, 4: 274–277.

Zhao Mingchao, Wang Kaihua, Fu Xinchu., 2009. The Dynamical Properties of 2-D Coupled Logistic Map and Applications to Stream Cipher. *Comm. On Appl. Math. and Comput.*, 23(1): 87–92.

Electronics, Information Technology and Intellectualization – Song & Kwak (Eds)
© *2015 Taylor & Francis Group, London, ISBN 978-1-138-02741-1*

Wavelength dependent birefringence in dual-core hybrid Photonic Crystal Fibre

Y. Liu & Q. Pan
Xi'an Communications Institute, Xi'an, China

X.G. Xie
Xi'an Communications Institute, Xi'an, China
Institute of Communications Engineering, Plaust, Nanjing, China

Y.L. Che
Xi'an Communications Institute, Xi'an, China

J.H. Li
Institute of Communications Engineering, Plaust, Nanjing, China

ABSTRACT: This paper investigates the birefringence mechanisms in dual-core hybrid Photonic Crystal Fibre (PCF). For this type of PCF, the birefringence is achieved by the coupling of the hybrid core mode and the cladding deficit mode. It is very different from those obtained by the asymmetry of the fibre structure or in another way by increasing the index differences between the two orthogonal polarization modes. It is therefore very easy to control to a certain wavelength range in order to match our need by simply adjusting the geometric structure parameters. The numerical results show that the birefringence theory of the proposed PCF is correct.

Keywords: Birefringence; Coupling; Hybrid photonic crystal fibre; Dual-core; Confinement loss

1 INTRODUCTION

In recent years, photonic crystal fibres have attracted much attention due to their unique properties which would be difficult to achieve in conventional fibres. Furthermore, it is well known that PCFs can be made highly birefringent. Many works have been reported so far. Generally speaking, these highly birefringent fibres can be divided into three categories. In the first category, the birefringence is introduced by the asymmetry of the fibre structure [1–26]. Typical designs of this structure can be achieved by the asymmetry of the fibre core region and the asymmetry of the fibre cladding, or the simultaneous asymmetry of the fibre core and cladding. The other way to get high birefringence is a stress-induced technique [27, 28]. The third way to achieve high birefringence is to infiltrate index-tunable liquids, polymers, or optofluids into the air holes of the PCFs to get tunable high birefringence [29–33]. In those PCFs, high birefringence can just be achieved in a numerical region, but the operating region was uncertain which is not easy to control.

In this paper, a wavelength dependent birefringence in a dual-core structure hybrid photonic crystal fibre is first proposed. For this type of PCF, the birefringence can be controlled in a certain range which would be needed. This type of birefringence is achieved by the coupling of the hybrid core mode and the cladding deficit mode. The value of the birefringence will increase sharply near the phase matching wavelength λ_p, and it will get to the top on λ_p. It remains almost zero when the wavelengths are apart from λ_p. This type of birefringence is very different from those obtained by the asymmetry of the fibre structure or in another way by increasing the index differences between the two orthogonal polarization modes. We can easily control its operating wavelength by changing the structure parameters of the proposed PCF. A comparison of two kinds of hybrid structure is also proposed to have a comprehensive understanding of the wavelength dependent birefringence in dual-core hybrid photonic crystal fibre.

2 GEOMETRIC AND NUMERICAL METHODS

In this paper, two kinds of hybrid structure were proposed. In order to have a visible comparison, the geometric parameters of the two hybrid PCFs were kept in uniform. However, the locations of the high index rods are different. The cross sections of the proposed PCFs are shown in Figures 1(a) and 1(b). The air holes and the doped rods in the hybrid DCF are arranged in a hexagonal pattern, which has a pitch Λ

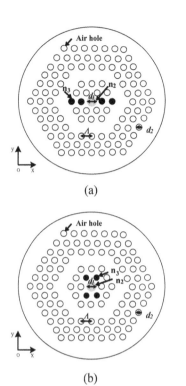

(a)

(b)

Figure 1. Geometric structure of the proposed PCF for structures A and B.

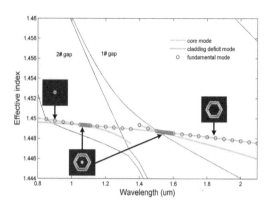

Figure 2. Birefringence mechanism of the proposed PCF.

from the difference in the real part of n_{eff} for the two polarizations as:

$$B = \left| n_{eff}^x - n_{eff}^y \right| \qquad (1)$$

The confinement loss [20] can be calculated from the imaginary part of the n_{eff} factor as:

Confinement loss

$$= \frac{40\pi}{\ln(10)\lambda} \cdot \mathrm{Im}[n_{eff}] \times 10^3 \quad (dB/km) \qquad (2)$$

3 ANALYSIS OF THE MECHANISM AND NUMERICAL RESULT

In this section, the mechanism of the birefringence of the dual-core hybrid photonic crystal fibre is analysed, and properties are discussed in a numerical way.

The proposed PCF has a special birefringence mechanism different from others. It is achieved by the coupling of the hybrid core mode and the cladding deficit mode in a certain wavelength range. The peak value of the birefringence is changed with the phase matching wavelength λ_p. Therefore, if we could find the evolution law of the λ_p, we can easily control it to match our needs.

The birefringence mechanism of the proposed PCF is shown in Figure 2. We can see that the coupling of the hybrid core mode and the cladding deficit mode occurrs when the hybrid core mode is close to the boundary of the band-gap. That is because the guiding mode in the core region is guided by both the index-guiding and band-gap guiding mechanism when the propagation wavelengths are within the band-gaps. However, nearing the edge of one of the band-gaps, the confinement ability of a band-gap to the guiding mode will become weak. Therefore, the mode confined in the core field will leak gradually to the cladding deficit area of the fibre. The coupling between the guiding mode in the edge of the band gap and the cladding deficit mode then occurs. The coupling changes the mode field distribution of the PCF, leading to an obvious difference between the two orthogonal

of 7.5 μm, the central deficit core is a positively doped region with a refractive index of 1.451 and the diameter d1 of 9 μm; the other four positively doped rods around the core have a refractive index of 1.49, and the diameter d2 of 3 μm, and the diameter of all the other air holes is also d2 = 3 μm, and the background index is 1.45. The third ring of air holes was absent to make a dual-core structure, in order to make sure that the coupling between the hybrid core mode and the cladding deficit mode occurs. For the structure A, the core has a higher refractive index than the cladding along the Y direction, the propagation light confined in the core in this direction can be guided by total internal reflection, which is the so called index-guiding mechanism. On the other hand, along the X direction, the cladding rods have a higher index than the doped core region, which ensures that the propagation light in this direction can be only guided by the photonic band-gap effective. This special structure ensures that the designed PCF has a hybrid lighting guide mechanism. For structure B, the hybrid light guiding mechanism is similar to structure A, only the index guiding direction and band-gap guiding direction are on the opposite.

To analyse the modal birefringence of the proposed PCF, the full vector finite-element method (FEM) with anisotropic perfectly matched layers (PML) was employed to calculate the mode field distribution as well as the complex modal effective indices of the proposed PCF. The modal birefringence can be found

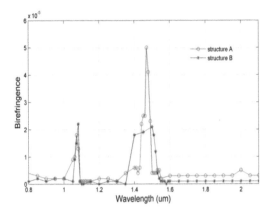

Figure 3. Birefringence curves of the proposed PCF of structure A and B.

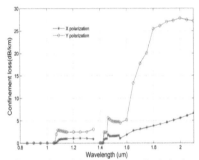

(a) Confinement loss of the two orthogonal polarization modes for structure A

(b) Confinement loss of the two orthogonal polarization modes for structure B

Figure 4. Confinement loss of the two orthogonal polarizations of the proposed PCF for structure A and B.

polarization modes. Because of the band-gap guiding mechanism of the hybrid core mode, the proposed fibre has multiple band-gaps. This, in turn, means that the coupling of those two modes will happen in multiple times. That is to say there will be multiple peak values of the birefringence. In our case, the coupling between the two modes takes place twice. Therefore, there are two peak values of the birefringence. The birefringence curves of the two different structures are shown in Figure 3. The value of the birefringence increases sharply near the phase matching wavelength λ_p, and it gets to the top on λ_p. The value remains almost zero when the wavelengths are apart from λ_p.

For different structures A and B, the phase matching wavelengths are different. It is because the coupling of the two modes of a dual-core structure PCF is affected by the first ring of the air holes [34], but the location of the doped rods for structure A and B is different, the mode field distribution is different when the coupling occurs. The difference between the two orthogonal polarized modes is different too. The evolution law of λ_p for the proposed PCF is proposed in our previous studies [35]. We can change the geometric structure parameters to adjust λ_p in order to match our needs. Because the birefringence is achieved by the coupling of the two modes, the value of the birefringence is not as high as that obtained by the asymmetry of the fibre structure, or in any other way by increasing the index differences between the two orthogonal polarization modes. However, the operating wavelength can be controlled in a certain range. The birefringence can be increased and controlled in a certain wavelength by destroying the symmetry of the fibre structure and keeping the dual-core structure.

Figure 4 shows the confinement loss of the two orthogonal fundamental polarized modes of the two different structures. We note that it remains almost zero before the coupling occurs. However, around the phase matching wavelength λ_p, the value of confinement loss increases quickly and then decreases. It is because the coupling leads to a leakage of the hybrid core mode to the cladding deficit mode. After the coupling, the cladding deficit mode will play a dominant role for propagating, it is a leakage mode compared to the hybrid core mode. Therefore, the confinement loss increases. For structure B, the confinement loss increases sharply at phase matching wavelength λ_p, which occurs because the energy exchanged in the coupling of the hybrid core mode and the cladding deficit mode is stronger than in structure A.

4 CONCLUSIONS

We investigate birefringence of a dual-core hybrid photonic crystal fibre. A comparison of two kinds of hybrid structure is also proposed. The proposed dual-core hybrid photonic crystal fibre has a special birefringence mechanism which is different from those obtained by the asymmetry of the fibre structure or in another way by increasing the index differences between the two orthogonal polarization modes. It is achieved by the coupling of the hybrid core mode and the cladding deficit mode. Because of the special birefringence mechanism, it can be controlled in a certain wavelength range to match our needs by adjusting the geometric structure parameters. The numerous results show that the birefringence theory of the proposed PCF is correct.

ACKNOWLEDGMENT

This work was supported by the Natural Science Foundation of Jiangsu Province, China (No. BK2012509).

REFERENCES

Ademgil, H., and Haxha, S., 2009. Ultrahigh-Birefringent Bending-Insensitive Nonlinear Photonic Crystal Fiber With Low Losses. *IEEE Journal of Quantum Electronics*. 45, (4):351–358.

Andrew Michie, John Canning, Katja Lyytikäinen et al., 2004. Temperature independent highly birefringent photonic crystal fibre. *Optics Express*. 12(21): 5160–5165.

Bin Hu, Min Lu, Weinan Li et al., 2010. High birefringent rhombic-hole photonic crystal fibers. *Applied Optics*. 49(31): 6098–6101.

Bhawana Dabas, Sinha, R.K., 2011. Design of highly birefringent chalcogenide glass PCF: A simplest design. *Optics Communications*. 284: 1186–1191.

Changming Xia, Guiyao Zhou, Ying Han, and Lantian Hou, 2011. Highly birefringent octagonal photonic crystal fibers with two zero-dispersion wavelengths. *Chinese Optics Letters*. 9(10): 100609.

Chunshu Zhang, Guiyun Kai, Zhi Wang et al., 2005. Tunable highly birefringent photonic bandgap fibers. *Optics Letters*. 30(20): 2703–2705.

Daru Chen and Genzhu Wu. Highly birefringent photonic crystal fiber based on a double-hole unit. *Applied Optics*, 2010, 49(9): 1682–1686.

Daru Chen and Linfang Shen, 2007. Highly Birefringent Elliptical-Hole Photonic Crystal Fibers With Double Defect. *Journal of Lightwave Technology*. 25: 2700–2705.

Felipe Beltrán-Mejía, Giancarlo Chesini, Enrique Silvestre et al., 2010. Ultrahigh-birefringent squeezed lattice photonic crystal fiber with rotated elliptical air holes. *Optics Letters*. 35(4): 544–546.

Jia-Hong Liou, Sheng-Shuo Huang, Chin-Ping Yu, 2010. Loss-reduced highly birefringent selectively liquid-filled photonic crystal fibers. *Optics Communications*. 283:971–974.

Jia-Hong Liou, Sheng-Shuo Huang, and Chin-Ping Yu, 2010. Birefringent selectively liquid-filled photonic crystal fibers. Photonic and Phononic Crystal Materials and Devices, SPIE Vol. 7609.

Jianhua Li, Rong Wang, Jingyuan Wang et al., Highly birefringent photonic crystal fiber with hybrid cladding structure. ACP2010, Vol.7986, December 8 2010.

Jian Liang, Maojin Yun, Weijin Kong et al., 2011. Highly birefringent photonic crystal fibers with flattened dispersion and low effective mode area. *Optik*. 122:2151–2154.

Ju, J., Jin, W. and Demokan, M. S. 2003. Properties of a Highly Birefringent Photonic Crystal Fiber. *IEEE Photonics Technology Letters*. 15(10): 1375–1377.

Kunimasa Saitoh, Masanori Koshiba, 2002. Photonic Bandgap Fibers With High Birefringence. *IEEE Photonics Technology Letters*. 14(9): 1291–1293.

Li Jianhua, Wang Rong, Wang Jingyuan, Liu Ying, 2011. Highly birefringent liquid-filled photonic crystal fibers with selectively filled in cladding. *Optical Engineering*. 50(2):025001-5.

Liang Wang and Dongxiao Yang, 2007. Highly birefringent elliptical-hole rectangularlattice photonic crystal fibers with modified air holes near the core. *Optics Express*. 15(14):8892–8897.

Lin An, Zheng Zheng, Zheng Li et al., 2009. Ultrahigh Birefringent Photonic Crystal Fiber With Ultralow Confinement Loss Using Four Airholes in the Core. *Journal of Lightwave Technology*. 27(15):3175–3180.

Liu, Y.C. and Lai, Y. 2005. Optical birefringence and polarization dependent loss of square- and rectangular-lattice holey fibers with elliptical air holes: numerical analysis. *Optics Express*. 13(1). 225–235.

Marcin Szpulak, Tadeusz Martynkien, and Waclaw Urbańczyk, 2003. Birefrigent photonic crystal holey fibers based on hexagonal lattice. ICTON2003, 333–336.

Ortigosa-Blanch, A., Knight, J. C., Wadsworth, W. J. et al., 2000. Highly birefringent photonic crystal fibers. *Optics Letters*. 25: 1325–1327.

Roberts, P. J., Williams, D. P., Sabert, H. et al., 2006. Design of low-loss and highly birefringent hollow-core photonic crystal fiber. *Optics Express*. 14(16):7329–7341.

Schreiber, T., Schultz, H., Schmidt, O. et al., 2005. Stress-induced birefringence in large-mode-area microstructured optical fibers. *Optics Express*. 13(10): 3637–3646.

Shah Alam, M., Kunimasa Saitoh, and Masanori Koshiba, 2005. High group birefringence in air-core photonic bandgap fibers. *Optics Letters*. 30(8):824–826.

Slawomir Ertman, Aleksandra Czaplal, Katarzyna Noweckal et al., 2007. Tunable highly-birefringent Photonic Liquid Crystal Fibers. Instrumentation and Measurement Technology Conference – IMTC Warsaw, Poland, May 1–3, 2007.

Soan Kim, Chul-Sik Kee, 2009. Dispersion properties of dual-core photonic-quasicrystal fiber. *Opt. Express* 17 (18):15885–15890.

Tadeusz Martynkien, Marcin Szpulak, Maciej Kieryk et al., 2005. Temperature sensitivity in birefringent photonic crystal fiber with triple defect. 17th International Conference on Optical Fibre Sensors, SPIE, Bellingham, WA, 5855: 912–915.

Theis P. Hansen, Jes Broeng, Stig E. B. Libori et al., 2001. Highly Birefringent Index-Guiding Photonic Crystal Fibers. *IEEE Photonics Technology Letters*. 13(6): 588–590.

Wang, A., George, A. K., Liu, J. F. and Knight, J. C., 2005. Highly birefringent lamellar core fiber. *Optics Express*. 13(16): 5988–5993.

Xiaoling Tan, Youfu Geng, Jun Zhou, 2011. A novel ultrahigh birefringent hole-assistant microstructured optical fiber with low confinement loss. Optics & Laser Technology. 43:1331–1334.

Xin Chen, Ming-Jun Li, Natesan Venkataraman et al., 2004. Highly birefringent hollow-core photonic bandgap fiber. *Optics Express*. 12(169): 3888–3893.

Yang Yue, Guiyun Kai, Zhi Wang et al., 2007. Highly birefringent elliptical-hole photonic crystal fiber with squeezed hexagonal lattice. *Optics Letters*. 32(5):469–471.

Ying Liu, Yuquan Li, Jingyuan Wang, et al., 2012. A novel hybrid photonic crystal dispersion compensating fiber with multiple windows. *Optics & Laser Technology*, 44(7):2076–2079.

Yuh-Sien Sun, Yuan-Fong Chau, Han-Hsuan Yeh et al. 2007, High birefringence photonic crystal fiber with a complex unit cell of asymmetric elliptical air hole cladding. *Applied Optics*. 46(22):5276–5281.

Zhongjiao He, 2009. Highly birefringent extruded elliptical-hole photonic crystal fibers with single defect and double defects. *Chinese Optics Letters*. 7(5):387–389.

Remote measurement and control system of nuclear radiation environmental accident emergency

Hongmei Zhong & Hong Yuan
Sichuan Institute of Nuclear Geology, Chengdu, Sichuan, China

G.Q. Zeng
Chengdu University of Technology, Chengdu, Sichuan, China

ABSTRACT: Nuclear accident emergency is different from the normal order and the normal work procedures. Its main task is to control, reduce and eliminate nuclear accidents that may cause damage and destruction, and to ensure public safety. Once a radiation accident happens, rescue personnel should carry a nuclear emergency monitoring terminal, immediately rush to the scene in order to conduct an investigation, transmit the collected and processed data of the scene to the monitoring command centre, keep abreast of the situation so that the centre is able to respond quickly to the emergency situation, and reduce losses to a minimum. This paper describes a remote control machinery tracked vehicle measurement and control system which uses wireless communication. The system, operated via a remote monitoring and control information system, without staff being present at the scene, would play a significant role in case of a nuclear accident emergency.

Keywords: Nuclear accident; Emergency disposal; measurement and control crawler

1   INTRODUCTION

A nuclear accident is a term which refers to an accident that has the potential to cause abnormal exposure to a supercritical accident and/or a serious leakage accident of radioactive substances to the general public[1]. A nuclear accident emergency is a term which refers to the urgent action that needs to be put in place in order to control or alleviate the nuclear accident and reduce the consequences of the nuclear accident, which is different from the normal order and the normal work procedures[2]. Once the radiation accident happens, rescue personnel should carry a nuclear emergency monitoring terminal, immediately rush to the scene in order to conduct the investigation, transmit the collected and processed data from the scence to the monitoring command centre, keep abreast of the situation so that the centre is able to respond quickly to the emergency situation, and reduce losses to a minimum[3]. However, traditional measuring or monitoring requires personnel for real-time supervision; the radiation will inevitably affect the personnel, which is a considerable health hazard[4].

To the nuclear accident emergency disposal, this paper researches a portable emergency monitoring and scheduling command system. This can be put into use immediately after entering the scene, and achieve continuous real-time monitoring of the radiation indicators (γ radiation dose rate, total α, total β) and the environmental indicators (temperature, humidity, VCO (volatile poisonous gases), atmospheric pressure, and the local storage of video and voice and by the wireless communication technology, realize high speed and reliable data transmission of field devices and remote scheduling command centre, that can transmit back the site actual situation in time, realize voice and video call to the scene. The equipment can be realized by using remote control, and there is no need to have personnel arriving at the scene of the accident, which can effectively ensure the safety of personnel. Command scheduling provides a good solution for nuclear accident emergency disposal.

2   TECHNICAL BACKGROUND

During the 11th five-year plan, nuclear power plants and technology have enjoyed great development and wide application in our country, helped by nuclear science itself and social psychology, especially after the great panic that nuclear and radiation accidents have caused to the public. Since then radiation safety has become a focus problem in the field of nuclear industry, and a promoter for the development of environmental radiation monitoring technology[5].

Abroad made big progress in radiation environment, they broke the original manual single detector monitoring mode, used more frequent automation and computerization technology to deal with monitoring data[5]. In terms of field monitoring, they are equipped with the vehicles and vehicle monitoring instruments for the relevant geographical information description. In terms of the radiation detector, in the United States a high pressure ionization chamber is commonly used whilst in Japan a NaI is used earlier on, instead of an ionization chamber later on. In European countries a proportional tube and Geiger tube are more commonly used. An ionization chamber has the characteristics of high sensitivity, low background, good energy response, good angular response and good linear features, whilst a G-M tube has the advantages of cheap, lightweight, easy maintenance, convenient data transmission and strong environmental adaptability.

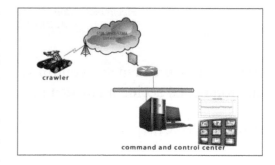

Figure 1. Structure of the system.

3 TECHNICAL SOLUTION

Remote measurement and control systems of nuclear radiation environmental accident emergency designs combine a digital control and data analysis system software and hardware The system includes two parts: first, is the small measurement and control crawler that is used for a variety of scene information gathering and can be remotely controlled. The information includes the scene radiation indicators (γ radiation dose rate, total α, and total β), environmental indicators (atmospheric pressure, temperature, humidity, VCO) and audio and video . Second, is the command and control centre located in the remote. The centre displays the real-time data that the crawler measures by a professional software, showing trend charts for various indicators, the audio and video of the scene, the spatial processing and analysis of the monitoring data, the remote call to the scene, and the control of the crawler route, direction of motion and data acquisition frequency. This greatly facilitates the rapid rescue and disposal of the accident. The structure of the system is shown in figure 1.

(1) Remote control terminal system

A remote measurement and control terminal system takes the form of a small numerical control crawler, mainly used for the data collection of radiation indicators (γ radiation dose rate, total α, total β), environmental indicators (temperature, humidity, VOC, atmospheric pressure), audio and video, that provides data support for the whole system. The structure of crawler includes three main parts, as shown in Figure 2.

① all kinds of detectors and sensors in front can quantify the relevant information into a standard value, including γ detector, α detector, β detector, temperature and humidity sensor, VOC sensor, atmospheric pressure sensor.

γ radiation measurement detection uses sodium iodide crystals as the front detector, that Na scintillator emits weak fluorescence, electron flow is formed after the photomultiplier multi-level amplifier, then charge-sensitive amplifier picks the electron and input processor interrupt port after amplification, the processor get the total measurement of γ particles by the

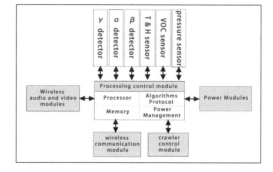

Figure 2. Structure of crawler.

pulse counting in the port. Total α particle radiation measurement based on semiconductor detector SI-PIN detector, adding reverse high pressure on both ends of the SI-PIN, when there is α launch into SI-PIN's PN, and SI-PIN exports detection signal, after through the modulation circuit sends processor to process. Because α has a large quality, some positive charges, low measurement and short penetration distance in the air, so in the system we design an α detection. Through forming an electric field pointing to the detection's centre, the α can be converged. By placing SI PIN detector in the centre of the detection, we can detect the total α. Total β detecting uses of mica Geiger counter tube detector. β and α have similar properties. β has some negative charges, light quality, low measurement in the air, so also need to design a detection.

Temperature and humidity monitoring use an AM2321 that integrates digital temperature and humidity sensor of humidity-sensitive capacitance. AM2321 is a containing composite sensor has been calibrated digital signal output of the temperature and humidity. The purpose of Volatile organic matter (VOC) measurement is to monitor the poisonous gas of organic matter in the air. In the system, we use the sensor for air quality sensor QS-01. QS-01 is a type of tin oxide semiconductor gas sensor; it has a high sensitivity to all types of air pollution with fast response, and can obtain excellent response characteristic under the conditions of low power consumption. Pressure measurement uses a MOTOROLA MPX200 silicon pressure sensor. The MPX200 pressure sensor is a type of silicon piezoresistive pressure sensor, outputs analogue voltage of high precision and high linearity, can directly get pressure after processor process. MPX200 data range is 0–200 kPa, and full range output voltage value is 60 mV.

② Processing control module has a control function of each detector and sensor, responds to user instructions, and completes the execution of instructions, includes a processor, memory, algorithm and protocol, power management system and other parts; the module is the centre of the whole equipment.

The processing control module's function is to realize the system scheduling for each part of the equipment and data processing. The system uses Atmel Company launched MEGA16 processor. The processor is responsible for the collection of the detector output signal, wireless remote control of tracked vehicles, power system management and so on.

③ For the equipment to work normally, it needs other modules, including mainly a communication module and a power module. The communication module includes caterpillar vehicles wireless remote communication, wireless audio and video collection and communication with local cable communication. Wireless remote communication realizes the wireless interaction between the scheduling command centre and the scene outside, including data transmission that the centre commands receive. Wireless audio and video collection communication part realizes audio and video playing using 1.2 GHz microwave that picks-up the audio and video from the PC in the control centre. Local cable communication exists mainly to meet the needs of the local store that stores various important information directly to EEPROM of the processor block internal.

The function of the wireless communication module is to realize data interaction between the monitoring device and caterpillar vehicles. In this system, we use an SI4432 radio frequency module, where we can configure the baud rate of the communication signal between the module and the processor, and the communication channel and the signal transmitted power. The reliable transmission distance can reach 1200 m. The wireless module is based on a 433 MHz wireless transparent data transmission module of GFSK modulation mode with a convenient interface and a stable performance.

Wireless remote control caterpillar vehicles drive using a high power brushless DC motor; the monitors can be far away using the PC software in order to realize remote control of caterpillar vehicles. The brushless DC motor adopts an MC33035 brushless DC motor controller and an MC33039 closed-loop brushless motor adapter for controlling the brushless motor drive, that can realize speed, direction, brake regulation of the motor, and is convenient for the movement of the remotely-operated tracked vehicles. The communication between processor and MC33035 motor

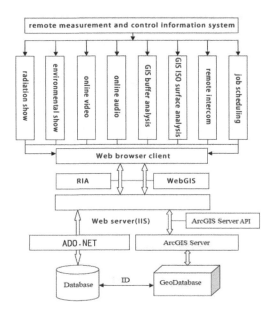

Figure 3. Structure of remote measurement and control information system.

driver uses optical coupling isolation which avoids any harm to the work of processor and monitoring circuit from the high power of the motor.

The power supply module involves power supply for each detector module, processing control module, wireless audio and video acquisition system, which uses lithium batteries and supplies power to all parts of the equipment after a voltage stabilizing treatment, which can realize power consumption management and enhance the system life. The system can realize the battery monitoring and the power supply switch for each part, as well as the management of power consumption.

(2) Remote measurement and control information system of the command and control centre

The remote measurement and control information system of the command and control centre adopts B/S mode to carry on the design and development, the structure is shown in Figure 3. The database at the bottom is the basis of the system operation, including the space data, scene gathering information data, system operation data, etc. The system provides web services on IIS. The spatial data uses an ArcGIS Server API to connect. The scene gathering information data and system operation data are connected by ADO.NET. Users can login in the browser of any computer that can access the system, then use all the functions provided by the system.

The system uses RIA (rich Internet application) in order to provide highly interactive, rich, smooth and beautiful graphics applications in the browser, and uses an ArcGIS Server to provide GIS function. The main functions of the system include radiation indicators show, environmental indicators show, online video, online audio, GIS buffer analysis, GIS ISO surface

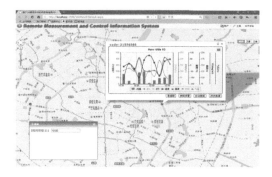

Figure 4. System's main interface.

analysis, remote intercom, job scheduling, etc. Indicators show uses a combining way for the proposed graph and data table, which update silently and automatically in the background. The system's main interface is shown in Figure 4 (using the simulated data).

4 CONCLUSION

The scene situation of the nuclear accident is very complicated, the temperature and humidity have changed dramatically, there is a considerable amount of dust, poor working conditions, and higher requirements on the stability of the equipment. In this study, the hardware system uses an AVR microprocessor MEGA16 with high stability, which has the characteristics of stable running, and strong function expansion. The acquisition of audio and video uses a microwave transmission method of 1.2 GHz, with clear image, long transmission distance, without being limited by third party operators, and many other features. In addition, the system uses a tracked vehicle that can be remotely controlled to load monitoring devices. The vehicle is driven by a brushless DC motor using the independent power supply that does not affect the power system of the measurement part. The hardware system uses a power supply of a wide range, whose input voltage is in the range of DC 7V to 30 V, and its output voltage can be kept constant. Using the power supply for the equipment can improve the service life of the equipment and obtain steady audio and video information. From a complete set of remote measurement and control information system uses GIS spatial analysis and static real-time refresh technology that realize the real-time monitoring of data and video, and data analysis of scientific and reasonable. The system can be effectively used in a nuclear accident emergency disposal and rescue operation.

ACKNOWLEDGEMENTS

Supported by Key Subject Laboratory of National High Technology Research and Development Program 863 (No. 2012AA063502), "Nuclear waste and environmental safety" National Key Laboratory Open Fund (No. 12zxnp06), Sichuan Province Science and Technology Plan Project (No. 2013GZX0169).

REFERENCES

China Conference. Fifteenth National Nuclear Science Nuclear Detection Technology Conference Proceedings and [C]. Guiyang: [Publisher Unknown], 2010.
Jiang Weihua, Zhang Xianmin. Nuclear Accident Emergency Radiation Monitoring System Design [J]. Application Research of Computers, 2005(4):180–182.
Liu Zhenjing. Radiation field data acquisition and wireless transmission processing [D]. Harbin: Harbin Engineering University, 2009.
People's Republic of China State Council. Nuclear Power Plant Emergency Management Regulations [S]. People's Republic of China State Council Order, 1993.
Tan Wei. The Improvement Design of an Environmental Radiation Monitoring Instrument Based on MCU [D]. Hengyang: University of South China, 2010.
Xing Lili. Reach of Long distance Radiation Emergency Inspection System [J]. Nuclear Electronics & Detection Technology, 2010, 3(30):446–450.

Research on superior mobile communications based on a fibre optic repeater system in a coal mine

Wei Li
Institute of Mining Engineering, Heilongjiang University of Science & Technology, Harbin, China

Hong-ming Kang, Chun-feng Jia & Xiao-lin Jiang
Institute of Electric & Information Engineering, Heilongjiang University of Science & Technology, Harbin, China

Wei-guo Zhao
Shuangyashan Coal Group Subsidary Corporation of Long May Group Corporation, Shuangyashan, China

ABSTRACT: In order to improve the effectiveness of mobile communications in a coal mine, the DSP automatic monitoring system has been put into practice. It introduces an FIR digital filter, which reduces noise by subtraction and so on, based on a fibre optic repeater system application platform. It makes the mine communication system more stable, voice quality better, and it provides new thinking on the best applications for mobile communications in a mine.

Keywords: fibre optic repeater system, digital signal process, noise suppression, and mine

1 INTRODUCTION

Mobile communications in a coal mine is a range of application for limited space, as underground is a special environment with narrow roadways, inflections and more branches, a large communication range and scattered service locations[1]. Electromagnetic waves cannot spread freely in the roadway and it can result in inefficient utilization. The ground mobile communication system is applied to the underground after transforming, which can result in a great waste of communication resources. This occurs because of the increase in the mechanical and electrical degrees of mine modernization. The artificial noise interference becomes more serious, affecting the normal work of the mine's communication system, revealing call quality deterioration and communication reliability reduction. It is a key part of solving underground mobile communication that enables the communication system to play effectively, adds to the communication stability and improves anti-interference performance and voice quality, etc.

2 THE COMPOSITION OF FIBRE OPTIC REPEATER SYSTEM IN MINE

A Fibre Optic Repeater is an effective way to extend cellular communication systems on the ground to mine. A Fibre optic repeater system is composed mainly of three parts which are the MHU, the ROU and optical fibre[2]; the ROU is located in the mine.

The MHU consists of duplex filter, automatic electric level control unit, monitoring unit, electro-optical conversion module etc. The ROU includes an optical receiver module, a low-noise amplifier, a power amplifier, an automatic level control module, a system monitoring unit, a duplex filter and so on. The MHU generally is placed in the room of the base station, which couples the RF signal directly from the BTS of a land cellular network. The RF signal will be filtered through the diplexer band pass filter and its level will be detected and adjusted to adapt for the demodulation request of an optical transceiver and then it will convert the electrical signals to optical signals by the optical transceiver and optical signals transmission to the ROU through a fibre optic. The ROU transforms the optical signal into an electrical signal, then it is amplified by the power module of the TX-link and finally fed to the antenna through the duplexer, covering the target area. A Fibre Optic Repeater system is found both in the form of a single link and a multi-link. The biggest advantage of the multi-link form is that it is suitable for a wide range of communications in mines. A multilink fibre optic repeater system consists of an MHU and many ROUS[3], its structure is shown in Figure 1.

3 AN AUTOMATIC MONITORING AND CONTROL UNIT BASED ON DSP SYSTEM

The complex environment in a coal mine and all kinds of severe interference always affects the normal

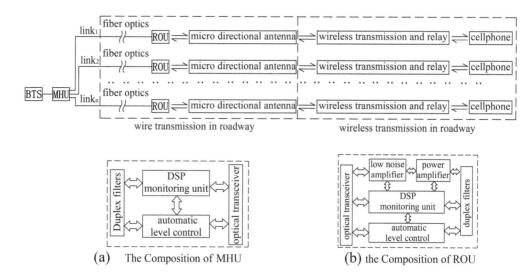

(a)   The Composition of MHU

(b) the Composition of ROU

Figure 1.   The Composition of Multilink Fibre-optic Repeater System.

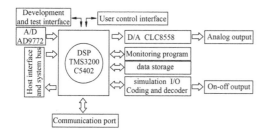

Figure 2.   Chart of Automatic Monitoring Units.

operation of the communication system in a mine. The fibre optic repeater system uses a DSP automatic monitoring unit to ensure reliable operation of the system. Its structure is shown in Figure 2.

DSP is in general a digital signal processing chip, with strong real-time performance, which can complete the processing of the external input signal in a short period of time, so that the introduction of the DSP for the system provides a high speed computing platform[4]. The main function of the monitoring unit is to monitor and record all kinds of parameters and status of the main system operation in real-time and pass the network management centre through a communications port, then control and adjust the fibre optic repeater system running state following the instructions of the network management centre. Monitoring tasks include: an amplifier switch of RX and TX, the attenuation values of each module, the output power level, keeping a gain balance of the RX and TX links, judging amplifier power of the system, standing waves and the second carrier frequency locking, etc.; to detect the important parameters such as access control and power whether it is normal or not, and making adjustments to any abnormal parameter in real-time. The underground environmental parameters can also be monitored by a DSP control unit for temperature, humidity, gas density and so on.

4   FIR DIGITAL FILTERS AND DSP IMPLEMENTATION

A digital filter is the basic application field of DSP and it is an important method of digital signal processing. In order to achieve the purpose of fast digital signal processing, DSP chips generally have a bus architecture for the program and data separately, a pile line operation function, a hardware multiplier in a single-cycle complete multiplies and an instruction set suitable for digital signal processing. It performs well in real-time, has more processing speed, and is more practical and flexible for the digital filters that are implemented on a DSP. For the purpose of anti-noise function and interference resistance in a mine, it must be designed to be the best digital filter to implement.

4.1   Direct structure of digital filter

The input and output of an FIR digital filter can be expressed as:

$$y(n) = \sum_{k=0}^{N-1} a_k x(n-k) \qquad (1)$$

$$H(z) = \sum_{k=0}^{N-1} h(k) z^{-k} \qquad (2)$$

There, $h(k)$ ($k = 0, 1, 2, \ldots, N-1$) is an impact response coefficients of the filter; $H(z)$ is the transfer function of the filter; N is the length of the filter, that is, the number of filter coefficients; $x(k)$ is the value of the input sample on time k; and $H(k)$ is the k grade tap coefficient. It can be seen that the FIR filter is structured with the collection of an adder and a multiplier, and for each sample, $h(k)$ must do N consecutive multiplication and ($N-1$) additions operation. FRI filter structure shown as Figure 3.

Hereinto, $x(n)$ is the input sequence, $y(n)$ is the output sequence.

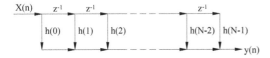

Figure 3.   Structure of N-order FIR filter.

Because of $h(z)$ is $(N-1)$ polynomials, it has $(N-1)$ zero-point on a flat surface, the origin $z = 0$ is $(N-1)$ order poles. Therefore, $h(z)$ is stable forever.

If the impulse response of the filter is: $h(0)$, $h(1)$, $h(2)$, ..., $h(N-1)$, $x(n)$ it indicates filter input at time n, whilst the output of the n time is:

$$y(n) = h(0)x(n-1) + \cdots + h(N-1)x[n-(N-1)] \qquad (3)$$

Because it has a characteristic of linear response, it is widely used in digital signal processing; the output expression of the length N for the linear phase FIR filter is as below:

$$y(n) = \sum_{k=0}^{N/2-1} h(k)[x(n-(N-1-k))] \qquad (n=0,1,2) \qquad (4)$$

4.2   Realization of a digital filter

The implementation of FIR filters using circular addresses and achieves centre point symmetry of the FIR filter with a circular addressing of TMS3200C5402 chip and FIR instructions. Established as a buffer, the input sequence is storage in two ring buffers. First, registered values of ring buffer size are set to $N/2$, auxiliary register AR4 will be specified at the top of the buffer 1 and will be secondary register AR5 which refers to the bottom of the buffer zone 2. A new sampling is stored in the buffer 1. First, the data at the top of buffer 1 should be moved to the bottom of buffer 2, then multiply accumulation should be conducted, that is: $h(0)\{x(0) + x(-N + 1)\}$, after each calculation, AR4 points to the next data unit of the buffer 1 and AR5 points to the next data unit of buffer 2.

Select 500 Hz and 2500 Hz sine signal as the input signal; the sampling frequency is 8000 Hz, the output waveform as shown in Figure 4 by using the filter, a is input, b for the output.

5   THE NOISE REDUCTION METHOD BY SUBTRACTING

As for underground mobile communication, it is another critical issue that enhances antinomies performance of communication systems and improves call comfort in order to resolve mobile communications underground[5]. To reduce the bandwidth or reduce interference, the comfort noise generation technology has become a key technology of VoIP[6]. The program of voice enhancement is often used in the voice transmission process and it uses the FEC coding techniques for data transmission, with a large quantities of data operation and detection. All of the above can be completed by the advanced DSP processor, which also includes noise reduction and voice coding.

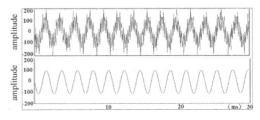

Figure 4.   Waveforms of FIR filter.

Figure 5.   Voice enhancement system of single channel.

5.1   Interference subtraction method of noise reduction

The noise characteristics can be estimated only when the background noise is left out, that is to say the single channel voice enhancement system (Fig. 5) must be no voice. Valid audio source and noise source can be collected effectively by Voice-activated detector (VAD), and noise reduction can be achieved by using a noise subtraction algorithm. A time-domain adaptive noise cancellation method can be used by generating a reference signal based on voice periodicity. Among them, the reference signal is formed from delaying a cycle of the main signal, which requires complex interval estimation algorithm.

First, we use FFT in a voice frame, the estimated noise amplitude spectrum subtraction, and inverse transformation to spectrum amplitude after the subtraction. Then, we use the original phase noise, and finally, we find the noise amplitude and short-time phase spectrum. Enhanced steps are completed on a frame by frame basis. This method that pollution voice decomposed into different frequency using band-pass filter group then each sub-band noise power is estimated during the period of free voice. Noise suppression can be obtained using the attenuation factor, of which the attenuation factor corresponds to instantaneous signal power of the estimated noise power ratio on each sub-band.

5.2   Spectral subtraction noise reduction technology

The majority noise reduction of communication is using DSP to complete, mainly using FFT to reduce noise. Among them, the spectrum subtraction provides an effective calculation method, by subtracting the noise spectrum from the noisy voice spectrum, which increases the voice, and reduces the noise. Noisy voice is segmented and set window, FFT of each data window is executed, and an amplitude spectrum is calculated. VAD is used to detect the input voice signal. In a voice segment, noise spectrum will be estimated, deposited in the buffer zone, and makes the data in buffer damping by the algorithm, thereby reducing the noise. During the non-voice period, there are two ways to generate output with the output of a fixed attenuation factor or set the output to 0. There is some residual noise (comfort noise) left during the period of non-voice frame. It can output high audio quality, because a voice of noise was shield locally during the voice frames, its value will be balanced by the noise of the same magnitude in the presence of non-voice segments[7−8]. The noise was amplified with the output set to 0, so it is best to reduce the noise by a fixed factor during non-voice period. It is a must to keep the balance between noise characteristic of amplitude and voice segment can be perceived, and the noise perceived at noise segment, so sound effects which are not needed, such as buzzing, clicks, jitter, and voice signal fuzzy can all be avoided.

It is necessary to set parameters and make data analysis before the algorithm comes into effect. Assuming the background noise within voice band is smooth and that the amplitude spectrum appears in front of the fixed voice, then enough time is provided to estimate the frequency spectrum with new amplitude value of background noise if the environment is changeful. For the noise algorithm to change slowly, we need to determine whether voice has terminated according to the parameters of VAD and to estimate the new influence on noise. We then, using the method of frequency spectrum subtraction, can suppress noise effectively.

Assuming the signal $s(n)$ is affected by the jamming signal $v(n)$ and causes loss, the signal which is contaminated with noise can expressed as:

$$\hat{s}(k) = |x(k)| - E|v(k)| \qquad (5)$$

If we carry out discrete Fourier transformation we can achieve:

$$|\hat{s}(k)| = |\hat{s}(k)| e^{j\theta x(k)} \qquad (6)$$

Assuming the $v(n)$ is a zero mean, and is not related to $S(n)$, the $S(K)$ estimation can expressed as:

$$s(k) = |x(k)| - E|v(k)| \qquad (7)$$

$E|v(k)|$ represents the expectations the noise spectrum appears in non-voice cycle.

Given estimation $s|(k)|$, then spectrum estimation can expressed as:

$$|s(k)| = |s(k)| e^{j\theta x(k)} \qquad (8)$$

Hereinto, $e^{j\theta x(k)} = \dfrac{x(k)}{|x(k)|}$

The $\theta x(k)$ is the measured phase of noise signal, using the phase of noise voice and can meet the demand of a specific target. So it achieves the reconstructed and processed signal by $s|(k)|$ and $\theta x(k)$, the estimator can expressed as:

$$\hat{s}(k) = [|x(k)| - E|v(k)|]\dfrac{x(k)}{|x(k)|} = H(k)x(k) \qquad (9)$$

Hereinto, $H(k) = 1 - \dfrac{E|v(k)|}{|x(k)|}$

The spectrum subtraction algorithm avoids the calculation of phases in equations. It can realize in floating point DSP hardware in order to reduce noise to get superior voice effects.

6 CONCLUSIONS

A multilink fibre optic repeater system, which is established in a mine, can solve the problem of wide-area scope mobile communication. The optimization measures such as a monitoring system, digital filtering and a noise restriction method by subtraction etc., can improve the performance of the communication system which has been applied to a mine, and can also improve the voice quality and the simulation level of comfort. To sum up, the integral system has the following advantages:

1) A multilink fibre optic repeater system is perfectly suitable for a wide range of communication in a mine and presents superior performance.
2) The DSP monitoring unit provides a reliable guarantee for the long-term stable operation of the fibre optic repeater system.
3) The digital filter has better filtering effectively and strongly. Anti-interference performance compare favourably with an analogue filter in application.
4) The technology of spectrum subtraction noise can increase the voice intensity and reduce the noise amplitude, which enhances voice quality and auditory effect.

REFERENCES

Huang Wei, Yao Shan-hua. The difficult technical problems of communication in coal mine underground and solving scheme [J]. Coal mine machinery, 2003, (12):33~36.

Jeong, K. S.; Cheong, J. M.; Park, T. H. et al., Performance analysis of DS-CDMA reverse link with fiber-optic repeaters, 2000 IEEE 51st, 2000, vol. 3: 2439~2443.

Li Wei, Sun Guang-yi etc. Research on the Signal Introducing System of GSM in Mine [J]. Mining research and development, 2006, 26 (6): 82–84.

Andrew Bateman, Iain Paterson-Stephens, The DSP Handbook: Algorithms, Application and Design Techniques, ISBN 7-111-12206-6.

Yao Shan-hua, Huang Wei. A new method to improve the performance of restraining noise of communication system in mine [J]. Coal mine safety, 2008, 11: 75–77.

Li Ruo-chun, Zhao Li. Research on comfortable noise Algorithm [J]. Computer simulation, 2009, 26(2): 341~344.

Wang Long. The research and design on voice enhancement system based on DSP [D]. Donghua University, 2011.1, pp. 15~20.

Lu Hong-wei. Research on noise reduction by voice union system based on DSP [D]. Nanjing Forestry University, 2009.6, pp. 21~24.

Electronics, Information Technology and Intellectualization – Song & Kwak (Eds)
© 2015 Taylor & Francis Group, London, ISBN 978-1-138-02741-1

A staircase modulation scheme for the MMC allowing fundamental switching frequency

Shuangwu Ni, Jianhui Su, Jinwei Li & Hang Gao
The Institute of Energy, Hefei University of Technology, Hefei, Anhui, China

Songlin Zhou
School of Electrical Engineering, Tongling University, Tongling, Anhui, China

ABSTRACT: In order to decrease switching action times of submodules and reduce converter power dissipation, this paper presents a new modulation method for the Modular Multilevel Converter (MMC). The proposed method is based on a staircase modulation. By the cooperation of different submodule switching signals, the submodule capacitor voltages can be balanced and the desired AC-side waveform can be obtained. The simulation results show that stable operation can be maintained at the fundamental switching frequency while successfully balancing the submodule capacitor voltages.

Keywords: MMC; DC transmission; switching frequency; voltage balancing; staircase modulation

1   INTRODUCTION

With the development of full-controlled power electronics and control technology, high voltage DC transmission technology based on voltage source converter (VSC-HVDC) has a giant application potential in the grid-connected field of renewable energy and the asynchronous AC grid interconnection [1–2]. The MMC topology is first presented by the German authors R. Marquart and A. Lesnicar. Compared with the traditional two-level or three-level inverter, it has less harmonic content in the output voltage, lower switching losses and a smaller AC-side filter, and can be easily extended to a high voltage level and a high power level due to the special structure of its modular composition [3–6]. Therefore, the MMC has recently become an important issue in domestic and international industry.

The balance control of submodule capacitor voltages is a very critical issue when the MMC is applied to the VSC-HVDC transmission. In the view of the recent literature reports in domestic and foreign research fields, there are two balance control methods for submodule capacitor voltages based on MMC. One approach is to monitor the submodule capacitor voltages and input controller to sort, and then insert or bypass the submodules, according to the corresponding arm current direction. The other is to use a phase-shifted carriers modulation technique [7–8], in order to achieve a balance of submodule capacitor voltages by increasing feedback control.

Numerous control strategies for MMC are mainly based on the sorting algorithm that was presented in the available literature, such as the traditional sorting method introduced in [9] and [10]. The method combined with voltage balancing algorithm of sorting capacitor voltages is a widely used modulation strategy because of its simple implementing process. However, the switching frequency for switching devices ranges from several hundreds to thousands hertz (Hz), which causes relatively high power loss and reduces system efficiency. Additionally, this method determines the insertion and bypassing of the submodules by current polarity and measured capacitor voltages, which introduces one issue of relative high and uncertain switching frequency for submodules. On the basis of the traditional sorting method, by using a keeping factor, the switching frequency of submodules is reduced in [11]. There is some degree of freedom in the choice of the keeping factor, so the method is non-quantitative. In the paper [12], the method with a maximum voltage deviation between the submodules is presented, but this method is only applicable to the steady-state system, so it is limited. The paper [13] proposed one type of sorting algorithm with variable sorting time of capacitor voltages, or the insertion and bypassing of submodules. The research based on phase-shifted carriers' modulation has been carried out for the capacitor voltage balancing control in [14] and [15].

This paper presents a new modulation method for MMC. By using this method MMC can be operated at the fundamental switching frequency without measuring the capacitor voltages or using any other forms of feedback control. This reduces the complexity of the system as the need for voltage measurements is

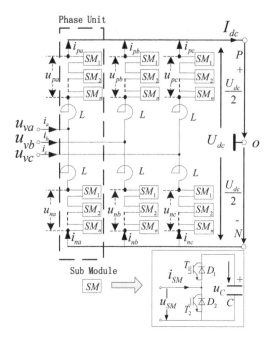

Figure 1. Three phase diagram of MMC and submodule Configuration.

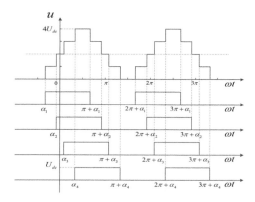

Figure 2. Disposition mode of five-level staircase wave.

reduced to fault-handling purposes or increased performance, and is not a necessity for normal operation. The proposed method is based on staircase modulation. By the cooperation between different submodule switching signals, the submodule capacitor voltages can be balanced and the desired AC-side waveform of can be obtained.

2 THE BASIC STRUCTURE OF MMC

A diagram of a three-phase MMC and a submodule is shown in Fig. 1. Each phase unit of the MMC consists of one upper arm and one lower arm. Each arm consists of N series-connected submodules and one arm inductor with the inductance L. The reactor is used to compensate for the imbalance of output voltage between the upper and lower arm and the DC-side [10].

Each submodule of MMC consists of two IGBTs and a capacitor with the capacitance C as shown in Fig. 1. There are two effectively working states of the switch in the working process, respectively insertion and bypassing. By selectively controlling the switch, the system can obtain the desired output power and voltage level [11].

3 STAIRCASE MODULATION WITH FUNDAMENTAL SWITCHING FREQUENCY

3.1 Splitting of staircase

A five level staircase can be split into four square pulses with the width π radians in a fundamental

frequency cycle as shown in Fig. 2. These pulses start at α_1, α_2, α_3, and α_4. Similarly, the splitting method is also applicable to n level staircase, and the starting angle is from α_1 to α_N.

In order to analyse the capacitor voltage variation of submodule k, the state signal s_k, defined as a square pulse with the width π radians, starts at α_j ($j = 1, 2, \ldots N$) radians, that is

$$s_k = \begin{cases} 1 & \alpha_j \leq \omega t \leq \pi + \alpha_j \\ 0 & 0 \leq \omega t < \alpha_j, \pi + \alpha_j < \omega t \leq 2\pi \end{cases} \quad (1)$$

To simplify the analysis, the square pulse s_k can be expressed as a Fourier series expansion, that is

$$s_k(t) = \frac{1}{2} + \frac{2}{\pi} \sum_{n=1}^{\infty} \frac{1}{2n-1} \sin\left[(2n-1)(\omega t - \alpha_j)\right] \quad (2)$$

3.2 Variation in capacitor voltage

The current flowing through the capacitor in submodule k is then given by the product of s_k, as defined in (2), and the arm current. The variation in capacitor voltages Δu_{pk_C} in the lower arm during one period T are then given by

$$\Delta u_{pk\_C} = \frac{1}{C} \int_0^T s_k i_p dt \quad (3)$$

Similarly, the variation of capacitor voltages Δu_{nk_C} in the upper arm is given by

$$\Delta u_{nk\_C} = \frac{1}{C} \int_0^T s_k i_n dt \quad (4)$$

This means that, since there are only odd-order harmonics in s_k and only even-order harmonics in i_p and i_n, the harmonic components will not contribute to any variation of capacitor voltages.

In the case when submodule k is situated in the lower arm, the variation of capacitor voltage can be calculated from (3) during one period T, that is

$$\Delta u_{ank\_C} = \frac{1}{C}\int_0^T s_k i_{an} dt$$
$$= \frac{1}{C}\left[\frac{I_{dc}}{6} - \frac{I}{2\pi}\cos(\alpha_j + \varphi)\right]T \quad (5)$$

By expanding the cosine term in (5), the variation in capacitor voltage during one period can be expressed as

$$\begin{cases} \Delta u_{ank\_C} = \Delta u_{ank\_CA} + \Delta u_{ank\_CR} \\ \Delta u_{ank\_CA} = \frac{1}{C}[\frac{I_{dc}}{6} - \frac{I}{2\pi}\cos(\alpha_j)\cos(\varphi)]T \\ \Delta u_{ank\_CR} = \frac{1}{C}[\frac{I}{2\pi}\sin(\alpha_j)\sin(\varphi)]T \end{cases} \quad (6)$$

Therefore, it is concluded that the variation in the capacitor voltage that occurs due to the active and reactive power transfer can be separated.

Consider that the dc-side input power is equal to the output power at the ac side, that is

$$\begin{cases} U_{dc}I_{dc} = \frac{3}{2}\left(\frac{1}{2}mU_{dc}\right)I\cos(\varphi) \\ I_{dc} = \frac{3}{4}mI\cos(\varphi) \end{cases} \quad (7)$$

The variation in capacitor voltages due to the active power flow Δu_{ank_CA} can then be expressed as a function of α_j and the load current. Substitute I_{dc} in (6) with (7) yields

$$\Delta u_{ank\_CA} = \frac{1}{C}\left[\frac{m}{8} - \frac{1}{2\pi}\cos(\alpha_j)\right]TI\cos(\varphi) \quad (8)$$

As the submodules in the upper and lower arms are inserted and bypassed in a complementary manner, every pulse in the lower arm is associated with a complementary pulse in the upper arm. The variation in the capacitor voltages of the complementary submodule in the upper arm can, then, be calculated by using the angle $\alpha_j + \pi$. It is then found that the variation in capacitor voltages of the submodule in the lower arm and its complementary submodule in the upper arm are equal during one period. Consequently, it is sufficient only to consider the lower arm when analysing the variation in capacitor voltages of the submodules.

4 REPETITION OF DIFFERENT STARTING ANGLE SWITCHING SIGNALS

4.1 *Balancing the submodule capacitor voltage*

The pulse pattern that is used to control submodule k can be defined by the vector p_k that contains the angles α. That is, a pulse pattern with n angles is defined by

$$\vec{p}_k = [\alpha_1 \quad \alpha_2 \quad \alpha_3 \quad \cdots \quad \alpha_N] \quad (9)$$

By shifting the pulse pattern, one fundamental frequency period in time between the submodules can be obtained and the same pulse pattern can be used to control all of the submodules in the arms. Thus, if the pulse pattern to submodule k is given by (9), the pulse pattern to submodule $k + 1$ is then given by

$$\vec{p}_{k+1} = [\alpha_N \quad \alpha_1 \quad \alpha_2 \quad \cdots \quad \alpha_{N-1}] \quad (10)$$

The fundamental frequency component h_{j_1f} that is imposed by a square pulse which started at α radians can be expressed as

$$h_{j\_1f} = \frac{2}{\pi}\left[\sin(\omega t)\cos(\alpha_j) - \cos(\omega t)\sin(\alpha_j)\right] \quad (11)$$

Thus, the total fundamental frequency component H_{1f} that is imposed in one arm can be expressed as

$$\begin{cases} H_{1f} = \frac{2}{\pi}(A+B) \\ A = \left[\sum_{j=1}^N \cos(\alpha_j)\right]\sin(\omega t) \\ B = -\left[\sum_{j=1}^N \sin(\alpha_j)\right]\cos(\omega t) \end{cases} \quad (12)$$

It is observed that the choice of α will influence both the phase and amplitude of the imposed fundamental frequency component. In order to produce the desired voltage waveform, the fundamental frequency component in (12) must be described by a cosine function. This is the case when the sum of all sine-components in (12) is zero. This gives the following constraint for the angles α_j

$$\sum_{j=1}^N \sin(\alpha_j) = 0 \quad (13)$$

It is evident that if the angles α_j are chosen in pairs that are symmetrically distributed around zero radians such that they are of equal magnitude but opposite signs, the sum in (13) will, in fact, become 0.

The modulation index m can then be expressed as the normalized amplitude of the fundamental frequency component in (12). Assuming that (13) is satisfied, the modulation index m

$$m = \left[\frac{2}{\pi}\sum_{j=1}^N \cos(\alpha_j)\right]\bigg/(N/2) = \frac{1}{N}\left[\frac{4}{\pi}\sum_{j=1}^N \cos(\alpha_j)\right] \quad (14)$$

The total variation in capacitor voltages Δu_{ank_C} that is transferred to a submodule that is controlled by the pulse pattern (9) can be expressed using (6), that is

$$
\begin{cases}
\displaystyle\sum_{j=1}^{N}\Delta u_{ank\_Cj} = \sum_{j=1}^{N}\Delta u_{ank\_CAj} + \sum_{j=1}^{N}\Delta u_{ank\_CRj} \\[2mm]
\displaystyle\sum_{j=1}^{N}\Delta u_{ank\_CAj} = \left[N\frac{m}{8} - \frac{1}{2\pi}\sum_{j=1}^{N}\cos(\alpha_j) \right] TI\cos(\theta) \\[2mm]
\displaystyle\sum_{j=1}^{N}\Delta u_{ank\_CRj} = \left[\frac{I}{2\pi}\sin(\varphi) \right] T\sum_{j=1}^{N}\sin(\alpha_j)
\end{cases}
\quad (15)
$$

Substituting the sum in (15) with (13) shows that the sum of all Δu_{ank_CRj} is zero. Also, if (13) is satisfied, the modulation index m is described by (14). Substituting the modulation index m in (15) with (14) shows that the sum of all Δu_{ank_CRj} is zero. Consequently, the total variation in capacitor voltages Δu_{ank_C} in (15) that is moved to or from a submodule over all N periods is simply zero since assuming that (13) is satisfied.

4.2 Calculation of starting angle

The relationship between the time t_{j_on} when submodule k is inserted and the angle α_j where the pulse is started is given by

$$
t_{j\_on} = \frac{\alpha_j}{\omega} \quad (16)
$$

The angle α_j should satisfy

$$
N\frac{1 + m\sin(\alpha_j)}{2} = j - \frac{1}{2} \quad (17)
$$

Solving (17) for α_j yields

$$
\alpha_j = \arcsin\frac{2j - 1 - N}{mN} \quad (18)
$$

Substituting α_j in (13) with (18) shows that the sum of all $\sin(\alpha_j)$ is zero, meaning that (13), and, thus, also (14) and (15) are always satisfied.

5 SIMULATIONS RESULTS

For simplicity, the simulated converter is connected to a constant direct voltage source and is operating in inverter mode, supplying a passive three-phase load as shown in Fig. 3.

A three-phase converter with 20 submodules per arm was simulated in MATLAB/Simulink. In the simulated converter, the specifications are chosen such that the DC-side voltage value is $U_{dc} = 200\,\mathrm{KV}$, the submodule capacitance is $C = 13\,\mathrm{mF}$, the arm inductance is $L_{arm} = 6\,\mathrm{mH}$, the connected inductance is $L_T = 3\,\mathrm{mH}$, the submodule voltage reference is $U_C = 200\,\mathrm{KV}$ and

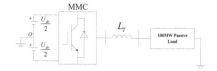

Figure 3.   Diagram of twenty one-level MMC supplying passive load.

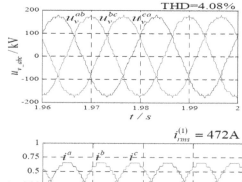

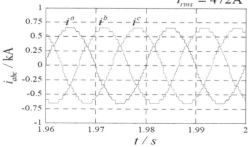

Figure 4.   AC output of twenty-one level MMC.

the fundamental frequency is $f_N = 50\,\mathrm{Hz}$. The AC-side line-to-line voltage and the three line currents are shown in Fig. 4.

The ac-side waveforms indicate good performance of the converter in Fig. 10. The THD of the simulated ac-side line-to-line voltage is small, which is only 4.08%. This is because one arm of MMC consists of twenty submodules and accordingly the number of voltage level is large.

The voltage ripple in the capacitors has a periodicity of 20 fundamental frequency periods, which is 0.4 s. Furthermore, the capacitor voltages in the different submodules are virtually the same, but shift 20 ms in time. Fig. 11 shows two complete periods of the capacitor voltage of the first submodule in the upper arm and all the submodules in the upper arm.

To confirm if the modulation method is applied to the situation where the balance of submodule capacitor voltages is broken suddenly, a large disturbance is introduced in two of the submodules in the upper arm of phase A at the time 1 s During this disturbance, one of the capacitors is charged to 13 kv and another capacitor is discharged to 7 kV. The capacitor voltages of these two submodules are shown in Fig. 6. It is observed that the capacitor voltages naturally converge back to their nominal values without the need for any additional control actions. This shows that the question of a sudden unbalance in the capacitor voltages can be

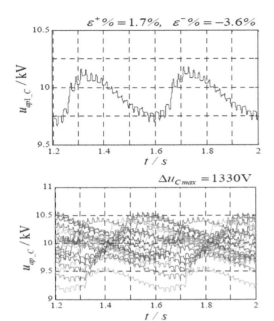

$$\varepsilon^+\% = 1.7\%, \quad \varepsilon^-\% = -3.6\%$$

$$\Delta u_{Cmax} = 1330V$$

Figure 5. Submodule capacitor voltages in phase A upper arm.

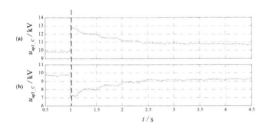

Figure 6. Waveforms of capacitor voltages under disturbance.

solved successfully by using the proposed modulation scheme.

6 CONCLUSION

In this paper, a new modulation scheme based on staircase modulation is proposed for MMC. In the modulation method, MMC can be operated at the fundamental switching frequency without measuring the capacitor voltages or using any other forms of feedback control, so the complexity of the system is reduced. Compared with traditional phase-shifted carriers modulation, by using such a modulation scheme, MMC has a lower switching losses due to the fundamental switching frequency. The simulation results show that stable operation can be maintained at the fundamental switching frequency while successfully balancing the submodule capacitor voltages and ensuring the desired AC-side waveforms.

REFERENCES

Ding Guanjun, Ding Ming, Tgang Guangfu, et al. 2009. Submodule capacitance parameter and voltage balancing scheme of a new multilevel VSC modular [J]. *Proceedings of the CSEE* 29(30): 1–6 (in Chinese).

Guan Minyuan, Xu Zheng. 2011. Optimized capacitor voltage balancing control for modular multilevel converter based VSC-HVDC system [J]. *Proceedings of the CSEE* 31(12): 9–14 (in Chinese).

Hagiwara M., Maeda R., and Akagi H. 2010. Theoretical analysis and control of the modular multilevel cascade converter based on double-starchopper-cells (MMCC-DSCC) in Proc. Int. *Power Electron. Conf.* pp. 2029–2036.

Konstantinou G. S. & Agelidis V. G. 2009. Performance evaluation of half-bridge cascaded multilevel converters operated with multicarrier sinu-soidal PWM techniques. *In Proc. 4th IEEE Conf. Ind. Electron. Appl.*, pp. 3399–3404.

Li Xiaoqian, Song Qiang, Liu Wenhua, et al. 2012. Capacitor voltage balancing control by using carrier phase-shift modulation of modular multilevel converters [J]. *Proceedings of the CSEE* 32(9): 49–56 (in Chinese).

Liu Jian-tao, Wang Zhi-hua, Wang Ke 2013. Comparative analysis of losses of voltage source converters with different structures [J]. *Power System Protection and Control* 41(6): 105–110.

Luo Yu, Song Qiang, Rao Hong et al. 2013. An Optimized Design Method of Cascade Number for Sub-modules in Modular Multilevel Converters [J]. *Automation of Electric Power* 37(4): 114–118.

Qin J C, Saeedifard M. 2010. Reduced switching-frequency voltage-balancing strategies for modular multilevel HVDC converters [J]. *IEEE Transactions on Power Delivery* 25(4): 2903–2912.

Rohner S., Bernet S., Hiller M., Sommer R. 2010. Modelling, simulation and analysis of a Modular Multilevel Converter for medium voltage applications[C]. *Industrial Technology (ICIT), IEEE International Conference on 2010*: 775–782.

Saeedifard M., Iravani R. 2010. Dynamic performance of a modular multilevel back-to-back HVDC system [J]. *IEEE Transactions on Power Delivery* 25(4): 2903–2912.

Song Pinggang, Li Yunfeng, Wang Lina, et al. 2013. A Modified Ladder Wave Modulation-Based Circulating Current Suppressing Strategy for Modular Multilevel Converter [J]. *Power System Technology* 37(4): 1012–1018.

Steffen Roher, Steffen Bernet. 2010. Modulation, Losses, and Semiconductor Requirements of Modular Multilevel Converters [J]. *Industrial Electronics* 57(8): 2633–2642.

Tu Qingrui, Xu Zheng, Zheng Xiang, et al. 2011. An optimized voltage balancing method for modular multilevel converter [J]. *Proceedings of the CSEE* 26(55): 15–20 (in Chinese).

Xin Yechun, Wang Chaobin, Li Guoqing, et al. 2014. An Improved Balance Control for Sub-Module Capacitor Voltage of Modular Multilevel Converter [J]. *Power System Technology* 38(5): 1291–1296.

Zhao Xin, Zhao Chengyong, Li Guangkai, et al. 2011. Submodule capacitance voltage balancing of modular multilevel converter based on carrier phase shifted SPWM technique [J]. *Proceedings of the CSEE* 31(21): 48–55 (in Chinese).

Electronics, Information Technology and Intellectualization – Song & Kwak (Eds)
© 2015 Taylor & Francis Group, London, ISBN 978-1-138-02741-1

The identification and control method of a multiple audio codec based on volume settings

X.G. Xie, F.J. Zhang, H.Y. Shi & Q. Gao
Qiongzhou University, SanYa, Hainan Province, China

Y.H. Han
DatangMobile Communication Co. Ltd, Beijing, China

ABSTRACT: This paper describes an automatic DT (Drive Test) system, Test Equipment (TE) is used for data acquisition and the testing process in a CS (Circuit Switched) domain. During the testing process, the processor needs to identify and control the multiple audio codes. This paper proposes a method to identify and control multiple audio codecs. After a codec power-up, a processor sets a certain volume value for each codec. When a device ID of codec changes, the processor can find the new device ID for each codec based on the volume setting, and then the processor can identify and control each codec independently. Datong Mobile designs the Test Equipment for automatic DT by using this method. It can perfectly identify and control the 10-channel audio codec independently in a Windows system, and can also completely achieve the voice test for four kinds of mobile communication networks at the same time. The equipment performance indicators meet or exceed the requirements for China Mobile Index.

Keywords: automatic DT; multiple audio codecs; volume setting

1 AUTOMATIC DT OVERVIEW

Network optimization of communication has been one of the major communication carriers. Due to its specificity, the network optimization of a mobile communication network is especially important. One of the main tasks of network optimization is the collection and analysis of the testing data. Automatic DT system is used for data acquisition and analysis at present, which replace the traditional network optimization. As the norm for China Mobile automatic DT equipment, it sets testing standards in the CS domain and PS (Packet Switch) domain for different kinds of standard terminals (GSM, GSM, CDMA, WCDMA, TD-SCDAM, TD-LTE, etc). Automatic DT equipment overall is as shown in Figure 1.

2 AUTOMATIC DT EQUIPMENT DEMAND

As can be seen in Figure 1, the automatic drive test system is a tool for network optimization, which achieves an automatic test by mounting test terminals onto a moving vehicle. Test equipment completes the data acquisition of different radio interfaces based on real-time pre-set testing program or instruction of automatic DT platform. Radio interface includes GSM, CDMA, WCDMA, TD-SCDMA, LTE, etc. Test

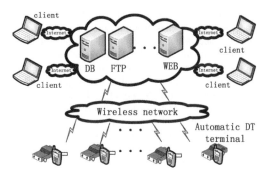

Figure 1. Automatic DT system block diagram.

equipment can automatically store test data in the memory card. Test data can also be returned back to a remote server through a modem module in real time or at an agreed time. The main components of the test equipment are shown in Figure 2.

3 THE CS DOMAIN SPEECH TEST PROCEDURE OF AUTOMATIC DT

CS domain test of Automatic DT is mainly composed of a voice assessment module and a data acquisition module. The voice assessment module is composed

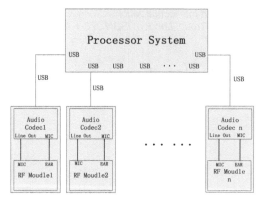

Figure 2. Main components of Test Equipment.

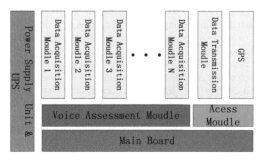

Figure 3. Voice test of Test Equipment connect in hardware.

of multiple audio codecs that can play and record. Each audio codec has an input interface and an output interface, which connects to the output interface and the input interface of the telephone module (RF module), respectively. Playback and recording is then achieved via software. Finally, the program calculates the speech scores through the PESQ (Perceptual evaluation of speech quality) algorithm. Hardware components are as shown in Figure 3. Automatic DT equipment includes a processor system, an audio codec and the RF modules (telephone modules). Audio codec is one relationship with the RF module. The audio codec and the RF module of the same group are in a slot, as is shown in Figure 3, assuming the number of slots are slot 1, slot 2, slot 3…slot n from left to right. Among the RF modules, two RF modules are in pairs. The RF module pair calls each other during the test, for example when audio codec 1 plays back, audio codec 2 is recording. Correspondingly, when audio codec 2 plays back, audio codec 1 is recording.

As shown in Figure 3, during the recording process, the processor system sends an audio data stream through the USB interface, after an audio codec 1 processing, audio signals output into MIC interface of RF module 1. Audio signals are then transformed from RF module 1 to RF module 2 by air. The audio signals are then exported from the EAR interface of RF module 2, and audio signals get in MIC interface of audio codec 2. After the chip processing, the audio

Table 1. Windows low-level audio WaveX API list for recording and playback.

Low-level audio WaveX API		
playback	WaveOutOpen	Open Play Equipment
	WaveOutPrepare Header	Prepare wave header
	WaveOutAdd Buffer	Prepare the data block
	WaveOutWrite	Data is written to the device and start playing
	WaveOutReset	Stop playing
	WaveOutClose	Close playback device
recording	WaveInOpen	Open the recording equipment
	WaveInPrepare Header	Prepare wave header
	WaveInAdd Buffer	Prepare the data block
	WaveInWrite	Start recording
	WaveInReset	Stop recording
	WaveInClose	Close recording equipment

data stream from the USB output interface into the processor system.

During the recording back process, the processor system sends an audio data stream through the USB interface after an audio codec 2 processing; audio signals output into MIC interface of RF module 2, then audio signals are transformed from RF module 2 to RF module 1 by air. The audio signals are then exported from the EAR interface of RF module 1, and the audio signals get in MIC interface of the audio codec 1. After the chip processing, the audio data stream transfers from the USB output interface to the processor system.

The recording process and the recording back process will generate audio files or an audio data stream. Then, the program calculates the speech scores by the PESQ algorithm. We can make an assessment of the quality of the communication network using these scores.

4 WINDOWS SOFTWARE PROCESSING

On the software processing, Windows provides a set of API (Application Programming Interface) functions to control the audio codec. Users can call these interfaces for playback and recording, as specified in Table 1.

5 PROBLEM POSING

In order to contrast the voice quality among different networks simultaneously, an automatic DT device needs to integrate multiple audio codecs that can play and record. The audio codec is related to the RF module. The audio codec and RF module of the same group are in a slot. The RF modules include a GSM module, a CDMA2000 module, a WCDMA module, a TD-SCDMA module, a LTE module and so on.

When testing using the API function, the processor systems usually use device ID to distinguish multiple audio codecs. It means that the API function must be based on device ID as the object to control the audio codec. As in a multi-codec system, device ID number of voice input or output is an important parameter to open or close the corresponding equipment. Call follows:

DeviceInNum:-waveInGetNumDevs (VOID), //Get sound input device ID number

DeviceOutNum:-waveOutGetNumDevs(VOID), // Get sound output device ID number

The range of the input channel is $0 \sim$ DeviceInNum-1, and the range of output channel is $0 \sim$ DeviceOutNum-1.

Thus, there comes a problem, as the Windows operating system is unable to distinguish, identify, or control a specific piece of codec in a multi-codec environment. Although the processor system can access some information of the codec (e.g. device ID), the system can only know which one is the specific object from the information which it has been given. The processor would not control recording and playback function for a particular piece of codec. The reason for this is as follows: we take the input device ID as an example: device ID set from 0 to the codec number minus 1,

However, one cannot determine the specific device ID number which corresponds to a codec. When the codec working condition changed, the device ID will change. The processor system will reset the device ID for the codec within its jurisdiction.

For example, when only codec 1 and codec 2 are powered up, the processor system will assign its device ID, called ID1 and ID2. If the codec 3 is also powered up, the processor system will then assign device ID again, called ID1, ID2 and ID3. For the device ID of audio codec 1, it may be any one among ID1, ID2 and ID3. Processor system cannot distinguish which device ID is used for audio codec 1. Therefore, a codec confusing phenomenon will occur, as the system cannot control the audio codec.

On the other hand, when a codec is powered down, the system will reassign the device ID. It also can cause a confusing phenomenon. In a word, because of these shortcomings in the course of the actual automatic DT, the processor system cannot accurately control the playback and recording of each audio code according to the device ID.

6 EXISTING SOLUTIONS

In response to these shortcomings, a common solution is that we add an EEPROM chip to each piece of the codec, which is used to store some uniquely distinguished information for each codec, such as slot ID.

Therefore, Windows API function command 'waveoutGetDevcaps ( )' can obtain the product name and other information, and at the same time obtain the codec device ID which corresponds to a specific slot information. Using this method of modifying the hardware design, the system can set up a corresponding table between codec equipment and its device ID. This method can meet the needs of a multi-codec system, and can also avoid codec confusion. However, the existing solution has the following disadvantages:

1) An increase in development costs. During the development period, increasing the EEPROM chips would increase the cost of the product. At the same time, the changes in hardware design often require more time.
2) The production process is cumbersome. It needs production staff to write specific information according to the demands in equipment production.
3) Flexibility. In actual use, due to the need for testing, people will often exchange codec among several slots. It can cause a situation where specific information in EEPROM and the actual slot information do not match. There will still be a cause for a codec confusing phenomenon.

7 THE IDENTIFICATION AND CONTROL METHOD OF MULTIPLE AUDIO CODECS BASED ON VOLUME SETTING

This paper proposes a method to identify and control a specific codec in a multi-codec environment, which can reduce implementation costs, and also avoid confusion in a multiple audio codec application system.

The specific technical solutions are as follows:

1) In the multiple audio codecs application system, we assign to each codec a volume value that can only distinguish each audio codec.
2) According to the volume value, which corresponds to each codec, the processor determines the device ID number of an audio codec used at that moment.
3) The processor sends commands in order to control the corresponding codec which is based on the obtained device ID.

In the implementation process of this paper, regarding the multiple audio codecs application system, we set specific volume value as the only distinguishing identification for each codec. Thus, when the operation state of the codec has, changed, the device ID can be obtained timely by checking the current codec volume value. Thus, we can send command accurately in order to control each codec used for automatic DT. This method avoids the error-prone production processes, reduces the development cost and also avoids the occurrence of codec confusion. Therefore, it would be a flexible and effective way used for automatic DT.

The method to identify and control multiple codecs is based on volume settings, whose specific implementation is as follows.

In order to reduce costs, and avoid codec confusion in multi-codecs application systems, we use Windows operating system features in the implementation process of this paper. During the first power up for each codec, we assign a volume value (volume mark) which is always effective for each codec. In the follow-up process, we can uniquely identify the corresponding codec through the volume value of each codec. This means that we can send commands accurately in order to control each codec successfully by using Windows API. The API functions are used to effectively control the recording or playback of the codec.

Windows system has a feature on the codec. After a certain set of specific volume value, regardless of how the device ID changed, the volume value of the codec is always valid. Windows operating system sound volume ranges from 0 to 65535. The default volume settings and the actual sound volume are at the level of 10^{+4}. As the volume correction value of $\pm10^{+2}$ levels for the user's perception and equipment related indicators are very small. We do not worry about the volume settings affecting the codec performance.

In a real multi-codec application process, the following two states are likely to occur, which will be the influencing factors to identify and control the codec:

a) Once a new codec powers up, the processor needs to re-assign device ID for the powered codec: Or/And, once a codec powers down. The processor also needs to re-assign device ID for the powered codec. In both cases, there may be one, or may be present simultaneously.

b) When each codec changes slot, we also need to power up or down the codec. Therefore, the processor needs to re-assign device ID for powered codec. There is volume value (volume mark) in the codec and therefore the system will recognize the codec.

We determine each codec device ID according to the volume value of each codec. We also set up a mapping relationship between the volume value and device ID of each codec, meaning that we set up a mapping list.

Following this, we give an example in order to introduce the operation on the actual application scene. Assuming that there are three codecs and three slots and that the codec 1 is in slot 1, the codec 2 is in slot 2, and the codec 3 is in slot 3, we can power up or down the three codecs independently.

First, we power up the codec 1 in slot 1, and when the system detects the codec 1, it will assign a device ID, called ID1. Through the device ID1, the processor would set the volume value of the codec1 to a certain value. Preferably, the volume can

Table 2.   The mapping relationship of 1 codec.

num	Slot	Codec	Device ID	Volume value
1	Slot 1	Codec 1	ID1	(50000 + 1)

Table 3.   The mapping relationship of 2 codecs.

num	Slot	Codec	Device ID	Volume value
1	Slot 1	Codec 1	ID2	(50000 + 1)
2	Slot 2	Codec 2	ID1	(50000 + 2)

be set for the identification on $(50000 + 1)$. In this, '+1' corresponds to the slot 1, as shows in Table 2.

Second, the system powers up codec 2. When the system detects that there are two codecs, it would assign another device ID. At this moment, there are two device IDs (0~DeviceInNum-1). To describe them easily, we call them ID1 and ID2. Owing to the operation system, the system would re-assign ID1 and ID2, which would change device ID assigned in the first step, the device ID of codec 1 in slot 1 may be ID1, also could be ID2. Also, the device ID of codec 2 in slot 2 may be ID2, also may be ID1. At the time, the System can inquire the volume value of each codec through the two device IDs that were assigned previously. (right now, the system does not know which codec corresponds to a specific device ID). For example, suppose the query to ID2 volume value is $(50000 + 1)$, we can determine that the device ID of codec 1 in slot 1 changes to ID2. Then, we can make sure that the device ID of the codec 2 in slot 2 is ID1. Using the device ID, the processor would set a certain volume value for the codec 2 in slot 2, such as $(50000 + 2)$. In this, '+2' corresponds to the slot 2. As shows in Table 3.

Finally, the system powers up codec 3 in slot 3, and when the system detects that there are three codecs, it would assign another device ID. At this moment, there are three device ID (0~DeviceInNum-1). They are called ID1, ID2 and ID3. The system would re-assign ID1, ID2 and ID3. It would change device ID assigned in the second step. Such as the current device ID of codec 1 in slot 1 may be one of ID1, ID2 and ID3. Similarly, the device ID of codec 2 in slot 2 and codec 3 in slot 3 are uncertain. At the time, the system can inquire the volume value of each codec through the three devices ID assigned before. For example, suppose the query to ID3 volume is $(50000+1)$, we can determine that the device ID of codec 1 in slot 1 change to ID3. The system requires that the volume value of ID1 is $(50000+2)$, and we can determine that the device ID of the codec 2 in slot 2 is the ID1. Furthermore, we can determine that the device ID of the codec 3 in slot 3 is the ID2. Through device ID, the processor would set a certain volume value for codec 3 in

Table 4. The mapping relationship of 3 codecs.

num	Slot	Codec	Device ID	Volume value
1	Slot 1	Codec 1	ID3	$(50000 + 1)$
2	Slot 2	Codec 2	ID1	$(50000 + 2)$
3	Slot 3	Codec 3	ID2	$(50000 + 3)$

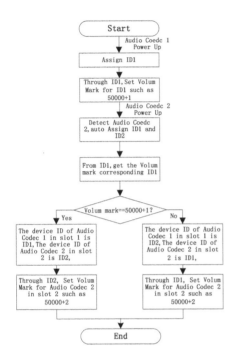

Figure 4. The identification control flow chart of two codecs based on volume settings.

slot 3, For example, $(50000 + 3)$. '+3' corresponds to slot 3, as shown in Table 4.

And so on, the road test system respectively provided a specific volume value for every codec. Through the volume value, we can know the current relationship between codec, slot and device ID. Using this mapping relationship table, the system can find the device ID of the codec in certain slot. Through the device ID, we can control the playback and recording for a certain codec in certain slot.

Power down process is similar. When a codec powers down, the system would automatically make device ID number minus one, and re-assigned the device ID. We obtain the corresponding volume value through the remaining device ID, and then we can set up the mapping relationship table among slot, codec, device ID and volume value. The flow chart is shown in Figure 4.

9 SUMMARY

This paper overall describes automatic DT equipment, test equipment composition of automatic DT, a multi-codec test program, and the voice testing process in CS domain of automatic DT. Then, we discuss the problem by using a multi-codec test program. We realise that we cannot identify and control the certain codec we want to. Having these reasons in mind, we propose a program for identification and control, which is based on volume setting in multi-codec applications.

From the above description of the theory and practical examples, we can see that the multi-codec identification and control method proposed in this paper, in a multiple codecs application system, the specific volume value distinction is the only identification of the codec. Thus, when the operation state of the card is changed, the changed device ID can be obtained timely by using the current codec's volume value. At the same time, we can obtain the mapping relationship between a slot number, a codec number, a volume value, and the device ID. Thus, we can send control commands accurately to each codec used for automatic DT. This method avoids the error-prone production processes, reduces the development cost, and also avoids the occurrence of codec confusion. Therefore, it will be a flexible and effective way used for automatic DT.

This method has been applied to the automatic DT product of Datang Mobile Communications Equipment Co., Ltd. We can identify and control the 10-channel USB audio codec independently in Windows system, and achieve the best results. Equipment performance indicators meet or exceed the requirements of China Mobile index. Products are sold to various provinces and cities nationwide, and occupy 1/4 of the automatic DT production market share.

REFERENCES

China Mobile Communication Co. Ltd. (2009). Technical Specification for Automatic Drivetest Equipment.
Cai, X.D. (2003). Application of automatic drivetest systems. *Wireless and Mobile Communications Commission Annual Conference Proceedings* 2003:246–248.
Han, Y.H., Xie, X.G. & Lei, H.J. (2012). Method and apparatus for identifying and controlling codec. *Patents. People's Republic of China Intellectual Property Office.* 1–16.
Sha, Q. (2008). Mobile DT data auto collection system based on embedded platform. *Journal of shanghai institute of technology.* Vol. 8, No. 2:149–152.
Zhao, W., Xie, X.G., Song, J.F. & Liu, C. (2012). Audio codec components, audio codec identification method and system. *Patents. People's Republic of China Intellectual Property Office.* 1–12.
Zhao, X.X. & Liu, L.F. (2013). Research on the data processing of the intelligence drive test system. *1994–2013 China Academic Journal Electronic Publishing House*: 442–446.

Fund Project: Hainan Natural Science Foundation (ID: 614237).
Qiongzhou University Youth Science Foundation (ID: QYQN201429).
Xie, X.G. (1982–), Male, lecturer, College of Electronics and Information Engineering, Qiongzhou University.

Electronics, Information Technology and Intellectualization – Song & Kwak (Eds)
© 2015 Taylor & Francis Group, London, ISBN 978-1-138-02741-1

The improved delay scheduling algorithm on Hadoop

Suzhen Wang & Bingrui Cao
Hebei University of Economics and Business, Shijiazhuang, China

ABSTRACT: According to the notion that the existing delay scheduling algorithm is lacking in consideration of the load of nodes, if the assignment data to be processed is set on one or several nodes, it will easily lead to an imbalance of the cluster load. In this paper, the original delay scheduling algorithm is improved. By detecting the load of nodes, the improved algorithm adjusts the waiting time of the job dynamically, and increases the number of copies of hot data block at the same time. The experimental results show that the improved delay scheduling algorithm, to a certain extent, improves the average response time of jobs.

1 INTRODUCTION

With the development of search engines, social networks and other data-intensive Internet applications, there is a trend for the information and data to grow explosively. Thus, it brings up the question of how to store and process more and more large data and information; the answer is cloud computing. As the commercial implementation of parallel computing, distributed computing and grid computing, cloud computing integrates multiple computing entities into a relatively strong cluster through the network. It has certain features including super-large scale, high reliability and high scalability, etc. Its core idea is forming a resource pool through unified scheduling and allocation of resources, which are allocated to each user according to his/her need.

Amongst numerous solutions for cloud computing, Hadoop is a particularly important cloud platform architecture. It is an open source for the implementation of the GFS data storage mode and the MapReduce distributed programming model, both of which are presented by Google. Its core technologies HDFS and MapReduce function as mass data storage and processing, respectively. Nowadays there are already many relatively successful applications on Hadoop, such as Yahoo, Facebook and Amazon, etc. The distributed storage of Hadoop greatly increases system parallel processing capabilities. Meanwhile, because when performing an assignment on the Hadoop platform, the computing resources and data resources it needs can be located in different physical nodes, thus causes the localization problem.

Currently used methods for localization problem is delay scheduling algorithm, namely the job waits for local computing resource nodes within time T, if the job gets nothing when the T time is out, then the job at the head of the queue will be scheduled, this method has greatly increased the efficiency of performing a job locally. Meanwhile for the reason that the job has been waiting for the local target node all the time, if the

assignment data to be processed is set on some certain nodes, it will increase the load of these nodes greatly and lead to imbalance of the cluster load easily, and then influence the execution efficiency.

In this paper, the delay scheduling algorithm is improved, specifically as follows: The first thing is to balance the load, namely to detect the load of the spare node while the job is waiting for a local target node. If the node load exceeds the threshold, then do not assign any task to this node temporarily. The second thing is to increase the number of copies of the hot data block, as the access frequency to different data block files is different. We can set different number of copies based on different heat of data block files. Using the above two ways, we can achieve the purpose of improving the operating efficiency of the Hadoop cluster.

2 THE EXISTING DELAY SCHEDULING ALGORITHM

A delay scheduling algorithm uses the strategy of "move computing" instead of "move data", namely migrating the computing task to a nearer location from the data. The basic idea of delay scheduling algorithm is as follows: With regards to the aspect of fairness, use the Max-Min fair scheduling algorithm to achieve the purpose of statistical multiplexing; with regards to the aspect of data localization, if there is no data of the job at the head of the queue to be processed at reaching current spare node, skip this job, schedule other jobs first and let the job at the head of the queue wait; if the waiting time of the job at the head of the queue exceeds the threshold, then schedule this job immediately rather than wait.

2.1 *Some basic related conceptions*

(1) Resource slot
In Hadoop, the computing resource is abstracted into the resource slot, and one task can run in one slot. Each compute node has a certain number of

resource slots. If the number is set as two, it means that there can be two existing tasks performing concurrently in the compute node.

(2) The heartbeat mechanism

Hadoop adopts a master-slave mode of resource management method, where the master node is responsible for all the resource management; the slave node sends the master node its status information on a regular basis, in order to report its current status.

(3) Job scheduler

Hadoop scheduler is divided into two levels, the first level is job scheduling, which means the job scheduler has to choose a ready job. The second level is task scheduling, meaning that the task scheduler has to select a ready task in the scheduled job.

(4) The waiting time threshold

When a spare node appears in the cluster, and if the job to be processed has no local task, then it has to wait, and schedule non-local tasks when the waiting time exceeds a certain value. The critical value of the waiting time is called the 'job waiting time threshold', denoted by 'T'.

2.2 Delay scheduling algorithm analysis

Delay scheduling algorithm adds to each job a private variable wait-time, which is used to record the time it takes a job while waiting for the local computing resource. Every time a job is to be scheduled, it is necessary to judge whether the schedule can be a local scheduling first. If it is not a local scheduling, it is necessary to check whether the value of the waiting time is bigger than the waiting time threshold T. If the value is higher than T, it means that the job has been waiting long enough, therefore it should no longer wait, and is ready to accept the non-local scheduling directly. The variable wait-time is set to 0 for a new timing. Otherwise continue to wait, if this is a local scheduling, and dispatch the localization task directly. The variable wait-time also needs to be set to 0 again for a new timing.

Algorithm 1: Delay scheduling algorithm
Input: The spare node n reached at present
for j in jobs do
  if j has unlaunched a task's data on node n then
    launch one local task on n;
    j.wait-time=0;
    return;
  else if j.wait-time<T then // T is the threshold
for waiting local computing resource
    j.wait-time++;
    continue;
  else
    launch one unlaunched j.task on n;
    j.waitime=0;
    return;
  end if
end for
return;

3 THE IMPROVED DELAY SCHEDULING ALGORITHM

Because the existing delay scheduling algorithm is lacking in consideration of the load balance and the hot data, in this paper, improvements are made in these two aspects to achieve the purpose of increasing efficiency of the Hadoop platform.

3.1 Define several concepts

(1) The node load

$$L_i=L_{cpu}*P_{cpu}+L_{mem}*P_{mem}+L_{net}*P_{net} \qquad (1)$$

In the above formula, L_i is defined as overall load of the node i, L_{cpu} as cpu utilization, L_{mem} as memory utilization, L_{net} as load condition of the current network, P_{cpu}, P_{mem}, P_{net} as the proportion of CPU, memory and network respectively, $P_{cpu} + P_{mem} + P_{net} = 1$. The specific value is set by the administrator according to the demand for load balance performance of the system. In this paper P_{cpu}, P_{mem}, P_{net} are set as 30%, 30% and 40% respectively.

(2) Average load of the cluster

If the Hadoop platform has n nodes, and the load of each node is L_i ($i = 1, 2, \ldots, n$) respectively, then the average load of the cluster is:

$$L_{avg}=(L_1+L_2+L_3+...+L_n)/n \qquad (2)$$

(3) Hot data block

The data block accessed frequently is called hot data block. And in this paper it is defined as follows:

$$num_{a[i]}^{advanced} = \begin{cases} num_{a[i]}^{now} + n2_{a[i]}, & nl_{a[i]}>n/2 \text{ and} \\ & n2_{a[i]}/nl_{a[i]}>0.5 \\ num_{a[i]}^{now}, & else \end{cases} \qquad (3)$$

In the above formula, the number of nodes in the cluster is n, the data block file is numbered in turn as a[i], i = 1, 2, 3 . . . m, $num_{a[i]}^{now}$ which is the number of the existing copies of data block. The default value is 3, and $num_{a[i]}^{advanced}$ is the number of the improved copies. The times of data block access is represented by $nl_{a[i]}$, times of non-local access by $n2_{a[i]}$, $n2_{a[i]} <= nl_{a[i]}$.

3.2 Description of algorithm process

The idea of an improved algorithm is when the job is waiting for scheduling of the local computing resource, meanwhile, adjust the job waiting time dynamically by certain rules according to both the speed at which the node free slot emerges and load condition of the node. Give full play to the performance of each node and improve the overall efficiency of the cluster. The specific steps are as follows:

a) Job queue sort. Determine the job priorities according to the Max-Min fair scheduling algorithm and sort out the jobs by priority going from high to low.

b) Free slot detection. Task Tracker sends status information to the master node every three seconds in order to report whether a free slot is available.

c) Local scheduling. When the free slot is available and the task which can be locally processed is also found in the job queue traversal, allocate the first task which can be locally processed to the queue for this compute node.

d) Compute the load of the nodes. When a free slot appears on the node i, but the task which can be locally processed is not found in the job queue, then detect the load condition L_i of this node.

e) Non-local scheduling, when all the jobs' waiting times do not exceed T. If $L_i < L_{avg}$ and $L_{avg} < L_{max}$, then select the job at the head of the queue to execute non-local scheduling, otherwise give up this scheduling. Here, L_{max} is the critical value of the load which shows whether the node is non-locally scheduled.

f) Extend the waiting time. When there exists job's waiting local scheduling exceeds T, either there's no local data resource. If $L_i > L_{max}$, then extend the waiting time, do not perform a non-local schedule task, or otherwise, perform a non-local schedule task.

g) Add copies of hot data block. The default number of data blocks in the cluster is 3; add the number of hot data block copies according to the rules set above.

Algorithm 2: Improved delay scheduling algorithm
Input: The spare node n reached at present
for j in jobs do
  if j has unlaunched task's data on node n then
    launch one local task on n;
    j.wait-time=0;
  else if j has unlanched task t then
    if j.wait-time>T and L_n>L_{max} then
      continue;
    elseif j.wait-time>T and L_n<L_{max} then
      launch t on n;
      j.wait-time=0;
      return;
    else if j.wait-time<T and L_n<L_{avg} and L_{avg}<L_{max} then
      launch t on n;
      j.wait-time=0;
      return;
    else
      continue;
    end if
  end if
end for
return;

Table 1. The set of job parameters.

Group Number	Jobs	Data Size	Tasks
1	15	64M	1
2	8	128M	2
3	6	256M	4
4	4	1024M	16
5	3	2048M	32

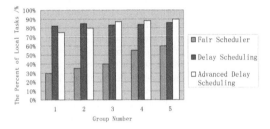

Figure 1. The comparison of the local tasks.

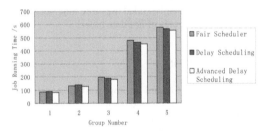

Figure 2. The comparison of the jobs' running time.

4 ANALYSIS OF EXPERIMENTAL RESULTS

4.1 Experimental environment

The Hadoop cluster environment has six nodes. One of them will run a job tracker as the name node, the other five will run a task tracker as the date node. The operating system is centos 6.3 final, the version of Hadoop is 1.2.1, the slot number per node is set to 2. The size of hdfs block uses the default 64MB. In order to compare the influence of scheduling algorithm to a different scale of jobs, the experiment uses five specific group jobs, as shown in Table 1.

4.2 Experimental result

Using fair scheduler, delay scheduling and the improved delay scheduling running, the results of the five group jobs are shown in Figure 1 and Figure 2.

4.3 Result analysis

Job localization analysis: Figure 1 compares the proportions of the local tasks in each algorithm. For small-scale jobs, the number of local tasks with improved delay scheduling algorithm is slightly lower than delay scheduling algorithm, but higher than fair scheduling algorithm. For large-scale jobs, the number of local tasks with improved delay scheduling algorithm is higher than delay scheduling and fair scheduling. It is because the improved algorithm adjusts the job's waiting local computing resources time dynamically, according to the load of nodes. When running small-scale jobs, the nodes are loaded lightly, the waiting time is shortened, and the non-local task is scheduled. In contrast, when running large-scale jobs, the nodes are loaded heavily, so the system prolongs the waiting time, and the proportion of the local tasks increases.

Job running time analysis: the job running time refers to the time between the job being submitted and the return of the final processed results. This is an important index, which reflects the interacting ability between the system and the users. The experimental results are shown in Figure 2, both for the small-scale jobs and large-scale jobs. The time using improved delay scheduling algorithm is always less than the delay scheduling algorithm and fair scheduling algorithm. This is because the improved delayed scheduling algorithm adjusts the waiting time of the local task dynamically, and at the same time, increases the number of copies of hot data block.

5 CONCLUSION

This paper puts forward an improved delay scheduling algorithm based on load balancing. The difference of this algorithm is that it thinks about the load of nodes when jobs are waiting for local scheduling, then the improved algorithm adjusts different jobs' waiting time according to the load of nodes dynamically, at the same time as increasing the number of copies of hot data block. Through constructing the Hadoop platform and doing experiments, it shows that the improved delay scheduling algorithm balances the node load, and reduces the average running time of jobs.

REFERENCES

Jin Jia-hui, Luo Jun-zhou, Song Aibo, Dong Fang. (2011). Adaptive delay scheduling algorithm based on data load analysis. *Journal on Communications*, 32:7.
Rajkumar Buyya, Chee Shin Yeo, Srikumar Venugopal, et al. (2009). Cloud computing and emerging IT platforms: vision, hype and reality for delivering computing as the 5th utility. *Future Generation Computer System*, 25(6): 599–616.
Tom White. Hadoop The Definitive Guide. (2011). *Tsinghua University Press*. Beijing.
Liu Peng. (2011). Cloud Computing (Second Edition). *Publishing House of Electronics Industry*: 1–9.
Hadoop.*http://hadoop.apache.org/*.
Ghemawat S, Gobioff H, LEUNG Shun-tak.(2003). *The Google File System. ACM*, 37:29–43.
Deng J, Ghemawa S. (2008). MapReduce: simplified data processing on large clusters.*Commum ACM*, 51(1):107–13.
Dean J, Ghemawat S. (2010). MapReduce: a flexible data processing tool. *Communications of the ACM*, 53(1):72–77.
Ralf L. (2008).Google's mapreduce programming model-revisited. *Science of Computer Programming*, 70(1):1–30.
Shvachko K, Kuang Hai-rong, Radia S. et al. (2010). The Hadoop Distributed File System. *MSST*: 1–10.
Amazon.Amazon elastic compute cloud. *http://aws.amazon.com/ec2*.
Fischer M J, Su Xueyuan, Yin Yitong. (2010). Assigning tasks for efficiency in Hadoop: extended abstract. *ACM Press*: 30–39. New York.
Zaharia M, Borthakur D, Sen Sarma J, et al. (2010). Delay scheduling: a simple technique for achieving locality and fairness in cluster scheduling. Proceedings of the EuroSys '10[C]. 265–278. Paris: France.

Electronics, Information Technology and Intellectualization – Song & Kwak (Eds)
© 2015 Taylor & Francis Group, London, ISBN 978-1-138-02741-1

An enhanced Kernel Fuzzy C-Means Algorithm based on bio-inspired computing methods

Yang Liu, Kunyuan Hu, Yunlong Zhu & Hanning Chen
Shenyang Institute of Automation, Chinese Academy of Sciences, Shenyang, China

ABSTRACT: In data analysis and data mining technique fields, one of the most widely used methods is clustering. Recently, one of the bio-inspired computing optimization algorithms called the Artificial Bee Colony (ABC) algorithm has been introduced, which has many characteristics, such as simple, robust, stochastic global optimization. In this paper, an enhanced Kernel Fuzzy C-Means Algorithm (KFCM) based on the ABC algorithm for data clustering is proposed. Compared with other popular bio-inspired computing optimization algorithms in data clustering, the results proved that the number of iterations is fewer, the convergence speed is faster and there is also a large improvement in the quality of clustering.

Keywords: Data clustering; bio-inspired computing optimization algorithms; kernel fuzzy c-means algorithm; Artifical Bee Colony

1   INTRODUCTION

Data clustering techniques are essential tools in data processing, which aim at the unsupervised learning of objects in different groups. The KFCM clustering algorithm is a popular centre-based clustering method, but it has disadvantages, such as depending on the initial state, sensitivity to initialization and noise data, and converging to local minima [1].

The bio-inspired computing optimization algorithm is an innovative artificial intelligence technique for solving complex optimization problems, which include Genetic Algorithms (GA), Particle Swarm Optimization (PSO), Ant Colony Optimization (ACO), Artificial Bee Colony (ABC), Bacterial Foraging Algorithms (BFA) etc. The Artificial Bee Colony (ABC) algorithm was first introduced by Karabog in Erciyes University of Turkey in 2005 [2]. Similar to the other stochastic optimization algorithms, the intelligent foraging behaviour of a honey bee swarm has been simulated in the ABC algorithm. When increasing the dimensionality of the search space, the ABC algorithm encompasses worse convergence result [3]. Therefore, a modified KFCM algorithm based on the ABC algorithm is put forward [4].

2   KFCM ALGORITHM

The KFCM algorithm partitions data objects into k clusters where the number of clusters, which can to a certain extent overcome the limitation of data intrinsic shape dependence, overcome sensitivity to

initialization and noise data and improve the algorithm robustness [5].

In order to minimize the following objective function, the KFCM algorithm partitions a given dataset $M = \{m_1, m_2, \ldots, m_n\} \in R^P$ into C fuzzy subsets, which is defined as Equation (1) [6]:

$$Perf(X,C) = \sum_{i=1}^{N} Min\{\| X_i - C_l \|^2 | l = 1,...,K\} \quad (1)$$

where C is a set of K clusters, the number of cluster centroids is denoted as K, $C = \{c_1, c_2, \ldots, c_k\}$. The profile data matrix is defined as $X_{n \times p}$ with n rows and p columns. The real-value p-dimensional profile vector characterizes each of its objects; p is the number of data points [7]. The similarity between object X_i and centre C_l is denoted by $\|X_i - C_l\|$ [8]. The Euclidean distance, which is the most used similarity metric in clustering procedure deriving from the Murkowski metric, and defined as Equation (2):

$$d(x_i, c_j) = \sqrt{\sum_{k=1}^{P} (x_{ik} - c_{jk})^2} \quad (2)$$

Equation (3) is used to recalculate the cluster centroid vectors after all data has been grouped.

$$c_j = \frac{1}{n_j} \sum_{\forall x_i \in c_j} x_i \quad (3)$$

where the number of data vectors is defined as n_j which belong to cluster j.

3 THE ENHANCED KFCM ALGORITHM BASED ON ARTIFICIAL BEE COLONY ALGORITHM

3.1 The traditional Artificial Bee Colony algorithm

The traditional ABC algorithm is a very simple, robust and population-based stochastic optimization algorithm, which is a new member of swarm intelligence (a branch of nature inspired algorithms focused on insect behaviour), which simulates the intelligent foraging behaviour of honey bee swarms. There are three groups of bees included in this: employed bees, onlookers and scouts. Employed bees and onlooker bees are primarily responsible for the exploitation of food sources, and the scouts are primarily responsible for the exploration of food sources [9]. Parameters such as its proximity to the nest, richness of energy and ease of extracting this energy decide the value of a food source. The original Artificial Bee Colony algorithm not only combines global searches, but also includes local searches, so the two aspects of the exploration and exploitation of the algorithm are able to achieve a better balance.

3.2 The enhanced kernel fuzzy c-means algorithm based on ABC [10][12]

The enhanced kernel fuzzy c-means algorithm based on ABC uses the capacity of global search in the ABC algorithm to seek optimal solution as initial clustering-centres of the KFCM algorithm, and then uses the KFCM algorithm to optimize initial clustering-centres, in order to get the global optimum [9].

In the proposed algorithm, a solution in K dimensional space is represented by each individual. The number of clusters is equal to the number of dimensions, a cluster centroid is represented by each component of an individual and a P-dimensional vector represents each cluster centroid. In the initialization phase, the maximum and minimum value of each component of the data set to be grouped were used as the proposed algorithm individuals' initialization range, and an initial solution was randomly generated in this range. The fitness function of individuals is calculated by Equation (1). The proposed algorithm generates a randomly distributed initial population of SN solutions (food source positions), where the size of employed bees or onlooker bees are denoted by SN. A bee denotes a cluster centre and a P-dimensional vector represents each solution x_i ($i = 1, 2, \ldots,$ SN). P is the number of optimization parameters, and a food source represents a possible solution which is to be optimized for the problem. The quality (fitness) of the associated solution corresponds to the nectar amount of a food source, which is calculated by Equation (1). A food source associated with the probability value is chosen by an artificial onlooker bee, p_i is calculated by Equation (4) [11]:

$$p_i = fit_i \Big/ \sum_{n=1}^{SN} fit_i \qquad (4)$$

Table 1. Main steps of the enhanced kernel fuzzy c-means algorithm based on ABC.

Step 1.	Cycle = 1
Step 2.	Initialize the food source positions x_i, $i = 1 \ldots$ SN
Step 3:	Evaluate the nectar amount (fitness fit_i) of food sources
	For data vector
	Calculate the Euclidean distance by using Eq. (2)
	Assign x_i to the closest centroid cluster c_j.
	Calculate the measure function using equation Eq. (1)
	EndFor
	Return value of the fitness function
Step 4:	Repeat
Step 5:	Employed Bees' Phase
	For each employed bee
	Produce new food source positions v_i by using Eq. (5)
	Calculate the value i fit
	Apply greedy selection mechanism
	EndFor
Step 6:	Calulate the probability values p_i for the solution by using Eq.(4)
Step 7:	Onlooker Bees' Phase
	For each onlooker bee
	Chooses a food source depending on p_i
	Produce new food source positions v_i
	Calculate the value i fit
	Apply greedy selection mechanism
	EndFor
Step 8:	Scout Bee Phase
	If there is an employed bee becomes scout
	Then replace it with a new random source positions
	by using Eq. (6)
Step 9:	Memorize the best solution achieved so far
Step 10:	Cycle = cycle + 1
Step 11:	Until cycle = Maximum Cycle Number

Table 2. Experimental data sets.

Data set name	Number of sample	Dimension	Class
Motorcycle	133	2	4
Iris	150	4	3
Wine	178	13	3
Contraceptive Method Choice	1473	10	3
Wisconsin breast cancer	683	9	2
Ripley's glass	214	9	6

where the fitness value of the solution i is represented by fit_i, the number of food sources is represented by SN. Using Equation (5) to produce a candidate food position from the old one in memory:

$$v_{ij} = x_{ij} + \phi_{ij}(x_{ij} - x_{kj}) \qquad (5)$$

116

Table 3. Comparison of intra-cluster distances for the three clustering algorithms.

Data set	Criteria	The proposed algorithm	KFCM	ABC
Motor	Average	2060.7	3012.3	2068.9
Cycle	Best	2060.6	2446.3	2060.6
	Worst	2062.4	4683.2	2126.7
	Std	0.325	439.06	19.118
Iris	Average	94.603	106.05	94.607
	Best	94.603	97.333	94.603
	Worst	94.603	120.45	94.644
	Std	0.00000000297	14.631	0.00777
Wine	Average	16294	18061	16298
	Best	16292	16555	16294
	Worst	16296	18563	16302
	Std	15.548	793.21	6.241
Contraceptive	Average	5693.8	5893.6	5695.4
Method	Best	5693.7	5842.2	5693.9
Choice	Worst	5693.9	5934.4	5698.6
	Std	0.0465	47.165	1.3824
Wisconsin	Average	2964.4	3251.2	2964.4
reast	Best	2964.4	2999.1	2964.4
cancer	Worst	2964.4	3521.5	2964.4
	Std	0.0000192	251.14	0.010731
Ripley's	Average	223.68	235.57	225.39
glass	Best	212.32	215.74	210.87
	Worst	246.27	255.38	253.20
	Std	7.798	12.471	12.685

where chosen k and j randomly, $k \in (1, 2, \ldots, SN)$ and $j \in (1, 2, \ldots, P)$, but k is not equal to i, choose φ_{ij} in the random number between $[-1, 1]$.

In the proposed algorithm, positions which cannot be improved further through a predetermined number of cycles (called limits) are provided, then assuming that the food source is to be abandoned, which is regarded as x_i, the scout discovers a new food source to be replaced with x_i. This operation can be defined as:

$$x_i^j = x_{min}^j + rand[0,1](x_{max}^j - x_{min}^j) \qquad (6)$$

Exploration and exploitation processes are carried out together and the global search performance of the algorithm depends on random search process performed by scouts, the neighbour solution production mechanism performed by employed and onlooker bees. Therefore, the enhanced kernel fuzzy c-means algorithm based on the ABC algorithm is an efficient optimization tool since it combines exploitative local search and explorative global search processes efficiently. The following algorithm is presented to show the steps of our proposed enhanced kernel fuzzy c-means algorithm based on the ABC method, which can be described as Table 1.

4 EXPERIMENT RESULTS

In order to evaluate the performance of the enhanced KFCM algorithm based on the ABC approach for clustering, the results of the KFCM, ABC, and the proposed algorithms are compared, and six different data

sets are used, which are selected from the UCI machine learning repository 23. The data characteristics are shown in Table 2 [10] [12].

There are three control parameters which are set as follows: SN = 100, MCN = 2000, limit = 30. The weighting exponent m is set to 2. For each data set, we have thirty tests for each algorithm, Table 3 summarizes the clustering results of each algorithm, also known as the 'cluster distance', and the Std represents the standard deviation.

In order to estimate the solution quality and convergence speed, there is a significant improvement in the performance, which is shown in Table 3. The average, best, and worst solutions of fitness from the thirty simulations, and standard deviation are presented. The comparative study of the proposed approaches with existing algorithms in the literature using the datasets from UCI Machine learning repository is satisfactory. It can be seen without a doubt that the results of the enhanced kernel fuzzy c-means algorithm based on ABC are superior to the other algorithms for all data sets, the search ability of the enhanced kernel fuzzy c-means algorithm based on ABC has being enhanced, and optimized speed is faster.

5 CONCLUSION

This paper proposes new data clustering based on the ABC algorithm. This algorithm searches cluster centres. Because the artificial bee colony algorithm has the ability of global search and local search, the advantage of the proposed algorithm can jump out of the local optimal solution. We tested the proposed algorithm on several well-known real data sets; the experimental data show that the proposed algorithm is superior to other algorithms for different data concerned on average value and standard deviations of fitness function. Additionally, the simulation result illustrates that the proposed optimization algorithm can be considered as a viable and efficient method to solve optimization problems, which is one of the main directions for future research.

ACKNOWLEDGMENT

This work was supported in part by the National Natural Science Foundation of China under Grant 61105067 and 61174164, the General Project of Education Department of Liaoning Province under Grant L2013446, the Engineering research centre of the IOT Information technology integration of Liaoning Province open-funded projects and Key Laboratory of Networked control System (CAS).

REFERENCES

Czajkowski, K., Fitzgerald, S., Foster, I, Kesselman, C. Grid Information Services for Distributed Resource Sharing. In: *10th IEEE International Symposium on High Performance Distributed Computing*, (2001) 181–184. IEEE Press, New York.

Dervis Karaboga and Celal Ozturk. A novel clustering approach: Artificial Bee Colony (ABC) algorithm. *Applied Soft Computing*, (2011) 652–657.

Dervis Karaboga and Basturk B. A powerful and efficient algorithm for numerical function optimization: Artificial Bee Colony algorithm. *Journal of Global Optimization* 39(2007) 459–471.

Dervis Karaboga and Bahriye Akay, A Comparative Study of Artificial Bee Colony Algorithm", *Applied Mathematics and Computation*, 214(2009) 108–132.

Izakian, H., Abraham, A. Fuzzy clustering using hybrid c-means and fuzzy particle swarm optimization. In: *Proceedings of the 2009 World Congress on Nature and Biologically Inspired Computing*, Coimbatore, (2009) 1690–1694.

Khang Siang Tan, and Nor Ashidi Mat Isa. Color image segmentation using histogram thresholding-Fuzzy C-means hybrid approach. *Pattern Recognition*, 44(1) (2011) 1–15.

Krishnamoorthi, M. and Natarajan, A.M. A Comparative Analysis of Enhanced Artificial Bee Colony Algorithms for Data Clustering. *2013 International Conference on Computer Communication and Informatics (ICCCI-2013)*, Jan. (2013) 04–06, Coimbatore, India.

Rui Xu and Donald Wunsch II. Survey of Clustering Algorithms. *IEEE Transactions on Neural Networks*, 16(2005) 645–678.

Wenping Zou. A Clustering Approach Using Cooperative Artificial Bee Colony Algorithm. Technical report, *Discrete Dynamics in Nature and Society* (2010).

Xiaoqiang Zhao and Shouming Zhang. An Improved KFCM Algorithm Based on Artificial Bee Colony. H. Deng et al. (Eds.) *AICI 2011, CCIS 237*, (2011) 190–198.

Yang Liu. An improved clustering method to evaluate teaching based on ABC–FCM. *World Transactions on Engineering and Technology Education*, 11(2) (2013).

Zhang, D.Q., Chen, S.C. A novel kernelized fuzzy C-means algorithm with application in medical image segmentation. *Artificial Intelligence Med*. 32(2004), 37–50.

Electronics, Information Technology and Intellectualization – Song & Kwak (Eds)
© 2015 Taylor & Francis Group, London, ISBN 978-1-138-02741-1

Based on low-carbon environment electric industry development ideas

Ning Li
Henan Engineering Consulting Centre 3rd Department, Zhengzhou, China

Yaning Zhu & Songfeng Tian
North China Electric Power University, Baoding, China

ABSTRACT: In the twenty-first century, climate changes and energy issues have been plaguing mankind. Reducing fossil energy consumption, CO_2 emissions and developing alternative sources are key elements in tackling climates and energy issues. Of all proven Chinese energy reserves, coal accounted for 94%, oil accounted for 5.4%, and natural gas accounted for 0.6%, the decision of Chinese energy structure dominated by coal, dependence up to 80%. Meanwhile, China has been ranked one of the two largest countries in the world with the largest CO_2 emissions, facing enormous pressure to sustainable development and energy security, as well as emission reduction pressure from the international community. Thus, this requires a change in the power structure, in order to achieve the development of low carbon economic development goals.

The concept of low carbon development in order to improve the energy efficiency, clean energy development as the core, mechanism of transformation of the mode of development and the innovation development as the key, with economic and social sustainable development as the goal, is the strategic orientation of Chinese economic and social development in the future. Low carbon development means low energy consumption, low pollution and low emissions based on the "three low" economic model. In essence, low carbon development means efficient use of energy, clean energy development and enhancing the integration of green GDP, and has at its core the energy technology innovation, emissions reduction technology innovation and industrial structure adjustment and institution innovation.

It will be a national core competitive ability and the comprehensive national strength important symbol that whether low carbon development mode can achieve. Low carbon development helps implement the scientific concept of development, alleviat environmental constraints in the development of resources, and is conducive to grabbing low carbon technology as representatives of emerging competitive industries.

Keywords: energy issues; low carbon; green GDP

1 THE CURRENT SITUATION OF THE DEVELOPMENT OF CHINA'S ENERGY AND CO_2 EMISSIONS' CHARACTERISTICS

1.1 *China's energy development*

The National Bureau of Statistics' data have shown that in 2008, the national total primary energy production was 2.6 billion tons of standard coal, which compared to 2007 has shown an increase of 5.2%. In this case, the crude oil output of 190 million tons has increased by 4.1%, natural gas production is 76.08 billion cubic metres, which has increased by 9.9%, coal production is 2.793 billion tons, which has increased by 4.1% and electric energy production is 3.46688 trillion kWh, which has increased by 5.6%. Compared with a national primary energy consumption of 2.85 billion tons of standard coal, which since 2007 has increased by 4.0%, crude oil consumption of 360 million tons, is up by 5.1%. Natural gas consumption is 80.7 billion cubic metres, which has increased by

10.1%, coal consumption is 2.74 billion tons, which is up by 10.1% and electric power consumption is 3.4502 trillion KWH, which is up by 5.6%[1].

The above data shows that our country's main energy resource is coal. The main way of consumption for burning has led to typical soot atmospheric pollution in our country[2]. Burning coal is the leading source of of the NO_x, SO_2, dust, soot and other pollutants[3][4][5]. In addition, coal products including polycyclic aromatic hydrocarbons etc., mark Yuan of poisonous and harmful pollutants, as well as arsenic, beryllium, lead, mercury, radioactive material. At the same time, the chlorine in the coal can be removed with the gas in flue[6][7].

1.2 *The characteristic of China's CO_2 emissions*

As a country that has high energy production and consumption, China's energy efficiency is low. Every ton of standard coal output efficiency is only equivalent to

28.6% of the US, 16.8% of the EU, 10.0% of Japan, whilst energy consumption per unit product is 2–3 times higher than in developed countries[8]; the high energy consumption per unit product is the problem in our country which needs to be solved urgently. Only through technical innovation, by changing the energy structure and reducing the energy consumption per unit product, can the benign development of the nation's economic low carbonization be realized.

Because of the standard unit of CO_2 produced by burning coal, the scalar is 1.7 times higher compared with natural gas emissions, and 1.3 times higher compared with oil emissions[9]. The coal naturally becomes the main source of CO_2 in our country, more than 85% of the total CO_2 emissions. With the development of the economy, the carbon dioxide emissions have increased from 3.38361 billion tons in 2000 to 6.89654 billion tons in 2008, and the growth is a 103.82% increase. Although carbon dioxide emissions are increasing year by year, the growth of carbon dioxide has been declining since 2003.

Of course, to the different population, developing countries think that by using indicators to measure the total carbon dioxide emissions of greenhouse gases emissions cannot embody the principle of fairness and justice. Therefore, per capita, a carbon dioxide emissions index is proposed as a standard measure, which is being accepted and used by more and more people. With the rapid development of the economy, China's per capita carbon emissions keep rising, from 2.7 tons/person in 2000 to 5.2 tons/person in 2008, showing a CO_2 emissions growth of 92.59%. But per capita emissions of carbon dioxide is at a low level compared with other developed countries.

Although carbon dioxide emissions are growing, carbon dioxide emissions intensity (ten thousand yuan GDP emissions of carbon dioxide) on the whole, shows a rapid decline. From the current price, 10,000 yuan GDP carbon dioxide emissions intensity decreased from 3.4 tons in 2000 to 2.3 tons in 2008, which is a 9.17% decline. From the stable price in 1978, 10,000 yuan GDP carbon dioxide emissions in 2000 decreased from 12.2 tons to 11.5 tons in 2008, which is a 5.74% drop. 10,000 yuan GDP carbon dioxide emissions tended to decline, and in 2003 (present price) and 2004 (constant) it appears to rebound, especially after the eleventh five year decline.

2 A GRID TO PROMOTE THE DEVELOPMENT OF A LOW CARBON ELECTRICITY INDUSTRY HUB

2.1 Smart grid analysis hub

Grid infrastructure is an important economic and social development, an important energy strategy relying on the layout, an important part of the energy industry chain and also an important part of the national integrated transport system. China's power grid enterprises in industry are, on the one hand, responsible for safe and efficient systems in order to ensure the livelihood of electricity, the support functions of the economic operation, hand charged with implementing the national macroeconomic policies, guidance, and promote energy saving electric power, electric power can be achieved sustainable development tasks for upstream and downstream enterprises have a key role in guiding. Therefore the power grid, by promoting a low carbon development trend of the industry, plays an important hub function.

2.2 Carbon connotation analysis of the smart grid

Despite the current world for smart grid have their own different set forth; but, in essence, are more consistent with the meaning: the advanced information, communication, and control technology as the basis for intelligent, digital, interactive, lean into features, with round, the whole process, the whole element of intelligent monitoring communications control, decision making and self healing ability of the power system. A smart grid covers improved grid technology content, asset utilization, reliability of the power supply and promotes energy conservation, in order to achieve optimal allocation of resources, and many other targets. Building a smart grid is an important way to achieve a low carbon power sector development, thus effectively promoting the vigorous development of a low carbon economy.

2.3 Building a smart grid is an important way to promote the development of the low carbon power sector

2.3.1 A smart grid to promote and protect the power structure of the low carbon development

Development of renewable energy is a necessary way to achieve low carbon development; a skeleton, strong grid and grid capacity are prerequisites for achieving large-scale development of renewable energy. However, because the power structure is dominated by thermal power, in contrast to accelerating the development of renewable energy, power peaking FM will become more prominent. The existing power grid operation control technology faces difficulties in taking into account large-scale renewable power and network access with a safe, stable and high-quality power supply.

A smart grid will run controlled large-scale renewable energy output forecasts, grid security and stability analysis, energy storage technologies and distribute power compatible with other key technologies in order to carry out in-depth research. The network will be developed in coordination with the plant operating mechanism of science optimizing the network structure, improving efficiency of conventional energy, guiding the construction of low carbon intensive power, raising the proportion of fossil energy alternatives and promoting the development of large hydropower and nuclear power, large-scale wind power, solar energy and other low carbon power, in order to achieve a low carbon power structure, resulting in huge carbon benefits.

2.3.2 *Smart grid transmission and distribution sectors to achieve low carbon*

Transmission and distribution aspects of the emissions reduction benefits can be attributed to the following two aspects:

1) Reduce transmission losses. In the transmission, power is often accompanied by loss of power equivalent to the consumption side additional fossil fuels, and indirectly, the CO_2 emissions. In part, CO_2 emissions are closely related to network planning, operation mode and transmission technology. A smart grid will build power at all levels of the coordinated development platform, integrate the application of new technologies, new materials and new technology, the full implementation of the state maintenance of transmission lines and the whole dough cycle management of critical transmission corridors and disaster-prone areas of the state parameters in real time monitoring and disaster warning, widespread adoption of flexible transmission technology to improve the level of transportation grid. To reduce transmission losses at the core, reducing transmission and distribution of CO_2 emissions.

2) Reduce SF6 emissions. As an excellent gas-insulated material, SF6 in high voltage electrical power is obtained in a wide range of applications, and SF6 in a high-voltage grid point has been widely used. However, the strength of SF6 is about 24,000 times the greenhouse effect of CO_2, and it is the one of the provisions of the six categories of greenhouse gases in the Kyoto Protocol. At present, China's annual production of SF6 is close to $1 \times 104 \, t$, and is expanding each year. SF6 circuit breakers used in the grid are about $2.5 \times 104 \, t$, although there is a serious part of circuit breaker overhaul leakage risks. If SF6 gas handling equipment is not properly maintained, resulting in emissions of greenhouse gases, the consequences will be very serious. Development of the smart grid promotes technological innovation, which can effectively reduce SF6 leakage and emissions in the production, installation, operation, maintenance and scrap processing stages, thereby reducing the transmission and distribution aspects of greenhouse gas emissions.

2.3.3 *A smart grid to guide the user to change their way of using electricity to reduce energy consumption*

In China, the proportion of energy in final energy consumption is more than 26%, by the end-user can change this model in order to improve energy efficiency; this will also have an important low carbon benefit.

A smart grid will promote technological innovation construction, appliances, transportation and other areas, in order to to improve the energy efficiency of the end-users. A smart meter management system as the basis for the measurement, the introduction of TOU energy trading and a two-way interactive mechanism to guide power users implementation of scientific and efficient conduct electricity, conduct research diagnosis of DSM energy efficiency technologies to reduce power system peak load, improve equipment utilization efficiency of the power grid, vigorously promote electric vehicles as an example of efficient energy use, improving end-use energy consumption the efficiency of energy use, would lead to huge carbon benefits.

3 SUMMARY

Global climate change is currently the focus of the international community. By addressing climate change against the background of reducing carbon dioxide emissions, a society of the future will move towards low carbon development. A low carbon economy is shifting away from industrial civilization to the ecological civilization of human society in the process of exploring a new model of economic development, which is the general trend of the global economy, and it is the inevitable choice for China's economic development. China is now in the fast-rising Economic development districts. China's accelerating industrialization, urbanization and modernization is at a stage of rapid growth in the field of energy demand. We could not stop large-scale infrastructure construction, "rich coal and less gas" in terms of resource, determines China's energy structure dominated by coal. Electric power is China's major energy source, whilst coal-fired thermal power plants account for more than 70% of the domestic total installed capacity, whose carbon dioxide emissions are high. The question of how to achieve low carbon development of the power industry in our country is the principle problem to be solved in developing a low carbon economy in China.

At present, the domestic capacity is dominated by thermal power generating units. Thermal power in terms of energy security is essential, clean, effective and a rational use of coal resources, which are practical for the low carbon development strategy of the electric power industry in China. Thus energy saving and emissions reduction in thermal power plants in service and supercritical/ultra-super critical coal-fired power generation technology are important for low carbon development in China's power industry policy. Improving energy conservation and emissions reduction in thermal power plants and developing new energy continue to change the existing power structure, and are also useful complements to low carbon development in China's power industry. Simultaneously, policies should also give new energy price subsidies, support the development of new energy industries and increase the intensity of research and development of new energy. Although carbon capture and storage technology can be a useful complement to power plant carbon dioxide emissions, and to a greater extent, to reduce carbon dioxide emissions from thermal power enterprises, however, due to carbon capture and storage technology is not perfect with higher costs, and carbon dioxide storage method is there is a big controversy, thus the current large-scale development and promoting encounter huge difficulty. The IGCC program also has some technical flaws, in the sense that the issue of the high cost of domestic large-scale promotion and the intelligent grid technology development has been rather slow. The new distributed energy sources, including wind and solar power, could also be useful complements to the Internet which is in urgent need of research and development, whilst in the meantime continuing their enhancement of the robustness of the network.

REFERENCES

Hu Y. H., Sun X., Zhang W. B., et al. Study on the environmental impact of coal [J]. China's energy, 2004, 26(1): 32–35

IPCC. Intergovernmental Panel for Climate Change Fourth Assessment Report [M]. Cambridge University Press, 2007

Liu C. W. Strategic Research of China's low-carbon energy development [D]. Jinan: Shandong University, 2010

Meij R., Winkel H. T. The emissions of heavy metals and persistent organic pollutants from modern coal-fired power stations [J]. Atmospheric Environment, 2007, V41 (40): 9262–9272

National Bureau of Statistics. Statistical Yearbook of China 2008 [M]. Beijing: China statistics press, 2009: 1–178

Tang X. S., Yan B., Huang Z., et al. Function of power grid to promote the development of low carbon economy [J]. Electric power technology economy, 2009, 21(6): 18–22

Wu J. S. The new cycle economics [M]. Beijing: Tsinghua University Press, 2005

Xu X. C., Chen C. H., Y., et al. Development of coal combustion pollution control for SO_2 and NO_x in China [J]. Fuel Processing Technology, 2000, V62 (2–3): 153–160

Zhao Y., Wang S. X., Nielsen C.P., et al. Establishment of a database of emissions factors for atmospheric pollutants from Chinese coal-fired power plants [J]. Atmospheric Environment, 2010, V44 (12): 1515–1523

Zhao Y., Wang S. X., Duan L., et al. Primary air pollutant emissions of coal-fired power plants in China: Current status and future prediction [J]. Atmospheric Environment, 2008, V42 (36): 8442–8452

Electronics, Information Technology and Intellectualization – Song & Kwak (Eds)
© 2015 Taylor & Francis Group, London, ISBN 978-1-138-02741-1

Melody composing algorithm for lyrics with meta-structures for melodic pieces

L.L. Liu & Y. Feng
Department of Cognitive Science, Xiamen University, Xiamen, China

ABSTRACT: An automatic melody composition technique for Chinese lyrics based on the collection of existing sample songs is presented in this paper. Some concepts involving meta-structure of a melodic piece and a song-template are proposed and used to implement an automatic melody composition system for Chinese lyrics according to user-specified genre, emotion, tempo, meter and mode. The final results show that the system can effectively achieve the goal of generating a variety of specified style of melody with given lyrics.

Keywords: algorithmic composition; Chinese lyrics; meta-structure of a melodic piece

1 INTRODUCTION

The area of automated composition refers to the process of using some formal process to make music with minimal human intervention (Alpen, 1995).

The research on composing algorithm has quite a long history, and there have been many successful systems. So far, the number of scholars involved in research on composing with lyrics, has been relatively small. The main technologies used are rule-based or stochastic models in existing automatic composing systems.

However, there exists one problem that the further increasing of the new style templates can only be operated by the developers, not by a user in the existing systems.

Also, as we know, the music content is related to human emotional expression. In recent years, there has been a lot of sentiment analysis based on music composition systems. The basic approach to such types of system used, is to set up a series of model tagging which is based on emotional psychology, and try to establish the correspondence among emotional tagging, musical concepts and a variety of music creation techniques (such as rhythm, tempo, pitch, tonality, harmony, counterpoint, etc.). The corresponding relations generally obtained through two channels are.

First, to establish the rule base which represents different musical styles directly, such as the experimental music system Inmamusys (2009), developed by Miguel Delgado, which works as a representative. It needs to be manually created for various compositional style models, otherwise a good deal of emotional mode combination and the corresponding model may change. Therefore, this method causes the system to lack of flexibility in its further expansion. Second is

to count the data that expresses different emotions by given musical pieces with certain attributes (such as rhythm). However, all kinds of survey data acquired are confined to the subjective judgment of a limited number of people. Therefore the investigation is not necessarily reliable and is time-consuming as well.

Based on the above-mentioned research issues, our work is mainly focused on the following two aspects.

1. Building up a friendly man-machine interface to expand the system database.

A friendly man-machine interactive interface is designed carry out a training phrase before composing in our system. Some simple and easy operations, for example collection and classification of existing songs with score and Chinese lyrics using existing composer software etc., which are dedicated to composing with Chinese lyrics, are required in the training phrase instead of manual programming or re-editing the knowledge base for knowledge improvement or expansion.

2. Making full use of the information of lyrics.

First, a concept of meta-structure of a melodic piece is proposed by converting the lyrics of the template into a poem so that seeking a solution to labelling the emotional attributes of the melody is feasible by reference to the mood and content of the lyrics the template song expressed. Thus, a database on meta-structure of a melodic piece with marked emotional attributes is built up.

Second, each sample song can be regarded as a sequence of meta-structure of a melodic piece after the poetic division. Based on the relationship between the adjacent meta-structure of a melodic piece (imitation, repetition, comparison, etc.), we put forward the concept of a song-template, and then construct a song template library. Hence, the problem of how to

Figure 1. An example of meta-structure.

generate the desired (or similar) style songs has been resolved.

Based on the above-mentioned points, a melody composition algorithm for Chinese lyrics is formed so that the machine can generate a new piece which expresses the prior pre-set emotion.

2 BASIC CONCEPT

Definition 1 (melodic piece): After the existing song's lyrics were written in poetry form, each row of Chinese character string is called a verse, and each melodic fragment corresponding to a verse in the song is called a melodic piece.

Definition 2 (Chinese character distribution): Let M be a melodic piece in a Chinese song. The number of Chinese characters of the verse, the positional relationship between each character and the notes above it in the score, is called Chinese character distribution of M.

Definition 3 (meta-structure of a melodic piece): Let M be a melodic piece. The Chinese characters distribution, sequences of the note duration, and of the intervals between the adjacent notes of M are collectively referred to as a meta-structure of the melodic piece (see Fig. 1), noted as M(n) for short, and where n is the number of Chinese characters of the verse in the Chinese character distribution of M.

Definition 4 (relationship between the two meta-structures of melodic pieces): Let M_A, and M_B be the two melodic pieces in a Chinese song. If there is a relationship between M_A, M_B satisfies R(j) described in Table 1, we deem that the relationship between the two of the corresponding meta-structures $M_A(n_A)$ and $M_B(n_B)$, is also defined as R(j) and vice versa. Where n_A and n_B, respectively, are the number of Chinese characters of the verses in the Chinese character distribution of M_A, M_B.

Definition 5 (song-template): Let V_1, V_2, ..., V_n, be a verse sequence–the poetic form converted from the lyrics of an existing song S. Sequence (1) is the corresponding meta-structure sequence. Each element $R_i(j_i)$ of the sequence (2) presents a relationship between $M_i(n_i)$ and $M_{i+1}(n_{i+1})$ in (1), i = 1, ..., L−1, which satisfies the relationship R(j) descried in Table 1. A song template is defined as the dual expression of (M_S, R_S), where, M_S, and R_S respectively are the sequence (1) and (2).

$$M_1(n_1), ..., M_L(n_L) \qquad (1)$$

$$R_1(j_1), ..., R_{L-1}(j_{L-1}) \qquad (2)$$

Table 1. Relationship of two meta-structures.

R(No)	Relationship	Description
R(1)	Repetition	1) $M_A = M_B$; 2) $n_A = n_B$; 3) Chinese character distribution of M_A is same as M_B.
R(2)	Rhythm repetition	1) Rhythm (M_A) = Rhythm (M_B); 2) $n_A = n_B$; 3) Chinese character distribution of M_A is same as M_B.
R(3)	Free sequence	1) M_A and M_B meet the condition of free sequence; 2) $n_A = n_B$; 3) Chinese character distribution of M_A is same as M_B.
R(4)	Strict sequence	1) M_A and M_B meet the condition of strict sequence; 2) $n_A = n_B$; 3) Chinese character distribution of M_A is same as M_B.
R (5–8)	Comparison A	Let $M_A = m_{A1}m_{A2}$, $M_A = m_{B1}m_{B2}$, 1) Duration of $m_{A2} \le$ Duration of m_{A1}; 2) Duration of $m_{B2} \le$ Duration of m_{B1}; 3) m_{A1} and m_{B1} satisfy R(j), $1 \le j \le 4$.
R (9–12)	Comparison B	Let $M_A = m_{A1}m_{A2}$, $M_A = m_{B1}m_{B2}$, 1) Duration of m_{A1} < Duration of m_{A2}; 2) Duration of m_{B1} < Duration of m_{B2}; 3) m_{A2} and m_{B2} satisfy R(j),$1 \le j \le 4$.
R(13)	Other Comparisons	Any relationship between M_A and M_B except R(1)~R(12)

*The two meta-structures that meet R(13) must be from the same song.

The first note of $M_{i+1}(n_{i+1})$ is called the i-th connection of template song(M_S and, R_S), meanwhile, $R_i(j_i)$ is the i-th connection relationship of template song (M_S, R_S).

Definition 6 (relationship with strong-R(j) connectivity): The relationship between $M_i(n_i)$ and $M_{i+1}(n_{i+1})$ satisfies the relation R(j) descried in Table 1, which is called the relationship with strong-R(j) connectivity.

Definition 7 (junction of the two meta-structures of a melodic piece): If $M_1(n_1)$ and $M_2(n_2)$ meet the relationship of strong connectivity, and $M_1 = m_{A1}m_{A2}$, $M_2 = m_{B1}m_{B2}$, where M_i is a melodic piece of $M_i(n_i)$ i = 1, 2, m_{A2}, m_{B1} each containing at least one note, then $m_{A2}m_{B1}$ is called the junction of $M_1(n_1)$ and $M_2(n_2)$.

Definition 8 (*Relationship with a week connectivity*): Let $M_1(n_1)$, $M_2(n_2)$, $M_3(n_3)$, $M_4(n_4)$ be four meta-structures of a melodic piece. $M_1(n_1)$, $M_2(n_2)$ is the relationship with strong-R(j) connectivity, but $M_1(n_1)$, $M_2(n_2)$ isn't. If the junction of $M_1(n_1)$, $M_2(n_2)$ is the same as the junction of $M_3(n_3)$, $M_4(n_4)$, i.e. $m_{A2}m_{B1}$, where, $M_1 = m_{A1}m_{A2}$, $M_2 = m_{B1}m_{B2}$, $M_3 = m_{C1}m_{A2,}$ $M_4 = m_{B1}m_{D2,}$ at the same time the sum of the duration of $m_{A2}m_{B1}$ should exceed a connection threshold H. The relationship between M_3 (n_3) and $M_4(n_4)$ is called the relationship with weak-R(j) connectivity.

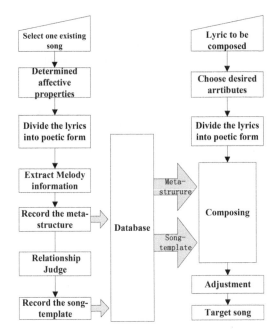

Figure 2. System flowchart.

It is not difficult to draw a conclusion that each relationship with a strong connectivity satisfies the requirement for a weak connectivity.

3 SYSTEM ARCHITECTURE

A song collection is required in order to make sure that there are enough sample songs with score and Chinese lyrics information for building up a database of meta-structure of a melodic piece and song-template. Each existing sample song is inputted and saved as the 'mus' file format on a computer by using a <Composer> tool which is develop by Wuhan FengYa company in China.

The system architecture involves two phrases, for example, training and composing (see Figure 2).

During the training phrase.

Determine affective properties. Trainer determines affective properties (lyrical, beautiful, bold, sadness, etc.), genre (carols, marches, etc.) of the sample song by their cognitive;

Divide lyrics into poetic form: Click on the button to read the lyrics of a sample, and, the trainer divides the lyrics into poetic form in the human-computer inter-action interface. Then the melody corresponding to each verse is called a melodic piece; thus the basis of dividing the meta-structure of melodic piece is formed.

Extract melody information. Data representation of 'mus' file has been obtained using early work. On this basis, the system extracts the song tempo, the speed, mode, each line of the lyrics, and the sequence of notes in each row for later use.

Record meta-structure. After each melodic piece is obtained, a computer records the Chinese charac-ter distribution (the number of characters, number of notes, sequence of words assigned), rhythm sequences,

pitch sequence and so on. In addition, the differential scale sequence and the differential interval sequence are calculated, recorded and used to assist the identifi-cation of relationships. The affective properties of the sample song are taking as the attribute of the melodic piece, thus completing the emotional labelling. Finally, to facilitate later the composition stage, the computer also needs to record relative abscissa, DoPlace value, etc. All of the above information combined together, got the meta-structure of a melodic piece. Each of the meta-structures are stored in the meta-structure database and are distinguished by a different ID.

Relationship judge. Now the sample song can be seen as a meta-structure ID sequence. Then system follow the relations shown in Table 1, in order to judge a relationship between the two adjacent meta-structures, and record it as the relation sequence, so that the machine can learn to develop a model, or a musical structure, which implies in the internal of the sample. This property, together with the total number of verses, will be very important at the composition stage.

Record song-template. The meta-structure ID sequence and relation sequence are very important attributes in the song-template database. Besides, the song name, meter, emotion, speed, genre, and mode also need to be recorded in the song-template database.

During the composing phrase.

Choose Attributes and Divide Lyrics. A user inputs lyrics in the composition interactive interface in accor-dance with the rhythm of poetry (verses are denoted as $V_1, V_2, ..., V_K$ rows; there is a total of k rows; the number of characters corresponding to row k is N_K). Select a desired emotion, rhythm and mode. Enter the name of the song to be composed. Click on the button to compose it.

Composing: The main idea of the composing algo-rithm is screening out a collection of template songs, a material collection of meta-structure in accordance with a user required emotion and rhythm. Then fol-lowing the development model of the selected song template, choose the appropriate meta-structure to link, after some adjustments, there comes a new song.

Composing section detail is shown in Table 2.

Adjustment. Systems need to make some adjust-ments in $P_1, P_2, ..., P_K$ in accordance with musi-cal knowledge, such as adjusting the time duration between each of the two bars, adjusting certain pitches, the position of each character of the lyrics and so on.

Target song. In this step, the program trans-lates the text document which recorded the adjusted $P_1, P_2, ..., P_K$ sequence information into 'mus' format. It automatically opens it with the software <Composer>, to provide flexibility when viewing the results of the system.

4 EXPERIMENTAL RESULTS

As a result generated by the system, an example of Gong mode, 2/4 beat, cheerful passionate song <Farewell to Cambridge> is shown in Figure 3.

Table 2. Composing section.

Initial Setting:
1. {S} ← Template songs in the database with a user requirement of the same emotion, rhythm and mode;
2. {M(n)} ← All meta-structures of melodic pieces which constitute {S};
3. $\{S\}_K$ ← A song template in {S} which has k verses;
4. $(M_S \ R_S)$ ← A selected generating template in $\{S\}_K$, where $M_S = M_1(n_1), ..., M_K(n_k)$, $R_S = R_1(j_1), ..., R_{k-1}(j_{k-1})$. Characters number of M_i is n_i;
5. $V_1, V_2, ..., V_K$ ← Sequence of poetic lyrics. Where characters number of V_i is N_i.

Composing Flows as:
if $\{S\}_K$ is not empty
    Match = Yes
    i = 1;
    $P_1 = \text{Random}(\{M(n)\})$;
while $i \le k$
  if $(N_i \ne n_i)$ or $(N_{i+1} \ne n_{i+1})$
  then
    $M_i(N_i) = T[M_i(n_i)]$, $M_{i+1}(N_{i+1}) = T[M_{i+1}(n_{i+1})]$
  end if
  if RelationJudge$[M(N_i), M(N_{i+1})]$ = weak−R(j)
  then
    $P_{i+1} = M_{i+1}$;
    i++ ;
  else
    Match = No;
  end if
if Match = Yes
then output $P_1, P_2, ..., P_K$

*T[M] is a transformation of words number of the meta-structure (increasing or decreasing within a threshold).

再别康桥

Figure 3. <Farewell to Cambridge>.

5 CONCLUSION

There must be some emotional set and formation of developing mode of the melody when composers engage in music creation. Through the research on this project, the computer simulates the process of this particular mode of thinking. This goal is achieved by dividing the sample songs into minute fragments, in order to obtain emotional tagging, and analyse the connecting relationship between verses through different types of samples. In the application, the melody composing for lyrics system has substantially improved the productivity of the music composition.

After we trained 100 songs in the song-template database, we invited three music majors to judge the system ability for generating songs. A total of 75 lyrics had been composed, and the system basically reached the level of generating target emotional songs. Furthermore, they suggested that the current speed of the song that the system has generated had a fixed value, and that it could vary it with different emotional requirements for the future improvements.

REFERENCES

Alpen A. 1995. Techniques for Algorithmic Composition of Music.

Hugo R. Goncalo Oliveira, et al. Tra-la-Lyrics: An approach to generate text based on rhythm, http://citeseerx.ist.psu.edu/viewdoc/download?doi=10.1.1.98.5209&rep=rep1&type=pdf.

Michael Edwards. Algorithmic Composition: Computational Thinking in Music, Communications of the ACM, Vol. 54 No. 7 (2011):58–67.

Miguel Delgado et al. Inmamusys: Intelligent multi-agent music system, Expert System with Applications, Vol. 37, No. 1 (2009), 4574–4580.

Yin Feng et al. Advances in Algorithmic Composition, Journal of Software, Vol. 17, No. 2 (2006):209–215.

Yin Feng et al. Harmonizing Melody with Meta-Structure of Piano Accompaniment Figure, Journal of Computer Science and Technology, Vol. 26, No. 6 (2011):1041–1060.

Electronics, Information Technology and Intellectualization – Song & Kwak (Eds)
© 2015 Taylor & Francis Group, London, ISBN 978-1-138-02741-1

Reappearing and interacting with real scenes system designs based on CAVE

Xiaoqing Duan, Yan Wang & Jian Ou
Harbin Institute of Technology, Harbin, Heilongjiang Province, China

ABSTRACT: With the continuous development of virtual reality technology, we hope that we can not only interact with the virtual world, but also communicate with the real world at the same time. To this end, we designed a system based on CAVE system, allowing people to interact with the world which maxed the virtual world and the real world together. This paper mostly describes this system from three parts: acquiring the scene, showing the scene and interacting with the scene, and discusses the significance and the field applicable to this system.

Keywords: virtual reality; maxed reality; CAVE system

1 INTRODUCTION

Virtual reality technology is a new field of computer technology, which has been developing as a combination of computer graphics technology, human-computer interaction technology, multimedia technology and simulation technology, and other technologies[1]. It is also the integrated application to the optics, mechanics, kinematics, mathematics and other subjects. Due to its characteristic, virtual reality technology is gradually been applied to medicine, entertainment, military, construction and other fields.

Tourism is a group of people conducting non-settlement travel for seeking spiritual pleasant feeling and all relations or phenomena which occur during the travel. In the international community, tourism is defined as travel activities for relaxation, entertainment, visiting friends or relatives, or business. With the development of virtual reality technology, the virtual tourism began to appear. Virtual tourism refers to the process in which a virtual environment for tourism is created by demonstrating the real tourism areas realistically through the network platform, then making the users do various "tourism activities," and getting emotional and rational knowledge/information which is related to the tourist attractions. IT uses the virtual reality technology in order to achieve three-dimensional simulation of tourist attractions and lets the user feel the virtual scene.

However, in virtual reality the user still interacts with the virtual world which is created by the computer; the user can only change and interact with the virtual objects, but cannot interact with the real world. To this end, with the further development of virtual reality technology, mixed reality has appeared. Mixed reality refers to different types of reality (mostly the real reality and virtual reality) mixed with each other.

This mix creates a new environment in which physical objects and virtual objects can coexist and interact with each other[2]. Mixed reality, speaking about a certain meaning, is the development of augmented reality technology and interactive media.

To this end, we hope to design a system that takes people into the mixed reality world using the CAVE system which has great immersive, allowing users to not only interact with the virtual world but also to interact with the real world.

2 SYSTEM DESIGN AND IMPLEMENTATION

2.1 Technical feasibility analysis of the system

Unity3D is a comprehensive game development tool developed by Unity Technologies, and it is a fully integrated professional game engine[3]. Unity3D has a convenient visual development environment[4], and supports a variety of scripting languages including C# and JavaScript[5]. Furthermore, it is compatible with a variety of operating systems. Therefore, developers can simply and intuitively develop a variety of application based on different types of platforms or hardware[6].

Arduino is an open-source electronics prototyping convenient platform, which itself includes both hardware and software parts[7]. It can be connected to the various sensors in order to sense the environment, and by controlling the lamp, various devices such as sensors and motors to change or affect the environment.

CAVE system[8] as a kind of virtual reality system, has a very good immersion, and it can let people interact naturally with the virtual world without using any wearable devices or sensors. CAVE system is a display device composed of multiple projection screen,

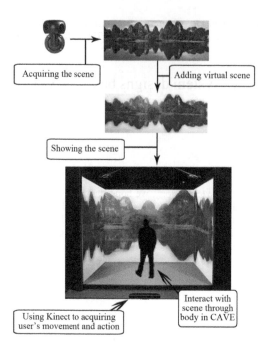

Figure 1. Process Schematic.

which screen forms a small room like a cave. Because of having a projection screen at each direction, the CAVE system not only has a strong immersive, but also becomes an ideal display device for displaying large scenes.

2.2 Design and implementation

The whole system is divided into three parts: acquiring the scene, showing the scene and interacting with the scene. Acquiring the scene includes the acquisition of a real physical world scene and creation of virtual world; showing the scene includes the screen used to show the scene and adjusting the rendered scene based on a viewpoint; interacting with the scene includes interactions with the real world scenario and the interaction with the virtual objects. The process schematic is shown on Figure 1.

2.2.1 Acquiring the scene
2.2.1.1 Acquiring the real scene
After obtaining the images of the scene by erecting a panoramic camera in the remote scene, the camera splits the image, and the adjusted images are transmitted through the network to the CAVE system. Also, it obtains sound information from the scene through a microphone, and transmits the information to the CAVE system through the network. And it obtains other scene information such as temperature, wind and other information in a similar way.

2.2.1.2 Adding virtual scene
By creating markers in the real scene, the computer can identify the spatial position information of the

camera, then it can use augmented reality technology in order to create digital models, and sound and sound effects to add to the scene. It can create a new environment named mixed reality in which physical reality and virtual reality coexisted.

2.2.2 Showing the scene
First of all, the hardware of the CAVE system needed to be constructed First, four pieces of glass were used to form a shape of a room as the projection surface. Then the projection film which was the same size as the glass would be glued onto the glass. Five projectors were placed around the room, which allowed a projector to completely cover the projection plane. Then each projector would be connected to a computer through data line. This computer was equipped with high-performance graphics card which could output six pictures simultaneously. Then one of the pictures would be displayed on a single monitor, to monitor the entire scene in all planes of the screen, and other five pictures which would be projected on each plane of the room were output by projector.

Then it needed to carry out the software. The Unity 3D was chosen as a basic tool. First, the skybox in Unity 3D was used to display the above-described mixed reality images. Then ed five cameras were positioned towards the top, bottom, left, right and at the front at the same position in space, in order to obtain the content of the scene that would be projected on respective planes. Then a plane before each camera was created, and the position of the plane was the same as the projection plane of the screen. After that, by adding scripts to the camera, the camera could turn the three-dimensional scene that it had into the two-dimensional projected images according to the position of projection plane. Then the images from the camera were shown on the newly created plane, and which were then used in order to make up a large plane. In order to control the shape of each plane, some control scripts were added to each plane to adjust the images when they were projected on each respective glass in CAVE. Then, the whole image was divided by using graphic card which would be projected on its projected screen. Finally, each image needed to be adjusted, in order to ensure that they were correct and continuous in the CAVE.

Because of the projection transformation which was used in the CAVE related to the viewpoint position, and different viewpoint will made different images. The location of the user's head in the CAVE system through Kinect, as the viewpoint position, needed to be established. After transforming and transferring the information into the CAVE system, the information would be used to adjust the viewpoint of projection transformation, in order to allow the users to get the right images.

2.2.3 Interacting with the scene
By adding a variety of sensors in the system, it allows users to fully and naturally interact with the scene and get a better experience.

It obtains the user's walking through the depth camera. Then the motion information is transformed into control signals and sent to the remote scene. Through the Arduino, the movement of the camera which was erected in remote scene, could be controlled. The user roaming in the scene could be realized The user only needs to use his/her hand to grip the objects and move them where he/she wants to move an object in the scene. If the object is a virtual object, then the system can directly implement it in the software. And if the object is an object in the remote scene, then the command would be sent to the remote robot, and the remote robot would complete the operation of gripping and moving.

Meanwhile, by adding the wind speed, temperature and other sensors to the system in order to restore the sensory information such as wind and temperature, it will not only allow the user to feel the scene more roundly and intuitively, but it would also increase the users' acceptance of this system.

3 SYSTEM ANALYSIS AND VALUE

3.1 *System analysis*

The system is based on the CAVE virtual reality system, adding the interaction with the real world, in order to achieve a simple mixed reality system. Because the most important tool which the system used was Unity 3D, which allows the system to have a good augment ability, therefore the system can be improved and some new functions in the future can be expediently added. For example, when people want to experience some special effects, they can be realized through a non-photorealistic rendering algorithm, by rendering picture to the desired effect. Therefore the system performance can be rich.

However, there are still many places that need to be improved. Some functions are also incomplete, and need a constant improvement.

3.2 *Application value*

Virtual reality has a very strong immersive feeling, and it pursues making users in the midst of a computer system completely and cutting users' links with the real world. In contrast, the mixed reality makes users keep the link with both the real and virtual world, and it lets people adjust themselves according to their own situation.

This is not only not the real world, nor is it a virtual world. It can help people have a different experience. In such an environment people can not only access information from remote real scene, but also get more information and more diverse experience through the virtual world mixed together.

Such a system can be applied to tourism, letting people experience the scenery of distant places. It can also be applied to an art exhibition hall or other places, in order to let people use a new way of feeling the atmosphere of art. This is good for spreading the cultural connotation which is expressed in the works of art.

ACKNOWLEDGEMENTS

This paper is an achievement of the 2012 annual national philosophy and social science funded project "technology of phenomenological horizon, the basic problem study" (project number: 12BZW017) achievements. It is an achievement Subject of the national science and technology support plan in 2012 "cultural industry science and technology innovation oriented common key technology research and application" (project number: 2012BAH81F01). And it is also an achievement for the subject of the national science and technology support plan in 2012 the "online interactive media production system" (project number: 2012BAH66F02).

REFERENCES

Li Min, Han Feng. 2010. Summary of virtual reality technology. *Software Guide* 9(6): 142–144.

Huang Mingfen. Art and mixed reality. 2009. *Journal of Southeast University (Philosophy and Social Sciences)* 10(6): 74–78.

Zhang Lili, Li Renyi, Li Xiaojing, Ma Jin, Hui Duoduo. 2014. Research on Communication Method between Unity3D and Database. *Computer Technology and Development* 24(3): 229–232.

Fu Yongming, Wang Yunjia. 2005. The Design and Implementation of a 3DGIS. *Geomatics & Spatial Information Technology* 28(1): 49–51.

Wu Chuanmin, Zhang shuai, Zheng Jinming. 2012. FPS game design and develop based on the Unity3D. *Journal of Sanming University* 29(2): 35–40.

Ling Dongwei, Li Xin, Zhu Meizheng. 2007. Design and implementation of virtual globe system based on GIS. *Computer Engineering and Design* 28(19): 4774–4777.

Cai Ruiyan. 2012. Principle and application of Arduino. *Electronic Design Engineering* 20(16): 155–157.

Cruz-Neira C, Sandin D J, DeFanti T A. 1993. Surround-screen projection-based virtual reality: the design and implementation of the CAVE. *Proceedings of the 20th annual conference on Computer graphics and interactive techniques*. ACM: 135–142.

Electronics, Information Technology and Intellectualization – Song & Kwak (Eds)
© *2015 Taylor & Francis Group, London, ISBN 978-1-138-02741-1*

A racing game's design with research into its impact on users' behaviour based on teleoperation

Xiaoqing Duan, Jian Ou & Yan Wang
Harbin Institute of Technology, Harbin, Heilongjiang Province, China

ABSTRACT: This paper attempts to explore through designing a teleoperating racing game and its positive impact on behaviour patterns for the players. Teleoperation has the real image, real-time feedback, and interactive features such as "present"; its technical system is made up of virtual reality and teleoperation, giving users a virtual telepresence. This paper first analysed the possibility of realizing racing games with teleoperation, and it used augmented reality technology to design game scenarios and game modes. On this basis, with the user experience at the centre, and using the teleoperation of remote operations and the present characteristics of the real-time feedback, combined with augmented reality, it put forward a racing game model based on teleoperation. Finally, it inquired about teleoperating racing game's impact on the user experience and the user driving behaviour patterns through the experiment. Teleoperating racing game linked the virtual world and the real world together; establishing that the behaviour patterns of the user in the virtual world and the real world can be harmonized.

Keywords: Teleoperation; teleoperating racing game; user experience; user behaviour patterns

1 INTRODUCTION

Racing game is a kind of electronic game in order to meet players' desire for speed and pursuit of passion. However, when the players in the racing game experience the thrill of speed and adventure, their patterns of behaviour in the virtual world will have a negative impact on the real social behaviour patterns. The remote science as a technical means, gradually matures in order to meet the needs of space scientific experiments[1]. As the technology maturing and hardware's development, telescience has begun to gradually come into people's lives. This paper proposes making teleoperation be used for racing game to create a new kind of racing game through analysing racing game's feature and teleoperation's characteristic, that is teleoperating racing game. Furthermore, it discusses the teleoperating racing game's impact on user's behaviour patterns. Comparing with the existing electronic games, a game developed with teleoperation is more realistic, and it brings a new way of thinking in finding the solution for the negative social impact brought by electronic games' virtual, weak normative and other characteristics.

2 TELEOPERATION'S CHARACTERISTICS AND PSYCHOLOGICAL EXPERIENCE OF THE GAME

2.1 *Teleoperation's characteristics*

The teleoperation system uses a variety of sensors that can get a variety of perceptual information acquisition, and the real-time transfer to the operating side, and then through to the corresponding device to restore the information, can make the operator to throw himself into it which has a strong immersion. This allows the operator to have a present experiment which feels like the operator operates at present. Teleoperation creates a field which is the same as the remote environment, drawing the player wholeheartedly into it, where they can fully interact, allowing the operator to throw himself to the created field, making a kind of "presence" interaction.

2.2 *Teleoperation system structure*

Virtual reality technology is a new field of computer technology, which is developing as a combination of computer graphics technology, human-computer interaction technology, multimedia technology and simulation technology, and other technologies[2]. Augmented Reality is an important branch of Virtual reality technology[3], and augmented reality technology is a technology that combines the computer-generated virtual graphics with the scenes of real-world captured through camera.

Using the delay problem which exists in teleoperation can effectively be solved using virtual reality technology, but there will still be computing systems errors accumulated through augmented reality technology. These errors can be effectively eliminated, thereby improving the accuracy and effectiveness of operations. The attention of the operator can be focused more effectively on improving the operator's psychological identity, thereby improving overall system efficiency. Although Augmented Reality technology still has many problems, with the development of hardware and technology, augmented reality teleoperation system will become the first choice for a variety of teleoperation systems with its own advantages.

2.3 Psychological experience of the teleoperation game

2.3.1 Senses' replication and transmission of teleoperation game

In the game based on teleoperation, a variety of sensory perception information of the remote environment can be acquired via various sensors installed on the remote machine, which can be transferred to the operating side through the network. Enables player more realistic and comprehensive to perceive the specific circumstances of the remote gaming environment, by using different hardware devices to send feedback to the players with different sensory information using a different way of perception. Moreover, it creates a full immersion, through copying and replicating the senses of remote gaming environment.

2.3.2 Virtual world and reality in teleoperation game

In the game based on teleoperation, there are no consequences because people are no longer operating racing cars which are created by virtual world of games, but a real racing car existing in the real world, then people are no longer driving the car the same way as they playing electronic game, excluding consequences to driving a vehicle. People's behaviour would have consequences in the real world, for example, vehicles would be damaged after the collision, it would become impossible to drive, and it would not be possible to get out of muddy water, r, which would make people once again return to the real world, from the lack of norms and limitations of the virtual world. Therefore, the virtual game world is not only the same as the reality with all kinds of objects, and the laws of motion and other rules, but it also has a lot of norms and constraints when people are doing something in society. All this makes the virtual world and the real world blend together and interconnect. It will not only be able to avoid the problem of personality disorder caused by the virtual world and the real world being too similar to each other although not the same, which in turn can make people's value judgments become confusing. People's behaviour norms and basic principles of action can be the same in both worlds, avoiding people may do action with the same patterns of behaviour which is unlimited and with no consequences which people are accustomed to in the virtual world.

3 RACING GAME DESIGN AND IMPLEMENTATION BASED ON TELEOPERATION

3.1 Define racing game based on teleoperation

Teleoperation based racing game is a new racing game developed using the technical principles of

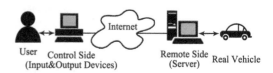

Figure 1. The process of racing game based on teleoperation.

teleoperation which is applied to the racing game. In the teleoperating racing game, players get visual information from the camera which is installed on a remote car, and can control the remote car's movements using the mouse and the keyboard to issue control instructions. In teleoperating racing game, players are driving a vehicle which is no longer a virtual vehicle generated by the computer, but a real vehicle in the real world. The movement of the vehicle is no longer required to be achieved by a computer simulation, so the motion of the vehicle can not only fully reflect the laws of reality, but can also bring special behavioural consequences which exist in the real world. It avoids the non-uniform gap created by different behaviour in the two worlds when players are getting relaxed through the game. The detail process is shown in Figure 1.

3.2 Technical feasibility analysis of racing game based on teleoperation

Arduino is an open-source electronics prototyping convenient platform, which includes both hardware and software parts[4]. It can be connected to the various sensors to sense the environment, and by controlling the lamp, various devices such as sensors and motors to change or affect the environment.

IP Camera, also known as network camera, in addition to its traditional image capture with the camera function, and it also has a digital compression controller and WEB-based operating system inside, making the video data obtained by the camera compressed and encrypted, sent to the end user through the network[5].

3.3 Teleoperation based racing game experience design

In the conventional racing game, the game scene and object models are constructed through modelling using computer graphics technology, where the material, properties, and various properties are calculated by means of the input data. However, in the teleoperating racing game, the game scene exists in the real world, and scene design needs to produce a physical model. Compared to electronic games, the teleoperating racing game needs more stringent model accuracy requirements, but it has no need to simulate other physical properties of the environment.

As the conventional racing game emphasizes the sense of speed more, the teleoperating racing game pays more attention to authenticity and close contact with the reality. Players can play with other players or computers as in the traditional racing game. They can also make use of the teleoperated car with which to explore the remote environment. Whilst they are exploring, they need to judge the situations of the car according to the visual screen, to control the car in order to avoid the emergence of barriers and so on. Meanwhile, we can set up different levels in the game by adjusting the remote scene and augmented reality, by adding puzzles and other elements to the game. This will not only make the game more entertaining, but will also make the game have more diverse patterns.

According to different real-life scenarios and the different content added by using the augmented reality technology, the teleoperation game's gameplay will be more varied, so that the teleoperating racing game is not only a racing game based on teleoperation, but also becomes a scalable game development platform, to which people can apply their new ideas. People can also add interesting gameplay to the platform as an application, or modify and improve the existing applications.

3.4 Teleoperating racing game implementation process

The teleoperating racing game presented here is just a prototype, and it has achieved some basic factors of the real teleoperating racing game. First, Teleoperating racing game must have a machine in the remote site, which is a real vehicle. We use a toy remote car as the vehicle and fixed the IP Camera on this toy car to get visual information. In addition, as teleoperating racing game also requires operating side for players to use, we use HTML and PHP to build the user interface and the server. This server can get the visual information from the server of IP Camera and show it to the users, and it also updates the commands which are used to control the movement of the vehicle according to the players' instruction input from the keyboard or mouse. These commands will be read through the network port and sent to the Arduino board by a C++ program. Then the Arduino board will check the commands and make an appropriate treatment to control the movement of the vehicle according to the processing program which burns on this board.

Up to now, the prototype of the teleoperation vehicle is completed, but it cannot be regarded as a teleoperating racing game. And because there are a variety of gameplays for the teleoperating racing game and the conditions of our laboratory, we decide to use this game mode in which we use the vehicle to explore the remote environment. We put the vehicle in the lab and use the objects of the lab to create various obstacles and road, and also set up some landmarks in some places. During the game, players need to explore this lab by driving the vehicle and finding out all of the markers to complete the mission of the game.

4 TELEOPERATING RACING GAME ON EXPERIMENTS AND ANALYSIS OF USER IMPACT

4.1 Teleoperating racing game testing and evaluation of user experience

4.1.1 The purpose of the experiment and prediction
The purpose of this experiment was to find out if the teleoperating racing game will have an impact on player's behaviour. Speculating the teleoperating racing game will make players behave more cautiously, and make their actions provided with certain social norms and restrictions of reality, making their behaviour and activities of the guidelines unity with the reality.

4.1.2 Experimental methods
We want to verify whether teleoperating racing game can influence the habits behaviour patterns or not, by analysing the experiment, in which the result was originally planned to be obtained by comparing the time and driving style when players driving through the same road after the players played electronic racing game and teleoperating racing game. However, because the date and the real car are difficult to get, and most of the volunteers participating in this experiment have no driving experience or have no driver's license, the original thought fails to be smoothly implemented, therefore, we plan to experiment with modifications and adjustments.

In order to be able to complete the experiment, we modify the experiment by comparing the time it takes players to complete the same level in the electronic racing game, and observing their driving habits and style to analyse whether the teleoperating racing game have an impact on players' behaviour, and if it does how it is affected. Comparison method and observation method are used in the experiment to verify the conjecture is correct.

4.2 Experiment process and data analysis

4.2.1 Experiment process
First, we selected a single racing game level of *Need for Speed: Shift 2 Unleashed*, to ensure that every experiment carries the same situation, avoiding that unexpected factors that will affect the players of the game. At this level, players needed to complete a lap on the road; every player participating in the experiment needed to complete this level in advance, in order to be familiar with the route on this level. Second, every player needed to complete the level several times; then, we recorded the time the player used and used video to record the process of racing through the routes. Third, the player needed to play 30 minutes teleoperating racing game, and to explore the lab, when they follow a certain route to move around the lab. They needed to avoid some of the obstacles on the way, and bypass some places. And after they finish the task, they can drive the car freely to explore the lab. Then, just like the second step, the player needed to complete the

level several times again, then we recorded the time the player used and the process of racing through the routes. We analysed the impact of teleoperating racing game on the players, by comparing the difference in time that players use to complete the game level, as well as the route when players driving the car to finish the level.

4.2.2 *Experiment data and analysis*

When the players are experimenting according to the above process, we need to observe the players' driving route as well as the driving style and record the time the players used. Record data would be shown in Appendix I.

We tested the data of this experiment with the Paired-Samples T Test and got the result that the correlation is 0.974(>0.8) and the Sig. (2-tailed) is 0.13(>0.01). We could have known that the two sets of the data are related and their difference is obvious.

By comparing the experiment data it can be found that the time players used to complete the level after driving teleoperating racing game increases a little. Similarly, after driving the teleoperating racing game, the players would reduce the speed before they would go into the corners as well as always trying to avoid the car away from the track on the whole process, and they would not keep increasing the speed like before. At the same time, their routes to finish the level become more stable and even more cautious.

4.3 *Teleoperating racing game's impact on the users' driving behaviour pattern*

Through comparative and analysis of the experiment data, and observation of the performance of the players during the experiment, we can get that teleoperating racing game has impact on the players to a certain extent. It makes the players' driving habits as well as behavioural patterns in the game tend to be more cautious. When driving the virtual vehicle to complete the level, the players would reduce the traveling speed of the car in the case of involuntary. Relative to previous blind pursuit of speed, they pay more attention to driving safety and stability.

When driving teleoperating racing game entering a corner, if the traveling too fast, the vehicle will not be able to successfully pass it. The vehicle may be stuck by some obstacle or even overturned, which led to the game cannot be finished. So players need to take account of the consequences of not safely pass. Therefore even in the electronic game, the players will take account of the consequence of excessive speed involuntarily, which can leave their own pattern of behaviour more realistic, taking into account the consequences.

Most of the players participating in the experiment have not played or rarely play *Need for Speed*. And comparing with the players who have not played this game, the players who have rarely played this game are more familiar with the operation mode of this game, and they would spend less time to complete the level,

and also the time that they spent for several times is also relatively stable. Most of the players do not have real driving experience, and whether such factors would affect the conclusions of the experiment need to be further verified. And for the same reason, when as far as possible to keep traveling safely, which great decrease the case of an accident situation, leads to the overall speed be improved, bringing about a better result than before. T situation happens in a relatively few cases, and only has little impact on the result of the experiment. That is why we did not consider this in the experiment. In addition, for some players, the constant repetition of the same level may lead to a better result because of being skilled which has also not being considered in this experiment. Those factors above mentioned may influence the result of the experiment to some extent, but their influence would be relatively small. The result of the experiment result may be slightly different because of these factors, but it will not affect the final conclusions of the experiment.

5 CONCLUSIONS

In this paper we analysed the features and shortcomings of the existing racing game, and used a remote operation feature with real-time feedback of presence feature of teleoperation, combined with augmented reality technology to design a racing game based on teleoperation, and verify the impact on users' behaviour patterns with teleoperation through experiment. The following conclusions:

1. Teleoperation has real image, real-time feedback and interaction in presence three properties, and it can constitute teleoperation system combining with virtual reality technologies. Teleoperation system is not only able to create a strong immersion, but also to improve the operator's psychological identity, and the efficiency of teleoperation.

2. Teleoperating racing games are different from the traditional racing games. In the teleoperating racing game, the player controls a real vehicle which exists in a real environment in a scene shown on the computer screen, which is composed of image information from the camera that is set on the vehicle and some virtual obstacles, checkpoint and other things created by augmented reality technology. Players need to control the vehicle in order to complete the game in the scene shown on the computer screen. Racing game based on teleoperation will not only be used as a racing game, but will also be used as a development platform, and widely used in other fields.

3. The teleoperating racing game links the virtual world and the real world, and has a positive impact on users' behaviour patterns. It tends to make the users' behaviour pattern more cautious and more realistic. It can also allow the users to take into account the consequences of actions when they in the course of activities, so that the players would

have the same behaviour pattern in the virtual world and the real world. The teleoperating racing game has practical significance.

In short, teleoperating racing games have the constraints of the real world but, compared with a traditional electronic racing game, it is more realistic. When players are playing this game they have fun but can still keep their behaviour and values identical to those of the real world, in line with social constraints and norms.

ACKNOWLEDGMENT

We gratefully acknowledge fund support of the Subproject for Support Project for National Ministry of Science and Technology (2012BAH81F01) and the Special Project for Hit 985 Social Science.

REFERENCES

Cai Ruiyan. 2012. Principle and application of Arduino. *Electronic Design Engineering* 20 (16): 155–157.

Li Henan. 2012. Video capture, storage and retrieval of webcam. *Computer CD Software and Applications* 23: 134–134.

Li Min, Han Feng. 2010. Summary of virtual reality technology. *Software Guide* 9(6): 142–144.

Ye Jinsong, Liu Lujiang. 2005. Away from science and technology and in the application of deep space exploration vision. *Chinese Society of Astronautics Technical Committee of deep space exploration Second Conference Proceedings* 6: 73–78.

Zhang Wei, Zhang Chunhua. Xu Wei. 2012. Augmented Reality Technology Research and Its Application. *Computer Programming Skills & Maintenance* 6: 029.

APPENDIX I

Table I.1.   Player 1.

Number of Completions	Time Spent (Before)	Time Spent (After)
0 (Familiar the level)	01:13'484	
1	01:15'359	01:11'792
2	01:12'425	01:16'039
3	01:09'570	01:16'110
Average time	01:12'451	01:14'647

Table I.2.   Player 2.

Number of Completions	Time Spent (Before)	Time Spent (After)
0 (Familiar the level)	01:17'818	
1	01:20'331	01:24'455
2	01:12'965	01:20'698
3	01:24'487	01:25'060
Average time	01:19'261	01:23'404

Table I.3.   Player 3.

Number of Completions	Time Spent (Before)	Time Spent (After)
0 (Familiar the level)	01:11'171	
1	01:14'413	01:14'447
2	01:12'256	01:12'572
3	01:10'921	01:11'405
Average time	01:12'530	01:12'808

Table I.4.   Player 4.

Number of Completions	Time Spent (Before)	Time Spent (After)
0	01:07'220	
1	01:06'201	01:06'718
2	01:05'851	01:06'059
3	01:05'616	01:06'177
Average time	01:05'893	01:06'318

Table I.5.   Player 5.

Number of Completions	Time Spent (Before)	Time Spent (After)
0 (Familiar the level)	01:06'766	
1	01:09'268	01:11'221
2	01:07'494	01:13'898
3	01:07'928	01:11'864
Average time	01:08'230	01:12'328

Table I.6.   Player 6.

Number of Completions	Time Spent (Before)	Time Spent (After)
0 (Familiar the level)	01:06'191	
1	01:06'593	01:07'539
2	01:03'059	01:05'239
3	01:00'850	01:04'290
Average time	01:03'501	01:05'389

Table I.7.   Player 7.

Number of Completions	Time Spent (Before)	Time Spent (After)
0 (Familiar the level)	01:08'698	
1	01:07'137	01:06'887
2	01:06'365	01:07'279
3	01:07'380	01:09'935
Average time	01:06'361	01:07'734

Electronics, Information Technology and Intellectualization – Song & Kwak (Eds)
© 2015 Taylor & Francis Group, London, ISBN 978-1-138-02741-1

A MIMU-based gait and pace detection system designed for human powered energy harvest devices

J.H. Bai & L.Y. Zhao
School of Instrument Science & Engineering, Southeast University, Nanjing, China

ABSTRACT: Energy harvesting from the human body is a field of promising research. To improve the efficiency of the harvester, there is a need to identify its motion accurately and control the harvester's time. This paper presents a human gait and pace detection system based on MIMU, which consists of a three-axis accelerometer, three-axis gyroscope, and the parameters of human gait detected by MIMU in real-time. An extended Kalman filter based on quatemions is designed for an attitude algorithm and getting gait information. Next the discrete Fourier transform is used to extract the frequency information and finally the k Nearest Neighbours classification algorithm is employed to identify the pace level. Through the testing and experimentation, it proves that this human gait and pace detection system is stable at working.

Keywords: detection of gait and pace; MIMU; quatemion; FFT; KNN

1 INTRODUCTION

Human motion analysis has important research significance and application value in medicine, bionics, computer and other fields. At present, video based detection and sensor based detection are both commonly used methods of movement acquisition (Lambrecht and Ama, 2014).

The video based detection method has many constraints, such as high cost, huge calculation time, and complicated recognition algorithm, which affect its application in human body posture measurement (Favre, 2009).

Posture detection based on MIMU which is small volume, light weight and low cost possesses the advantages of high real time capability, not affected by environmental constraints, therefore this technology is widely noted (Zekavat and Buehrer, 2012).

Although other studies have shown that the human pace can be classified by using the single-axis MEMS accelerometer (Liu, 2010), considering the information on the gait (the angle and the angular velocity of the joint) is essential for the harvester these existing methods are unbefitting for this research. In order to obtain both the gait information and the pace information, this paper presents a human gait and pace detection system based on MIMU. MPU6050 is applied as the microchip of MIMU, which consists of three-axis accelerometer and three-axis gyroscope. STM32F103TB is applied as the microprocessor chip, obtaining the raw data of MPU6050 using IIC bus in real-time, and fusing the raw data using extended Kalman filter (Qin, 2009), completing attitude calculation, extracting the frequency information using the discrete Fourier transform, and identifying the pace by the K nearest neighbour classification algorithm (Liu, 2007).

2 DETECTION SYSTEM DESIGN

The regularity of leg movement is distinct when people walk normally. Hip, knee and ankle are used in normally walk; gait and pace information can be acquired by these joints. This paper decides to fix the detection system on the outside of the thigh, and detect the movement of the hip joint. Through this method, the angle and swing velocity of the hip joint can be obtained directly, and ultimately gait and pace information can be obtained through certain data processing.

2.1 Hardware design

The MPU6050 Motion Processing Unit designed by Invensense, is applied as the microchip of MIMU. The MPU6050 has an embedded three-axis MEMS gyroscope, a three-axis MEMS accelerometer, and a Digital Motion Processor (DMP) hardware accelerator engine with an auxiliary IIC port that interfaces to third party digital sensors such as magnetometers. For precision tracking of both fast and slow motions, the parts feature a user-programmable gyroscope full-scale range of ± 250, ± 500, ± 1000, and $\pm 2000°$/sec and a user-programmable accelerometer full-scale range of $\pm 2\,g$, $\pm 4\,g$, $\pm 8\,g$, and $\pm 16\,g$.

The STM32F103TB is applied as microprocessor chips, which incorporates the high-performance ARM Cortex-M3 32-bit RISC core operating at a

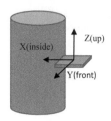

Figure 1. Detection system fixation schematic diagram.

72 MHz frequency, high-speed embedded memories (flash memory up to 128 kbytes and SRAM up to 20 kbytes). The STM32F103TB communicates with the MPU6050 through IIC bus at 400 kHz. The STM32F103TB communicates with the computer through RS-232 in debugging.

2.2 System fixation

Figure 1 shows the information on fixing the detection system on the outside of the thigh, ensuring the roll angle is 0° when the body is upright.

3 ATTITUDE CALCULATION

3.1 The EKF model based on quaternion

Euler angles algorithm, direction cosine algorithm, and quaternion algorithm are classical methods for attitude calculation (Zekavat, 2012). As a type of practical algorithm, quaternion algorithm can, by contrast, not only determine the whole attitudes because of avoiding angle degradation of Euler equation, but can also be calculated simply, and operated easily comparing with direction cosine algorithm (Yang, 2009).

R_b^n is defined as the rotation matrix rotating vectors in the NED (North-East-Down) earth fixed frame to the body fixed frame:

$$R_b^n(\vec{q}) = \begin{bmatrix} R_{11} & R_{12} & R_{13} \\ R_{21} & R_{22} & R_{23} \\ R_{31} & R_{32} & R_{33} \end{bmatrix} \quad (1)$$

where

$R_{11} = q_0^2 + q_1^2 - q_2^2 - q_3^2$

$R_{12} = 2(q_1q_2 + q_0q_3)$

$R_{13} = 2(q_1q_3 - q_0q_2)$

$R_{21} = 2(q_1q_2 - q_0q_3)$

$R_{22} = q_0^2 - q_1^2 + q_2^2 - q_3^2$

$R_{23} = 2(q_2q_3 + q_0q_1)$

$R_{31} = 2(q_1q_3 + q_0q_2)$

$R_{32} = 2(q_2q_3 - q_0q_1)$

$R_{33} = q_0^2 - q_1^2 - q_2^2 + q_3^2$

The quaternion differential equation relating the body axis angular velocity to the unit quaternion rates can be represented as a formula 2.

$$\dot{\vec{q}} = 1/2 \times \mathbf{\Omega}(\vec{q})\vec{\omega} \quad (2)$$

where

$$\mathbf{\Omega}(\vec{q}) = \begin{bmatrix} -q_1 & -q_2 & -q_3 \\ q_0 & -q_3 & q_2 \\ q_3 & q_0 & -q_1 \\ -q_2 & q_1 & q_0 \end{bmatrix}$$

The quaternion differential equation also can be represented as a formula 3.

$$\dot{\vec{q}} = 1/2 \times \mathbf{\Psi}(\vec{\omega})\vec{q} \quad (3)$$

where

$$\mathbf{\Psi}(\vec{\omega}) = \begin{bmatrix} 0 & -\omega_x & -\omega_y & -\omega_z \\ \omega_x & 0 & \omega_z & -\omega_y \\ \omega_y & -\omega_z & 0 & \omega_x \\ \omega_z & \omega_y & -\omega_x & 0 \end{bmatrix}$$

The state vector $\vec{x}$ consists of the attitude (unit quaternion) $\vec{q}$, and rate gyro biases $\vec{b}_\omega$.

$$\vec{x} = \begin{bmatrix} q_1 & q_2 & q_3 & q_4 & b_{\omega x} & b_{\omega y} & b_{\omega z} \end{bmatrix}^T$$

The EKF state equations as a function of the states are given as follows:

$$\dot{\vec{x}} = \vec{f}(\vec{x}, \vec{u}, \vec{w}) = \begin{bmatrix} \dot{\vec{q}} & \dot{\vec{b}}_\omega \end{bmatrix}^T = \begin{bmatrix} \frac{1}{2}\mathbf{\Omega}(\vec{q})\vec{\omega} & \vec{w}_b \end{bmatrix}^T \quad (4)$$

The measurement vector $\vec{z}$ is established by acceleration vector in the body fixed frame:

$$\vec{z} = \begin{bmatrix} a_x & a_y & a_z \end{bmatrix}^T$$

The output vector $\vec{y}$ to be used in the correction steps is given as follows:

$$\vec{y} = \vec{h}(\vec{x}) = R_b^n(\vec{q})\begin{bmatrix} 0 & 0 & g \end{bmatrix}^T \quad (5)$$

To implement the EKF it is necessary to linearize the state equations and the output equations:

$$\mathbf{F} = \frac{\partial \vec{f}}{\partial \vec{x}} = \begin{bmatrix} 1/2 \times \mathbf{\Psi}(\vec{\omega}) & 1/2 \times \mathbf{\Omega}(\vec{q}) \\ & \mathbf{0}_{3*7} \end{bmatrix} \quad (6)$$

$$\mathbf{G} = \frac{\partial \vec{f}}{\partial \vec{w}} = \begin{bmatrix} 1/2 \times \mathbf{\Omega}(\vec{q}) & \mathbf{0}_{4*6} \\ \mathbf{0}_{3*6} & \mathbf{I}_{3*3} \end{bmatrix} \quad (7)$$

$$\mathbf{H} = \frac{\partial \vec{h}}{\partial \vec{x}} = 2[\mathbf{H}_{bq} \quad \mathbf{0}_{3*3}] \quad (8)$$

where

$$\mathbf{H}_{bq} = \begin{bmatrix} -q_2 g & q_3 g & -q_0 g & q_1 g \\ q_1 g & q_0 g & q_3 g & q_2 g \\ q_0 g & -q_1 g & -q_2 g & q_3 g \end{bmatrix}$$

The EKF algorithm is a discrete time algorithm, thus the state equations and the measurement equations can be simplified as follows:

$$\vec{x}_k = \mathbf{A}\vec{x}_{k-1} + \mathbf{\Gamma}\vec{w}_{k-1} \tag{9}$$

$$\vec{z}_k = \mathbf{H}\vec{x}_k + \vec{v}_k \tag{10}$$

The process noise vector, $\vec{w}_k$, is assumed to be white and to have a covariance matrix, Q. The measurement noise, $\vec{v}_k$, is assumed to be white and have a covariance matrix, R:

$$p(\vec{w}) \sim N(0, \mathbf{Q})$$

$$p(\vec{v}) \sim N(0, \mathbf{R})$$

Model established in this paper is a continuous time model. Therefore, we must approximate using discrete time model for implementation. Formulas for approximating are given as follows:

$$\mathbf{A} = \mathbf{I} + \mathbf{F}T \tag{11}$$

$$\mathbf{\Gamma} = \mathbf{G}T \tag{12}$$

3.2 Attitude updating

3.2.1 Prediction step

Assuming that update interval is T, the fourth order Runge-Kutta method is applied in order to optimize the system state; recursive form is given as follows:

$$K_1 = 1/2 \times \mathbf{\Psi}(\vec{\omega}_{k-1})\vec{q}_{k-1}$$
$$K_2 = 1/2 \times \mathbf{\Psi}(\vec{\omega}_{k-1/2})(\vec{q}_{k-1} + T/2 \times K_1)$$
$$K_3 = 1/2 \times \mathbf{\Psi}(\vec{\omega}_{k-1/2})(\vec{q}_{k-1} + T/2 \times K_2) \tag{13}$$
$$K_4 = 1/2 \times \mathbf{\Psi}(\vec{\omega}_{k-1/2})(\vec{q}_{k-1} + TK_3)$$
$$\vec{q}_k = \vec{q}_{k-1} + T/6 \times (K_1 + 2K_2 + 2K_3 + K_4)$$

In order to ensure that the norm of quaternion is 1, thus avoiding the error of graduation, the following formula is applied to normalize the quaternion:

$$\vec{q}_k = \vec{q}_k / \sqrt{q_0^2 + q_1^2 + q_2^2 + q_3^2} \tag{14}$$

F and **G** are calculated according to formula 5 and 6.

State estimate error covariance matrix is calculated as follows:

$$P_k = \mathbf{A}P_{k-1}\mathbf{A} + \mathbf{\Gamma}\mathbf{Q}\mathbf{\Gamma}^T$$

3.2.2 Correction step

H is calculated according to formula 7.
$\vec{y}_k$ is calculated according to formula 4.
Kalman gain matrix is calculated as follows:

$$\mathbf{K} = P_k \mathbf{H}^T (\mathbf{H}P_k\mathbf{H}^T + \mathbf{R})^{-1}$$

The state vector is corrected as follows:

$$\vec{x}_k = \vec{x}_k + \mathbf{K}(\vec{z}_k - \vec{y}_k)$$

State error covariance matrix is corrected as follows:

$$P_k = P_k - \mathbf{K}\mathbf{H}P_k$$

Quaternion normalization is applied again.
The angle of pitch is calculated as follows:

$$\theta = -\arcsin(2q_0 q_2 - 2q_1 q_3)$$

The angle of roll is calculated as follows:

$$\gamma = \arctan(2q_2 q_3 + 2q_0 q_1)/(1 - 2q_1^2 - 2q_2^2))$$

4 PACE ANALYSIS

In reality, everyone's leg and gait are different, it is hard to get accurate information about the pace directly in the case of the limited sensor information. Related research shows that pace is increased mostly by changing the stride frequency and step length, thus eigenvectors in this paper consists of the amplitude and the frequency extracted from the roll angle (Ferreira, 2009). The K nearest neighbour algorithm is applied finally to identify three walking patterns of low speed walking, medium speed walking, and high speed walking (Davrondzhon, 2007).

4.1 Frequency extraction based on DFT

The roll angle is an approximate periodic signal in normal walking; the information of stride frequency can be extracted from power spectra implemented by discrete Fourier transform (DFT).
$\gamma(n)$, n = 1, 2, ..., N, consists of the most recent N roll angles calculated using the attitude algorithm.
The power spectra, X(k) is calculated the following formula.

$$X(k) = 1/N \times |FFT(\gamma(n))|^2, k=1, 2, ..., N.$$

As the sequence of real numbers, the power spectra, X(k), conjugates symmetric occasionally using k = n/2, thus $X(k) = X(N-k)$, k = 1, 2, ..., N/2−1. With that side, the first half of X(k) can express all the frequency information of $\gamma(n)$.

Assuming sampling rate, f_s, has been known, the discrete frequency, f_k, can be calculated by the following formula:

$$f_k = (k-1) \times f_s/N, \quad k = 1,2, \dots N/2 - 1 \quad (15)$$

Finding out the max of the X(k) except X(1), the frequency of $\gamma(n)$ can be extracted using formula (15).

The main defect in ordinary periodogram was that the frequency resolution was low, because data unobservable was consider to be zero. According to the average periodogram method (Bartlett method), this paper puts forward a new method implemented by averaging frequencies calculated according to different sampling interval length. $\gamma(n_i)$, $i = 1, 2, \dots, 10$. $n_i = 1, 2, \dots, 100 + i*10$.

f_i can be calculated using formula (15), then the frequency of $\gamma(n)$ can be averaged by this following formula.

$$f = \frac{1}{10} \sum_{i=1}^{10} f_i$$

4.2 Pace identification based on KNN

As a mature algorithm in theory, K-Nearest Neighbour (KNN) algorithm was first put forward by Cover and Hart in 1968 [11]. The concrete steps of KNN implemented in this paper were shown below.

First, let the tester walk on the treadmill using low speed (3 km/s), medium speed (4 km/s), and high speed (5 km/s) respectively, then as the sample information, the amplitude and the frequency of the roll angle measured at different speed, should be stored.

Second, when it is time to deal with the new unknown speed, according to the principle of KNN, the distance between the unclassified data and the simple data should be calculated, then K simples possessing the minimum distance should be found out.

Finally, the category of the unclassified data should be distinguished the same as the dominant simple.

The amplitude sample of the roll angle, $A_j(i)$, and the frequency sample of the roll angle, $f_j(i)$, are assumed to have been known. The distance between the unclassified data, (A_0, f_0), and the simple data is calculated by the Minkowski distance formula:

$$d_j(i) = \sqrt{w_1(A_j - A_0)^2 + w_2(f_j - f_0)^2}$$

$j = 1, 2, 3. i = 1, 2 \dots N_j$. N_j is the amount of simples acquired in different speed. w_1 is the weight factor of the amplitude, and w_2 is the weight factor of the frequency.

5 RESULT AND ANALYSIS

Figure 2, 3, and 4 show the information about the roll angle acquired at different speed.

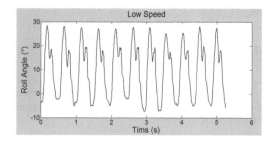

Figure 2. The roll angle information at low walking speed.

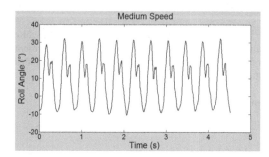

Figure 3. The roll angle information at medium walking speed.

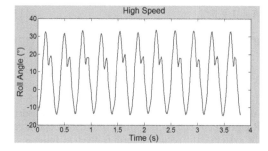

Figure 4. The roll angle information at high walking speed.

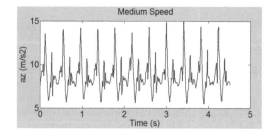

Figure 5. Z-axis acceleration in medium walking speed.

The Z-axis acceleration measured by MPU6050 at medium walking speed is shown in figure 5. It is evident from figure 5 that the vibration disturbance seriously impacts the accelerometer.

The X-axis angular rate measured by MPU6050 at medium walking speed is shown in figure 6. It is

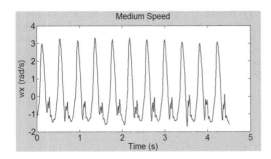

Figure 6. X-axis angular rate in medium walking speed.

Chart 1. The sample space of KNN.

Parameters	1	2	3	4	5	6
A_L	32.2	32.7	32.4	32.0	30.1	32.3
f_L	2.13	1.92	2.02	2.08	2.01	1.89
A_M	37.2	40.9	38.1	41.0	41.2	40.2
f_M	2.38	2.43	2.33	2.56	2.43	2.50
A_H	44.8	46.8	46.4	47.1	46.7	47.5
f_H	2.78	2.94	3.03	2.86	2.78	2.94

evident from figure 6 that the vibration disturbance slightly impacts the gyroscope.

Using a set of accelerometer and gyroscope, pace information may be obtained using the treatments for noise debasing based on the signal processing method, but the gait information cannot be obtained, which is important for controlling the harvester. Therefore, the process of fusing the raw data by extended Kalman filter is significant, not only the vibration interference whilst walking is filtered, but also the real-time gait information is obtained.

According to the method recommended in the pace analysis section, the sample space composed of the amplitude and frequency information of the roll angle gathered in different walking speed, is shown in chart 1.

According to the KNN algorithm, 27 groups of data are selected randomly, the result of pace identification is shown as figure 7. Red dots represent the unclassified data, blue dots represent the simple data of low walking speed, green dots represent the simple data of medium walking speed, and black dots represent the simple data of high walking speed. It shows that the method of utilizing the amplitude and frequency information to constitute the sample space of KNN can identify three different pace classes effectively.

6 CONCLUSIONS

Experimental results show that the process of fusing three-axis acceleration and three-axis angular rate using extended Kalman filter eliminates the vibration interference whilst walking effectively, and calculates the gait information in real time.

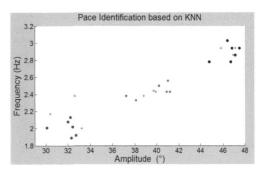

Figure 7. Pace identification based on KNN.

The accuracy of pace identification depends not only on the accuracy of frequency extraction, but also on the reliability of the classification algorithm. This paper utilizes the amplitude and frequency information to constitute the sample space of KNN algorithm and how it can identify three different pace classes effectively. Over all, the human gait and pace detection system designed in this paper furnishes the information of the pace and the gait effectively and has some practical value for improving the efficiency of the harvester.

ACKNOWLEDGMENTS

This work has been supported in part by National Natural Science Foundation of China (No. 61101163) and the Nature Science Foundation of Jiangsu Province of China (No. BK2012739).

REFERENCES

Favre, J., Aissaoui, R., Jolles, B.M., De Guise, J.A. & Aminian, K. 2009. Functional calibration procedure for 3D knee joint angle description using inertial sensors. *Journal of Biomechanics*, 42(14): 2330–2335.

Ferreira, Joao P., Crisostomo, M.M. & Coimbra, A.P. 2009. Human Gait Acquisition and Characterization. *Instrumentation and Measurement*, 58(9): 2979–2988.

Gafurov Davrondzhon, Snekkenes Einar & Bours Patrick. 2007. Gait Authentication and Identification using Wearable Accelerometer Sensor. *2007 IEEE Workshop on Automatic Identification Advanced Technologies Proceedings*: 220–225.

Kantardzic, M. 2003. *Data Mining: Concepts, Models, Methods, and Algorithms*. Wiley-IEEE Press.

Liu Yan, Li Yue-e & Hou Jian. 2010. Gait Recognition Based on MEMS Accelerometer. *2010 10th International Conference on Signal Processing Proceedings*: 1697–1681.

Liu Rong, Zhou Jianzhong, Liu Ming & Hou Xiangfeng. 2007. A Wearable Acceleration Sensor System for Gait Recognition. *2007 Second IEEE Conference on Industrial Electronics and Applications*: 2654–2659.

Qin Wei, Yuan Weizheng & Chang Honglong. 2009. Fuzzy Adaptive Extended Kalman Filter for Miniature Attitude and Heading Reference System. *2009 4th IEEE International Conference on Nano/Micro Engineered and Molecular Systems*: 1026–1030.

Raziel Riemer & Amir Shapiro. 2011. *Biomechanical energy harvesting from human motion: theory, state of the art, design guidelines, and future directions*. http://www.jneuroengrehab.com/content/8/1/22.

Stefan Lambrecht & Antonio J. del-Ama. 2014. Human Movement Analysis with Inertial Sensors. *Biosystems & Biorobotics*: 305–328.

Yang Jinxian & Zhang Ying. 2009. Research of Micro Integrated Navigation System Based on MIMU. *2009 International Conference on Industrial Mechatronics and Automation*: 147–150.

Zekavat R. & Buehrer R. 2012. Handbook of Position Location: Theory, Practice and Advances. *Wiley-IEEE Press*.

Electronics, Information Technology and Intellectualization – Song & Kwak (Eds)
© 2015 Taylor & Francis Group, London, ISBN 978-1-138-02741-1

The contextual expression of fresco art in the information society

Yan Zhang
Wuhan Textile University, College of Art and Design

ABSTRACT: This article first thoroughly explains the importance of context of fresco art in the information society. Secondly, it analyses the process of artistic language forming technology elements. Thirdly, it proposes the idea that information technology is the driving force behind fresco art, especially in the interaction between art language and modern technology, the exchange of materials, and the transformation of graphics. Fourthly, it states that the key to a strong visual impact is the rhythm that digital technology creates in artworks. And finally, it claims to sublimate the value of fresco art, the creation of new contexts for fresco art, and that the sense of humanity is essential.

Keywords: digital technology; context of the fresco art

1 INTRODUCTION

The way people live nowadays has changed enormously under the influence of the information society. Many traditional art forms are overthrown, while digital technology is showing its power to the world and impacts many other art forms as well. Fresco is challenged under these circumstance. If the art form does not change but follows the original pathway, it will lose its energy because of the changes in the aesthetic standards among both individuals and society. Fresco is both environmental and public and thus, the form and the expression need to satisfy people's spiritual needs as well as to display artistic expression on modern architecture. Modern architecture is developing at an unprecedented speed. As a result, fresco will have more diversified forms and more profound connotations. Therefore, discussing how to express fresco art within the information society has a positive impact on the development and revolution of this art form.

2 TECHNOLOGY ELEMENT OF ART LANGUAGE IN THE INTERNET ERA

The Internet age is a milestone. It has incomparable advantages over industrial era. It completely changed the way people live. With its influence spreading across the world both spiritually and physically, the prevalence of the Internet's existence is truly amazing. It is hard to imagine art without digital technology nowadays and digital technology is changing art with an irresistible force. Especially in the production of modern movies, one can be immersed in unbelievably life-like scenes through the use of visual technology, three-dimensional animation, and photography-editing software. Moreover, digital technology can also lead to digital painting, online music, and more. Thus, information technology is the driving force for changes in the form of artistic expression.

In fact, information technology is not merely a tool; more importantly, it provides artists and designers with more inspiration for continuing their trade. In this era of information, these creators will doubtlessly have broader views and more international ideas due to their unlimited platform and space for creation. Though heavily influenced by original forms of culture and art, traditional and modern ideologies will inevitably collide and technological elements will continue to impact the art language, especially the art form. Thus, this art form is not isolated, but it is a carrier for content. Art works are only valuable when the art form and content combine together. Therefore, only the transformation in ideology can make ideas vivid, and culture and art are more competitive in an open environment. Rigorous scientific attitude and meticulous attitude are the prerequisites for the introduction of information technology; this will open up new opportunities for the development of fresco art language.

Fresco art language and expression will undeniably generate unparalleled performances and artistic effects if technology elements are added. For example, virtual reality technology can make re-represent space; interactive art can make the interactivities between people and art a reality; audio-visual media technology helps achieve audible art works; computer graphics and image processing technology create a whole new artistic image. Once those technologies are added into fresco art, they will bring people keen shocks. Therefore, from the perspective of Kansei engineering, human sensory perceptions, including hearing, sight, smell, taste, and touch, can be used holistically through the use of technology. Especially when people's emotion and culture combine through language, material, and expression, a whole new perspective of fresco art will be recognized, and thus will create a more shocking effect using new media.

3 INTERACTION AND TRANSFORMATION BETWEEN THE LANGUAGES OF SCIENCE AND ART

By using computer graphics and image processing technology, artists and designers can achieve the transformation between realistic graphics, and apply them to fresco art. In terms of information collection, the use of computer technology can maximize the amount of information collected. The creators are able to collect the most relative and valuable information and combine them together. Generally speaking, after the draft of the fresco is created, it is still possible to add on a variety of effects through computer technology, such as changes in the composition, layout, location arrangements, as well as rearranging the tones until the best fit for the theme is selected. This builds a bridge between technology elements and the art language.

In fact, the real connection between technology elements and the fresco art language depends on the rich imagination and inspiration from the artists. Artists have to be observant and skilful to adept possibilities onto the creation and change of the fresco art.

The interaction and transformation of the fresco art is a current focus of the Internet era. With improving quality of the education and rapid pace of the urban construction, people's aesthetic standard is changing. Not only the fresco art ought to be appropriate for the environment around it, but also people are developing higher standards for the manifestation of art. People are wishing for more distinguishing arts. Whether the fresco art would meet people's needs is still worth studying even though this art form has a distinct characteristic. Rainer Marialatzke, the famous fresco artist, made some influential exploration on this question through the transformation amongst different art forms. Mr. Marialatzke worked indefatigably from the year 1986 to 1989, and turned the Chateau Thal from Kettenis, Belgium into the castle from the masterpiece of *One Thousand and One Nights*. His work was breath-taking and expressed the art to the extreme. His distinctive means of artistic expression helped him achieve the transition from a baroque style to a science-fiction style fresco art and create the new manifestation for modern fresco art in terms of the transition from two dimensions to multi-dimensions. He did not only use the most advanced informational technology at the time, but also retained the beauty of the original castle.

Another focus is the interaction and transformation of the material used in fresco art. The visual space of fresco art can be presented by different materials, and different choices of the material for the computer output printing materials can definitely create

various visual spaces with novel ideas. In the era of industrial civilization, many artists, such as Robert Rauschenberg, often take advantage of existing pictures to recreate. Yet in the age of information, the most characteristic method is to use digital technology and the Internet information to create fresco art, within which the most typical way is to transit between the materials of fresco-painting language and non-fresco-painting language, because this will create new visual spaces that bring the fresco language to a new level and extend the culture of fresco art. Therefore, it is necessary to grasp the notion of art reprocessing, and it is also crucial to understand the characteristics of the materials. They are the basis for the interaction and transformation of the material used in the fresco art.

4 THE RHYTHM OF THE INTERNET LANGUAGE IN FRESCO ART

Computer technology has a profound impact on modern fresco art. It leads art to the literary excellence to some extent. After applying distinct types of digital technologies to fresco art, people are fascinated by computer graphics, trompe l'oeil, and LED decorations and are surprised how unpredictable art is. These technologies have not only brought the visual feast to people, but have also provoked people into thinking about the rhythm of fresco art. And it is this rhythm that composed a unique artistic style for modern fresco art. This rhythm also produces a powerful artistic tension and force field, and gives people unprecedented shocks spiritually.

For instance, In the Optical Valley station of the Line 2 subway in Wuhan, to highlight features of the Optical Valley, the language of science and technology is integrated into fresco art – LED light was hidden behind the artwork and the rhythm of light and colour brought the artwork to life. This fresco artwork used some relatively concise graphics, which corresponded to people's fast pace and aesthetical standard nowadays; the hollow form perfectly combines the lighting technology. Imagine if there is no LED light behind fresco art, the expression would surely be tasteless.

The reason Chinese poems are so classic is that they contain extremely affluent artistic concepts, which are the product of rich imaginations from the great poets. So to speak, modern fresco art needs to fortify its influence by using exaggeration and rhythm. This new art form does not encourage artists to completely abandon the original style of fresco art; it stresses the rhythm and innovative ideas combined with modern information technology, and thus creates more novel expressions.

One needs to feel and comprehend artworks in different angles and perspectives in order to combine the information technology and fresco art. Artists can use contrast such as stableness and movement, reality and illusion, concentration and looseness – to arrange the piece, and achieve a lively artwork, and thus, give fresco art a stronger visual impact.

5 A NEW CONTEXT CREATED BY INFORMATIONAL TECHNOLOGY

Contextual performance is not easy to obtain in any form of art. It is generated from the product of the wisdom of great artists and the long-term accumulation. Such contextual performance of Chinese traditional painting is a necessary result of the long-term artistic practice. The discussion on how to express a new context so that fresco art will integrate well with information technology, needs to be made on different levels. First of all, the expression of the context is based on the new artistic language, which needs to integrate science and technology and later to refine it so that it is symbolic. Also, a new view on time and space needs to be built, so an artwork with different aesthetic value can be created. Second, the continuous innovation could be achieved using the modern audio-visual media technology which introduces new dynamic visual effects symbols. It can also be reached by applying "image reading" from new media, and form a new visual experience. Especially, the virtual digital technology serves as the carrier, and thus strengthens the artistic expression of modern fresco art. Finally, informational technology improves people's aesthetic standards, and leads people to find harmony amongst artwork, display space, and aesthetic beauty. Moreover, the sense of humanity is spread and promoted on this unique platform. The powerful software featuring the information age are the tools and platforms for modern fresco art.

In fact, artists of high self-discipline never give up the traditional art forms, yet they also absorb the latest technologies. For example, although the famous German artist Rainer Marialatzke is a genius trompe l'oeil master, he never forgot to go back to the classic fresco artworks to dig their values. He had a profound understanding of fresco art. He was especially recognized for implying video technology into fresco art, and made a revolutionary contribution to the inheritance and development of the classical fresco art.

Rainer Marialatzke's classic work is embodied beyond borders, and he has thus become a master spreading cultures. He created fresco art for the President of United Arab Emirates, and for this piece he added LED effect to the art. The beams of car light are emitting down to the Broadway; the windows of skyscrapers shimmer as if meteors crossed the sky. People are stunned by this twelve by four metre artwork called *the Night of Manhattan*. The reason he could create this artwork is closely related to the fact that he mastered computer technologies. Just as Rainer said, "Computer technology plays a truly crucial role in fresco art. If Leonardo da Vinci lived in our age, I believe he will use computer too."

In conclusion, Information technology has a very powerful advantage, and this advantage is continuously proved by the database. For instance, Chinese Dunhuang Art Institute is working with Wuhan University to construct a platform to exhibit Dunhuang's art. This platform will make visitors feel like they are on the scene by using audio, light, electricity, colours, and so on. They will present a historical Dunhuang, and meanwhile add a sense of humanity to it. This will allow the visitors to understand the new context under the support of digital technology. Therefore, this is the essence of the blend of traditional and modern art.

REFERENCES

Oil Painting Teaching Material Art Studio written by Zhang Yuan, published by Peking University Press, 2007. (12):414–416.

The Plight Of Environmental Art Pattern And The Way Out written by Lu Shizhu and Zhang Jinghui, published by Hundred Schools in Arts, 2004.(6):140–142.

Mural new elements – neon lights written by Xing Wenwen, published by Art and Design, the third edition in 2012.

The analysis of digital design's application in textile design

Bin Feng
Lecturer, Nanning, China

Conglie Cai
Professor, Wuhan, China

ABSTRACT: With digital technology as the representative of the modern high-tech science and technology, it not only makes rapid changes but also brings revolution on modern manufacturing industry. The traditional manufacturing industry is using digital technology to transform it and bring it into a new era. Textile industry which is marked with textile has a close relationship with people. In the market economy, its role is increasingly outstanding, which has both material and spiritual functions. Therefore, how to import digital textile design, combine it with modern design and make modern textile design more personalized and humane is an important study subject of the digital textile design. Focusing on textile design, this paper expounds textile design and digital design theories in simple terms, carries out research at different levels, proposes the upgrade of digital design-led textile design and demonstrates from lowering costs and reducing intermediate waste links. Meanwhile, this paper describes the digital dimensional, three-dimensional software of textile design, which makes digital design more spatial, integral, and full of material texture to enhance the beauty of art. According to the theoretical research findings, digital design has a broad space for the development in textile design, which has a long-term and practical significance in the realization of low-carbon economy, green design, and environmental protection of industry. Digital technology is becoming more and more important, which not only brings rapid changes but also a revolution to the modern manufacturing industry.

1   INTRODUCTION

Textile design is an art design activity, which is designed to meet the human living space environment. If the architectural space design and environmental design are called hard environmental design, then textile design is a soft environmental design. Compared with hard environment, the soft environment plays a more active role in improving environmental quality and creating indoor atmosphere. According to the index coefficient which foreign countries evaluate living space, the soft environment undoubtedly plays a decisive role, which is a mark of measuring the family living quality. Therefore, the artistic creation of textile design requires the designers' active thinking, keen observation and the ability to control the market and release human creativity in full. To fully release human creativity, we must bring people's initiative into full play. Human subjectivity is not arbitrary, therefore the design must follow the laws of textile art. Textile design must be guided by theory to make its shapes diversified and colourful.

1.1   *The key to artistic textile design*

Aesthetic Principles have theoretical significance for textile design. It is a scientific art theory that is summarized and extracted through long practice in human artistic activity. Therefore, how to correctly apply the laws of formal beauty will have a significant role on textile design's artistic effects, which will be reflected in a positive or consumer active.

1.2   *Formal beauty led by the form of material texture*

Needless to say, the material is the basis for textile design. The development and progress of textile design is inseparable from material. Thus, different materials mark different levels of science and technology development, symbolizing the face of social productive forces and the times. In addition, the characteristics of the material and texture give people visual and tactile sensor, thus producing physical and psychological sense. Based on the texture of the material factors, and choosing similar material when matching textile products, has the advantage to get the desired artistic effect and coordinate the relationship between materials.

2   DIGITAL DESIGN THEORY

The computer and graphics equipment which is used in helping designers is called CAD for short. In the field of engineering and product design, the computer can help designers in computing, information storage, mapping and so on. In the field of design, computers make a lot of calculations, analysis and comparison to different computer programs that are usually used, in order to determine the optimal solution.

Compared with the traditional jacquard patterns, art design software in computer-aided design shows a strong performance capability. It has a rich variety of drawing tools, such as brush, pencil, airbrush, markers, rulers, etc. Besides, there are a variety of images, graphics, collages, editing, mixing and mask synthesis tools and a variety of paper materials, as well as diverse special filter effects tools. They can produce diverse product image styling and rich infinite colour effects, showing the difficult performance of the illusory conception and a seamless combination of the two. The real and indescribable surreal way works well with the help of computers and art design software.

The graphic design is inseparable from the ingenious applications of graphic art design software. The design shape of the Jacquard product is on a two-dimensional plane. The application of jacquard fabric design would involve a three-dimensional space environment and the three-dimensional design software. Therefore, to successfully design a jacquard pattern, we should first learn and become proficient in the use of graphic art design software, in order to quickly and accurately unfold our numerous design ideas. Designers must not only learn concepts methods and techniques of images, and graphics software design, but also be able to skilfully and accurately apply software function to the design of jacquard fabric, making the software serve jacquard fabric design and realize the art design intent.

3 THE UPGRADE OF TEXTILE DESIGN LED BY DIGITAL DESIGN

Different from traditional interior textile designs, modern digital textile design products update their speed and rhythm faster. In the major cities of industrialized countries, a variety of new textile design products appear abundantly in the shopping malls, and spread popular trends in different media, especially in professional journals. Major cities in China, and even smaller ones are popular destinations for textiles. It tells people that the value of textile products is trendy, whose classic brand value is brought about tremendous changes through digital design.

3.1 Cost reduction is inevitable in digital design

The advantage of digital design has now been recognized, primarily reflecting on the characteristics and advantages of digital design from the technical aspects. Generally speaking, new and high technology is new and high gold content. However, it often faces many challenges in the course of implementation, in which the cost is an issue. Any business which wants to obtain good returns pays attention to technological innovations, business potential and efforts to reduce costs.

The cost includes raw materials, energy consumption, equipment depreciation costs, labour costs and so on., The decline in product costs should be based on the premise of the product quality assurance, rather than cutting corners, which otherwise could seriously affect the interests of consumers, and enterprises to despair.

3.2 Digital design can maximise savings

Digital design can be called a green design method, which can be expressed in a paperless manner. Therefore, it reduces the emission of waste and pollution of water for the earth and society. For the traditional hand-painted method, designers need to waste a lot of paper, paint due to the limitations of tools and materials. Coupled with water diluting the process of drawing and the washing of colour palette and brush wash, large amounts of sewage produces and constitutes the pollution of water bodies.

Therefore, the traditional method of hand-drawn form will do harm to nature and society directly and indirectly, because the increased wastes and wastewater are not suitable for human living and working environment. Although these wastes and wastewater are produced locally, it is not needed by green living environment of human on the whole. Therefore, in foreign industrial countries, digital design products are produced in volume using digital printing equipment and digital weaving technology, which produce digitized design pattern directly. Such products are highly competitive in the international textile market, thanks to the added value of advanced digital technology, its products also increase. By the same token, companies no longer use traditional design methods and tools, therefore saving a lot of funds annually, eliminating spending on paper, paint and in other ways.

3.3 Reducing the intermediate waste

The patterns and styles designed by traditional hand-painted method tend to go through the sampling process to examine the effects of quality, which is a complex process. Because of the restrictions of dyes and proofing, the sampling cannot achieve the design effect. Two methods are often applied as follows: the proactive approach is to experiment again and again using dye tests and processes until the same effect is achieved as the original design, whilst the other method is consumption which requires the designer to modify the original design, thus the contradiction arising.

4 THE APPLICATION OF SOFTWARE TO THE DIGITAL TEXTILE DESIGN

4.1 The application of plane software to the textile design

The foundation of textile design is fabric, and fabric patterns and colours play an important role in decoration and landscaping for the overall effect of textile design. It is the artistic charm of patterns and colours that makes the effect of it paid special attention in textile design. In fabric pattern design, the choice of subject matter is very important. Although the range of subjects is wide, there are more floral patterns in textile design. This is because people often use flowers to symbolize women and children in order to imply beauty and vitality. Of course, for the subject matter, there are geometric patterns, abstract patterns, landscape designs, classical styles, as well as other patterns full of human spirit. But regardless of the performance of the theme, people must meditate the whole idea so as to produce artistic beauty.

The software used in fabric design is Photoshop, which has a strong performance, and is easy to operate and perform. What is more, textile CAD professional software can also be used for fabric pattern designs.

4.2 *The application of three-dimensional software to the textile design*

Dimensional software has a good functional role in the indoor environment and space. There are a variety of methods in practical application. In order to use 3DSMAX dimensional software to design, it must first be modelled, shape good interior space environment, and the function of change is very powerful. Map after the completion of modelling, and then place the plane fabric pattern in specific location and environment. This software has its advantages in building the scene. Textile CAD design software can not only design graphic patterns, but also simulate three-dimensional space on an existing two-dimensional image, especially to fully demonstrate the effects of soft material and texture of the three-dimensional space, even the cloth texture effects can appear and be easily operated. Therefore, choosing which software is used should be based on need. However, the basic principle is that the overall textile design should be performed well. Meanwhile in the design, the local part should be fully performed, so the relationship between local and local, local and whole should be handled harmoniously.

5 THE INNOVATION OF DIGITAL DESIGN

Firstly, digital textile design should be understood from the concept of systematic engineering knowledge. It cannot be simply understood as the updates and changes of design tools. Simple awareness and understanding are impossible to realize digital design in real sense. Because digital design cannot be completed due to the use of computer tools to complete; if that is the case, it's too easy. But the fact is not the case, because the digital design is a large systematic engineering, which is composed of different subsystems. It relates to the depth and breadth of a company's decision-making, management, design department, technology department, market research. Whether the information collection and forecasts are correct or not plays an important part in the analysis and decision-making of market potential, technological innovation capability of enterprises, creativity of the design department. Each link and sub-system must be coordinated and assume their respective responsibilities. Otherwise, a problem occurs, which will cause the entire system to a standstill.

Secondly, grasp the digital design, particularly in the design department. It is important to select a sector director, whose main task is to design and manage. The business management of the designer is mainly done by the artistic director. Therefore, the supervisor should have a strong sense of responsibility and good organizational skills, to be able to fully mobilize the enthusiasm of each designer's activity and creativity. Meanwhile, he should have a high level of professional competence and aesthetic awareness. Otherwise, he could work to suppress some designers' work. In fact, there is a lot of creative work which is caused by the artistic director's negligence. Then, the designer is likely to change jobs due to this repression. But this is a big loss to the business. Sometimes, due to these mistakes, such talents in other companies will overthrow the enterprise. Fundamentally speaking, cybernetics is critical in digital design. Not only the design department, but also all the sectors must be proficient in using cybernetics. In terms of an enterprise's CEO, a successful entrepreneur must be an expert in grasping the cybernetics.

6 CONCLUSION

The application of digital design to textile design represents the trend and future of digital technology. It offers a new space for the development of digital technology, showing that traditional art design must be combined with digital design. Thus, it is a brand new issue worth researching. The development of modern textile design presents the diversity and novelty of artistic styles. Novelty is more significant and prominent, leading overall design to turn to versatile, multi-purpose design. The effect of the material's texture is more aesthetic, visual comforting and beautiful. People will pay attention to the culture of the textile product while enjoying it.

Thus, digital design will continuously innovate based on new technologies and theories, accelerating the combination of digital design and textile design and reflecting the superb technology integration innovation. Digital design systematically solves the problems in techniques, crafts, equipment, and production, making China's textile products truly competitive in the international market, which is positive for enhancing the overall level of China's textile industry. Textile design, led by an advanced digital technology and based on a solid art design, must make the art flowers blooming more colourful.

REFERENCES

Brarnett, Jonathan. 1982. *An introduction to Urban Design.*
  Mel, Scott. 1969. *American City Planing.*
CALLIES, D. & Freilich. R. *Cases and Materials on Land Use.*

Curtis, W. 1996. *Modern Architecture Since 1900.* Phaidon Press.
Emanvel, M. *Contemporary Architects.* MacMillan Press.

Electronics, Information Technology and Intellectualization – Song & Kwak (Eds)
© 2015 Taylor & Francis Group, London, ISBN 978-1-138-02741-1

An investigation on smartphone user behaviour and dependency

H.I. Chen & Y.C. Chen
Chinese Culture University, Taipei, Taiwan

Ching-Sung Lee
Fu-Jen Catholic University, Taipei, Taiwan

ABSTRACT: The present paper aims to analyse how categorized groups of smartphone users differ from one another in user behaviour and dependency. To collect data for analysis, the researcher adopts the questionnaire survey method. The research instruments include Smartphone use behaviour scale and Smartphone dependency scale. In total, 639 valid questionnaires were collected for data analysis. Data analyses show that smartphone users on average manifest moderate smartphone dependency. In contrast to younger users who are found to have higher degree of dependency, older users are less dependent on smartphones. Smartphone dependency is therefore in inverse proportion to the user's age. Younger users have a tendency to feel anxious where there is a need for interpersonal relationships, which results in their dependency on smartphones. It is suggested that young people pay more attention to interpersonal relationships in the real world and are less dependent on virtual online relationships.

1 INTRODUCTION

1.1 *Smartphone usage*

In its 2014 report, International Telecommunication Union (ITU) points out that the number of mobile phone users worldwide will reach a record high of 7 billion by the end of 2014, with the total market penetration rate being at 96% (ITU 2014). Indeed, smartphone with its high mobility and portability is more competitive than PC. As a result, the smartphone has enjoyed a higher use rate than PCs. According to a survey of American adolescents (12–17 year olds) conducted by Internet & American Life Project in September 2012, 95% of the surveyed adolescents are Internet users. The survey also found that 25% of them use smartphones as their interface for surfing the Internet. Specifically, more than one third of American adolescents were smartphone users (Pewinternet, 2013).

1.2 *Smartphone dependency*

By mobile phone dependency we mean the user's cognitive or behavioural dependency on a mobile phone. This phenomenon is similar to the psychological mechanism of addiction. It is frequently found that once a mobile phone fails to function, runs out of power, or receives no signal, its user may feel anxious or even become panicked, thus affecting his or her daily life (SecurEnvoy, 2012). Young people value peer relationships. Group or peer members regard the mobile phone as an important means of communication. In order to maintain interpersonal relationships with their group members, young people tend to use mobile phones more frequently. This phenomenon leads to overuse of mobile phones and aggravates the risk of phone dependency (Boasea & Kobayashi 2008; Igarashi, Motoyoshi, Takai & Yoshida 2008; Walsh, White, Cox & Young 2011).

In the advent of the digitalized world, interaction between people no longer needs face-to-face contact. Interpersonal links are made possible through the Internet, computer, or mobile phone. Interpersonal communication modes have become much faster and more convenient. With the popularity of smartphones, PC-based communication modes have gradually been replaced by mobile phones. Internet phone oriented communication has therefore become trendy (Lu, Watanabe, Liu, Uji, Shono & Kitamura 2011; Reid & Reid 2007).

In the report "2013 Our Mobile Planet: Taiwan," Google pointed out that Taiwan's smartphone popularity rate reached a record high of 51% in the first quarter of 2013. Of all users, 81% carry a smartphone wherever they go. Taiwan ranks No.1 in smartphone dependency (Google, 2013). Smartphones function as a micro-computers. People communicate with others by calling, text messaging and emailing. Additional functions include surfing the Internet, renewing community websites and providing online shopping. Using smartphones to communicate with others has become an integral part of our daily life. In view of this phenomenon, users' dependency behaviour and anxiety are questions worth our investigation. Therefore, this paper aims to investigate how categorized groups of users are related to smartphone user behaviour and dependency.

2 METHOD

2.1 Samples and sampling method

Smartphone users are the targeted respondents of the present research. The population approximately consists of 9.16 million Taiwanese people. At 95% confidence level and 4% sampling error, the effective samples are statistically rendered at 600 (Krejcie & Morgan 1970). Upon taking return rate and invalid questionnaires into consideration, the researcher decides to solicit answers from 800 respondents, who are randomly selected, as formal samples. Some paper-pencil questionnaire surveys are conducted at public sites, including the Taipei Railroad Station Square and several fast food restaurants, for on-site answers. Others are conducted on the Internet, where respondents answer the questionnaire and send it back. In all, a total of 639 valid returned questionnaires are collected and the answers serve as the source for data analysis. At 95% confidence level, the empirical sampling error of the present study is 3.88%.

2.2 Instruments

Smartphone user behaviour scale respondents' demographic backgrounds include: sex, age, educational background, and profession. Smartphone user behaviour includes: the type of smartphone Internet plan, use of social networking sites, use of social application and the number of good friends. In total, nine questions (items) were identified by the researcher.

Smartphone Dependency: the researcher developed a smartphone dependency scale by referring to the MPIQ (Mobile Phone Involvement Questionnaire) developed by Walsh et al. (2010), the Self-perception of Text Message Dependency developed by Igarashi et al. and other related papers. The questionnaire, which consists of 29 items, measures the users' dependency on a smartphone. The format of each item goes from strongly disagree through disagree, partially agree, agree, to strongly agree, all of which, in order of strength, respectively assume 1, 2, 3, 4 and 5 points. The total score of each returned questionnaire ranges between 29 and 145 points. The higher the total score, the more smartphone dependency there is.

2.3 Analytical technique

The SPSS 20.0 is implemented to process statistical analysis of the data. It undergoes exploratory factor analysis of "Smartphone dependency scale". It establishes the common factors in the scale by deleting items with factor loading less than 0.5 and maintaining items with factor loading more than 0.5. Thus, the construct validity of the scale is generated for further reliability analysis. Cronbach's α coefficient is then used to test internal consistency. An independent samples t test and one-way ANOVA are used to examine how categorized users differ from each other in mobile phone dependency.

3 RESULTS

3.1 Description of smartphone users

Of all the respondents, 48.83% are males and 51.17% are females. Most respondents are between 18 and 25 years old. Only 5.16% of the respondents are older than 56. This is due to the fact that young people have easy access to digital devices and are likelier to use a smartphone. With respect to the educational background, more than half of the respondents are college graduates (65.73%). As far as professional people are concerned, 35.37% are government service people. Concerning styles of Internet surfing, 51.96% prefer 3G-without limit. Presently, smartphone users prefer to surf the Internet at any time and in any place, without any restraint on time or download volume.

As regards experience in using social network, more than half of the respondents (56.49%) use mobile phones to connect to social networks every day. Most smartphone users visit social networks via their smartphone.

As to experience in using social apps, more than 70% of the respondents (70.42%) use a social app every day. Most smartphone users make the best use of the social app functions of a smartphone in place of voice calling or text messaging of a traditional mobile phone. Daily users of social apps outnumber those of social networking sites (70.42% vs. 56.49 %) This shows that smartphone users communicate with others via social apps more frequently. Social apps are more frequently used because they provide a simple and ready-for-use interface. Additional functions include text messaging, stickers providing and voice calling, and photograph and file transmitting. Because of its convenience and practicality, all age groups can accept and use it.

3.2 Types of Internet phone plans and their effect on dependency

Five constructs, including compulsive behaviour (F = 25.71), social connections (F = 4.57), withdrawal (F = 12.06), significant behaviour (F = 7.46), and friendship connection urge (F = 9.24), respectively display significant difference. Concerning the construct of compulsive behaviour, 3G-without limit is significantly different from 3G-with limit and WiFi. As to the construct of social connections, 3G-without limit is significantly different from 3G-with limit and Wifi. With respect to the construct of withdrawal, 3G-without limit is significantly different from 3G-with limit and Wifi. As far as the construct of significant behaviour is concerned, 3G-without limit is

Table 1. One-way ANOVA of Dependency on types of Internet Plans (n = 639).

Smartphone Dependency	F value	P value	Scheffe's comparison
Compulsive behaviour	25.71***	.000	A > B.C
Social connections	4.57*	.011	A > B
Interpersonal interaction happiness	0.15	.860	
Withdrawal	12.06***	.000	A > B.C
Significant behaviour	7.46**	.001	A > B.C
Friendship connection urge	9.24***	.000	A > B.C

Note 1: * p < 0.05; ** p < 0.01; *** p < 0.001
Note 2: A is 3G-without limit, B is 3G-with limit, C is WiFi

Table 2. Independent sample t tests of dependency on social network sites (n = 639).

Smartphone dependency		Mean	SD	T value
Compulsive behaviour				
	Yes	3.46	0.82	4.31***
	No	3.17	0.83	
Social connections				
	Yes	3.19	0.81	7.92***
	No	2.67	0.83	
Interpersonal interaction happiness				
	Yes	3.37	0.70	5.19***
	No	3.06	0.79	
Withdrawal				
	Yes	3.56	0.90	4.35***
	No	3.25	0.89	
Significant behaviour				
	Yes	3.50	0.90	4.31***
	No	3.19	0.90	
Friendship connection impulsive				
	Yes	2.63	0.85	4.52***
	No	2.33	0.80	

Note: * p < 0.05; *** p < 0.001

Table 3. Independent sample t tests of numbers of app friends & the effect on Dependency (n = 639).

Smartphone Dependency		Mean	SD	T value
Compulsive behaviour				
	100-minus	3.16	0.82	−5.54***
	101-plus	3.52	0.82	
Social connections				
	100-minus	2.78	0.85	−5.49***
	101-plus	3.14	0.83	
Interpersonal interaction happiness				
	100-minus	3.17	0.77	−1.82
	101-plus	3.28	0.76	
Withdrawal				
	100-minus	3.27	0.89	−4.42***
	101-plus	3.59	0.90	
Significant behaviour				
	100-minus	3.23	0.92	−3.63*
	101-plus	3.49	0.90	
Friendship connection urge				
	100-minus	2.34	0.80	−4.81***
	101-plus	2.66	0.85	

Note: * p < 0.05; ** p < 0.01; *** p < 0.001

significantly different from 3G-with limit and WiFi. As regards the construct of friendship connection urge, 3G-without limit is significantly different from 3G-with limit and Wifi (see Table 1).

Compared with other two types of Internet surfing, 3G-without limit enables users to surf the Internet at any time in any place without any restraint on download volume. This satisfies users in their desire to surf the Internet freely. This also heightens the use frequency of mobile phone and the interpersonal interactions of mobile phone users. However, mobile phone dependency arises accordingly.

3.3 Social network sites on dependency

All six constructs, including compulsive behaviour (t = 4.3), social connections (t = 7.92), interpersonal interaction happiness (t = 5.19), withdrawal (t = 4.35), significant behaviour (t = 4.31), and friendship connection urge (t = 4.52), show significant difference. Obviously, those who have access to social network sites have a higher dependency than those who do not (see Table 2).

3.4 Number of app friends and the effect on dependency

Of the 6 constructs, 5, including compulsive behaviour (t = −5.54, p < 0.05), social connections (t = −5.49, p < 0.05), withdrawal (t = −4.42, p < 0.05), significant behaviour (t = −3.63, p < 0.05), and friendship connection urge (t = −4.81, p < 0.05), display significant difference. The 101-plus group is more dependent on mobile phone than the 100-minus group. The construct of interpersonal interaction happiness shows no significant difference (see Table 3).

4 CONCLUSION

The statistical analysis of the present study reveals that all respondents, on average, have moderate mobile phone dependency. Younger users are found to have higher mobile phone dependency. Of the six constructs, five show significant difference. The only exception is interpersonal interaction happiness. Mobile phone dependency is in inverse proportion to the age. This finding is consistent with that of Lu's study (2011). Young people have higher mobile phone dependency and are have tendency to feel anxious and fearful in social gatherings. IPhone is a virtual electronic platform where young people can fully express their own ideas without having to be mindful of others' viewpoints or judgment of them. Interacting with others via smartphones may release them of the anxiety inherent in face-to-face interaction. They are allowed to have more time to think how to express what they would like to say. The mobile phone, indeed, is an ideal means of communication for those who tend to feel anxious in social life. Over-dependency on virtual online communication, however, inevitably leads to over-dependency on mobile phones. It is advisable that young people value interpersonal relationships and acquire skills necessary for interpersonal interaction. They need to develop ability in face-to-face communication.

REFERENCES

Butt, S., & Phillips, J. G. 2008. Personality and self reported mobile phone use. Computers in Human Behavior, 24(2), 346–360.

Goole. 2013. Our Mobile Plane. Retrieved January 22, 2014, from http://services.google.com/fh/files/misc/omp-2013-tw-local.pdf

Igarashi, T., Motoyoshi, T., Takai, J., & Yoshida, T. (2008). No mobile, no life: Self-perception and text-message dependency among Japanese high school students. Computers in Human Behavior, 24(5), 2311–2324.

Krejcie, R. V., & Morgan, D. W. (1970). Determining sample size for research activities. Educational and Psychological Measurement, 30, 607–610.

Lu, X., Watanabe, J., Liu, Q., Uji, M., Shono, M., & Kitamura, T. (2011). Internet and mobile phone text-messaging dependency: Factor structure and correlation with dysphoric mood among Japanese adults. Computers in Human Behavior, 27(5), 1702–1709.

Pewinternet (2013). Teens and Technology 2013. Retrieved January 22, 2014, from http://www.pewinternet.org/Reports/2013/Teens-and-Tech.aspx

Reid, D. J., & Reid, F. J. (2007). Text or talk? Social anxiety, loneliness, and divergent preferences for cell phone use. Cyberpsychology & Behavior, 10(3), 424–435.

SecurEnvoy (2012). 66% of the population suffer from Nomophobia the fear of being without their phone. Retrieved March 15, 2013, from http://www.securenvoy.com/blog/2012/02/16/66-of-the-population-suffer-from-nomophobia-the-fear-of-being-without-their-phone/

Walsh, S. P., White, K. M., & Young, R. M. (2010). Needing to connect: The effect of self and others on young people's involvement with their mobile phones. Australian Journal of Psychology, 62(4), 194–203.

Walsh, S.P., White, K. M., Cox, S., & Young, R. M. (2011). Keeping in constant touch: The predictors of young Australians' mobile phone involvement. Computers in Human Behavior, 27(1), 333–342.

Electronics, Information Technology and Intellectualization – Song & Kwak (Eds)
© 2015 Taylor & Francis Group, London, ISBN 978-1-138-02741-1

The secondary development of CMACAST management system in Shenyang City and counties

L.Y. Sun
Meteorological bureau of Shenyang, Shengyang, China

J.W. Wu
Institute for Atmospheric Environment (IAE), CMA, Shengyang, China

ABSTRACT: The system presented in this paper used Oracle DataBase, Tomcat JSP containers. The database layer used JDBC interaction with the underlying database. The logic layer packages were in Java class. Secondary development was based on the CMACast system, and realized the province, city, county three class large capacity information broadcast.

Keywords: Secondary development; Database; CMACast

1   BACKGROUND

1.1   *CMACast*

China meteorological satellite data broadcasting system (CMACast) used DVB-S2 satellite data broadcasting standards and C band communication satellite transponder to build a new generation of meteorological satellite data broadcast system (CMACast), instead of PCVSAT, FENGYUNCast, DVBS three set of broadcast system, which greatly increased the variety and quantity of meteorological data broadcasting, improved the timeliness and reliability of data distribution, and constituted the global earth observation system of information dissemination with USA GEONETCast and European centre EUMETCast, distribution of the WMO global data exchange and FY satellite products to the Asia Pacific broadcasting. However, in the city and county broadcasting, problems of delay, overflow error, manual transmission capacity is insufficient, there was need to adapt to development in city and counties, thus a secondary development based on CMACAST, to improve the applicability of the system.

1.2   *The main problem of CMACast*

There is one set CMACast in Shenyang City and seven other sets in other counties of Shenyang. All of the eight satellite stations completed the installation in 2011. The CMA Cast satellite radio system was officially put into operation in 2012,

The whole area had 8 sets of CMA Cast satellite broadcasting system, signal-to-noise ratio reached 7.0. Software were upgrade in the county.

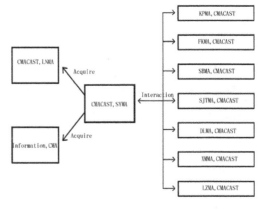

Figure 1.   Sketch map of system data interaction.

Up to now, the CMA Cast satellite broadcasting system has received data over 200G. With the business development in place, the following problems in the system were gradually found:

– Satellite data was not complete.
– To a large amount of data on the file server, received more than the rated capacity, data was not deleted automatically.
– The Micaps data processing overflow memory server caused the server to crash during data processing.
– The ability of the district level technical personnel using the SUSI LINUX system, is poor.

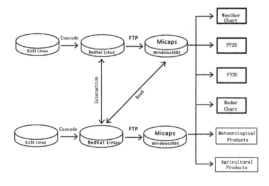

Figure 2. Sketch map of system structure.

2 DESIGN AND FUNCTION

2.1 System design

In view of the above, shenyang weather service network center two satellite radio system. Through repeated experiments, the establishment of "daily business data receiving schedule", to transform the Linux system into Redhat Linux system. Modify the program based on Redhat Linux system, realize the data real-time transmission, the main push send business data storage; Implement the data missing warning function and automatic compensation function.

2.2 Main functions

– The total area of eight sets of CMAcast system business operation.
– The implementation of the eight sets of CMAcast system for cold backup function area, and the function of remote debugging.
– Automatic data deletion and automatic data backup function.
– The implementation of Micaps data processing server timing start function, and releasing the cache function.
– Early warning function for data.

3 DEVELOPMENT AND IMPLEMENTATION

The system uses Oracle DataBase 11G XE as a database, and Tomcat as a JSP container. The database layer uses the JDBC to interact with the underlying database, the logic layer packages are in Java class, whilst the presentation layer is implemented by JSP.

3.1 User

The operator uses the authentication of the software system; system uses the scheduled tasks to achieve control of operation timing; system uses the concept of cloud.

The operator roles include: online user registration, username and password setting; node increase and decrease and node information modifying.

3.2 System features

The system features are as follows: a logic layer and presentation layer separation, the three- layer structure for server use, business logic and database development which are independent from each other performed in parallel, which can greatly reduce the development time. At the same time, this method reduces the coupling degree between the three layers, future modification, layer extension, can only have a small influence on other layers.

3.3 System structure

3.3.1 The role of module design
ddition to the user operation in the whole system by designing a robot users, the role of the different operating encapsulation into separate classes. Design an abstract parent class, the concrete role classes inherit it. A subclass of the common methods are divided into two kinds of design. These methods design directly in the parent class, subclass inherits; In another implementation method in each subclass

3.3.2 The connection with the database
Designed a class and were responsible for connecting the database, when the connection to the database after the success, the class will be able to return a reliable database connection object for other.

3.3.3 The log records
In order to debug the server information records, class design will be responsible for the necessary information for the local hard disk log file.

3.3.4 Assisted transaction processing
The class design specialises in processing some auxiliary affairs, such as strings of transcoding job.

4 CONCLUSION

System used in a three-tier architecture, including data layer, logic layer and display. Logic layer implemented a simple logging, can record any information to the log file, convenient server program tracking and debugging, at the same time can record some important transaction information. The server USES the database, transaction logic, the structure of the user interface independent of each other.

Electronics, Information Technology and Intellectualization – Song & Kwak (Eds)
© *2015 Taylor & Francis Group, London, ISBN 978-1-138-02741-1*

A study on operational standardization of urban delivery vehicles

Yanchang Lv
Shandong Jiaotong University, Jinan, P.R. China

ABSTRACT: This paper aims to help people who are interested in the distribution industry, to understand vehicles selection and operation related to urban delivery. In order to be compatible with customer and authority needs, some laws and regulations coupled with the characteristics of the cities were introduced. The delivery vehicles should have the same basic conditions and operational norms, delivery services and quality standards. Drivers should comply with laws and regulations, which should meet the diverse needs of the urban freight market. Therefore, the study mainly focuses on urban delivery operation which is in line with the norms, standards and service quality, and in this way, the existing deficiencies and shortcomings could be revised.

1 THE INTRODUCTION OF URBAN CARGO DELIVERY

Urban cargo delivery, usually known as Urban Delivery, within a reasonable range, serves urban and suburban customers in a city, in order to deliver cargoes such as groceries, chilled products, medicine, household appliances and so on. The services include distribution processing, packaging, sorting, delivering, handling, loading and unloading. It focuses on "the last kilometre" to finish the cargo movement from distribution centre to customer.

2 THE SELECTION OF URBAN DELIVERY VEHICLES IN CHINA

2.1 *The basic conditions of urban delivery vehicles*

As the vehicles are used in cities to deliver cargoes, they should meet basic conditions according to the laws and regulations of most cities in China. Some issues should be considered, such as air pollution, noise, safety and city environment; for example, in the city of Shanghai there are the following basic conditions:

a. Vehicles used for urban delivery should meet the regulations of GB1589-2004t, (the profile size, and axial load limits for road vehicles), the vehicles used for chilled goods delivery should meet Regulation JTJ019-2008 (the technical specifications for cold trucking).
b. The technical degree of vehicles should reach the standard of JT/T198-2004 (the technological hierarchies and assessment requirements for operational vehicles).
c. Vehicle emissions should meet the standard of GB17691-2005 (the Chinese limits and measurement methods for exhaust compression ignition and gas fuelled positive ignition vehicle engines).
d. A cargo van body should be made of full grate plate material or window plate structure.
e. A cargo van should, at the most, weigh 800 kg (the design of double row seats in a cab is allowed), as for lighter enclosed lorries, the limited design weight should be 450 kg (the seats in a cab should be only in one row), and 800 kg < loaded cargo ≤ 2000 kg.
f. There is a rigid isolation device between the cab of the lorry and the rear cargo van; the cargo van should be fully enclosed without glass windows.

2.2 *The technical specifications of urban delivery vehicles*

The van is usually connected with the driving cab, for which there are no nationwide regulations in China. Several cities have pioneered regulation schemes, for example, in Shanghai, the technical specifications for urban delivery vehicles are stated below:

a. The vehicle body should be stamped by using one iron sheet as a whole.
b. The power/mass engine ratio (kW/t) should be more than 20.
c. The minimum turning radius (M) ≤6.5. A power steering system should be supplied.
d. The climbing gradient should be higher than 30% per cent (16.5°).
e. The vehicle should have independent front suspension.
f. Disc brakes and hydraulic equipment should apply to all vehicles. The hydraulic brake which can be used is the total mass ≤4500 kg, while air brakes

or disc brakes should be required if the total mass is between 4500 kg and 8000 kg.

2.3 *The safety of a delivery vehicle when it is in motion*

Potential risks to the transportation process can be caused by human error. Only by conforming to relevant regulations can the job be done safely, punctually and correctly. The following are the transportation process regulations:

a. Vehicles should meet the requirements of the approved payload and it is strictly forbidden to go against relevant State regulations and overload the vehicles during the transportation process. Van-type lorries cannot be used to carry passengers.
b. Drivers or transport contractors are responsible for supervising the whole process of freight delivery. Cargoes should be kept intact during the process of cargo handling, verification, compartment sealing, delivery and acceptance of delivery.
c. Given China's record on actual conditions and technical specifications of distribution vehicles, it may not go smoothly. However, standards can only be achieved when regulations are set; the logistics industry will become more and more standardised through state supervision and local corporations.

3 THE OPERATIONAL STANDARDIZATION OF URBAN DELIVERY

3.1 *Complying with city regulations*

When freight delivery is in progress, the freight vans should be driven along specified routes, there should also be restriction times for loading and unloading on designated curb sides, the emissions and noise levels should meet the standard stipulated by local authorities.

3.2 *Standard accessory equipment should apply to all delivery vans*

The urban delivery van should have a tail lift to load and unload automatically; the pallets should be standardized in categories of 1200 mm × 800 mm, 1100 mm × 1100 mm, and 1200 mm × 1000 mm; some standard boxes and trolleys can be used when handling cargoes.

3.3 *The information system should be standardized*

Among supply chain companies, the ERP system is used to link every joint corporation even outside the supply chain which is, in this case, based on bar codes (EAN) or RFID information. A system should be established according to unified protocol. Some standard information standards are as follows: EDI (GB/T15191-1997, GB/T16833-1997 and

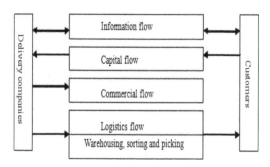

Figure 1. Service delivery based on customer satisfaction.

GB/TTI 6703-1996), for groceries delivery the UCS (Uniform Communication Standards) is now being used, most merchants are using VICS (Voluntary Inter-Industry Standards Committee), for inventories the WINS (Warehouse Information Network Standards) is popular now and for transport and delivery companies, the TDCC (Transportation Data Coordinating Committee) is widely used.

3.4 *The business flow of urban delivery services*

Service delivery is a clearly defined customer-based business model that offers a reliable and readily accessible system. It defines the quality and availability of its product and describes how and when it is delivered.

Service delivery lets the customer know what to expect from a service, and it helps them understand the scope of a service and how it operates.

The business flow of urban delivery is made up of three procedures. These are cargo receiving, tallying and delivery, and every procedure has its own content.

a. Cargo receiving. Before receiving cargoes from suppliers, the cargo information should be inputted into the information system by using a bar code or RFID, enabling the cargo to be received and handled automatically at the DC warehouse.
b. Cargo tallying. Through the information system, the quality and quantity can be easily made clear. This will benefit the customers satisfaction.
c. Cargo delivery. This is the core issue of the delivery. Cargoes should be delivered in a timely manner and make customers satisfied.

3.5 *The operational standardization of urban delivery*

The aim of urban service delivery is to meet the customers' needs. Four types of flows should be dealt with properly; these are information flow, capital flow, commercial flow and logistics flow, see Figure 1.

a. Information flow always exists among all steps and participants, and it is mutual. Usually it is realized through the Internet. Customers can place orders via the Internet, and trace the order process through online information according to the orders' numbers.

b. Capital flow is from the customer to the delivery company and most capital is paid electronically via the Internet.

c. Commercial flow is there to take the cargoes from the delivery company to the customers. Cargoes are delivered to the locations according to a timetable. Logistics flow includes many procedures such as warehousing, handling, loading and unloading by using standard equipment to operate efficiently.

4 CONCLUSION

Urban cargo delivery is related to many elements and issues, some are from government authorities, some are from inner companies, and some are from other companies such as manufactures and suppliers and information providers. Efforts should be united in order to standardize the delivery vehicle models and their operational flow.

REFERENCES

Wang Yang. Truck and Logistics [J].2011.1.5(58-62)
BESTUFS 3rd Workshop. Vehicle related constraints, Problems and Trends in Urban Freight Distribution. 2011.1.17
Jim York. UK Urban Delivery Vehicle Requirements. www.exel.com
Transferability of urban logistics concepts and practices from a world-wide perspective. 2010.3.31

Electronics, Information Technology and Intellectualization – Song & Kwak (Eds)
© 2015 Taylor & Francis Group, London, ISBN 978-1-138-02741-1

Fuzzy clustering based on intuitionistic fuzzy similarity relations

Juan Li & Jing Zhang
Qiongtai Teachers' College, Haikou, China

Yongliang Jiang
Hainan Normal University, Haikou, China

ABSTRACT: In this paper, the concepts of isomorphism and homomorphism of intuitionistic fuzzy sets (relations) are introduced. After that, the concept of classified isomorphism of the intuitionistic fuzzy similarity relation is given. We prove that an intuitionistic fuzzy similarity relation is classified isomorphic to its transitive closure. Finally, a new fuzzy clustering algorithm based on the intuitionistic fuzzy similarity relation is proposed, and an example is illustrated.

Keywords: Intuitionistic Fuzzy Set (IFS); Intuitionistic Fuzzy Relation (IFR); Classified isomorphism; Fuzzy clustering

1 INTRODUCTION

In 1986, the Intuitionistic Fuzzy Set[2] (IFS) was proposed by K. Atanassov which considered the degree of membership and the degree of non-membership of the element synchronously, provided more choices for the attribute description of an object, and had a stronger ability to express uncertainty than an ordinary fuzzy set[1] introduced by Zadeh in 1965. It had gained extensive attention from academic circles and the circles of engineering and technology.

As an important application, fuzzy clustering based on fuzzy sets was studied by many researchers and several typical clustering algorithms were introduced. The familiar algorithms are the transitive closure method based on fuzzy equivalent relations[5], the maximal tree algorithm[6] based on fuzzy graphs, and FCM algorithm[7]. However, there are rarely clustering algorithms based on IFSs. In this paper, we will discuss the clustering algorithm based on Intuitionistic Fuzzy Similarity Relations (IFSR).

Firstly, the concepts of isomorphism and homomorphism of IFSs (IFRs) are introduced, and then the concept of classified isomorphism of the IFSR is given. Secondly, we find an equivalent relation by the $\langle \lambda, \mu \rangle$ − cut set, by which we prove that an IFSR is classified isomorphic to its transitive closure. Finally, we propose a new fuzzy clustering algorithm based on the IFSR and an example is illustrated.

2 PRELIMINARIES

Definition 2.1 [2] Let X be a non-empty set, an IFS A on

X is an object of the form $A = \{\langle x, \mu_A(x), \nu_A(x)\rangle \mid x \in X\}$, where $\mu_A : X \to [0, 1]$ and $\nu_A : X \to [0, 1]$ define the degree of membership and the degree of non-membership of the element $x \in A$ respectively, with $0 \le \mu_A(x) + \nu_A(x) \le 1$ for any $x \in X$. $\langle \mu_A(x), \nu_A(x)\rangle$ is called the intuitionistic fuzzy value of A, denoted by $IFV_A(x)$. The set of the whole IFSs on X is denoted as $IFS(X)$.

Definition 2.2 [3] An IFR is an intuitionistic fuzzy subset of $X \times Y$ and is an expression R given by $R = \{\langle \mu_R(x, y), \nu_R(x, y)\rangle \mid x \in X, y \in Y\}$, where $\mu_R : X \times Y \to [0, 1]$ and $\nu_R : X \times Y \to [0, 1]$ satisfy $0 \le \mu_R(x, y) + \nu_R(x, y) \le 1$ for any $(x, y) \in X \times Y$. The whole IFRs of $X \times Y$ is denoted as $IFS[X \times Y]$.

Definition 2.3 [4] Let R be an IFR on X.

(1) reflexive i.e. $\forall x \in X, \mu_R(x, x) = 1, \nu_R(x, x) = 0$;
(2) symmetric, i.e. $\forall x, y \in X$,

$$\mu_R(x, y) = \mu_R(y, x), \nu_R(x, y) = \nu_R(y, x);$$

(3) transitive, i.e. $\forall x, y \in X$,

$$\mu_R(x, y) \ge \bigvee_{z \in X}(\mu_R(x, z) \wedge \mu_R(z, y))$$

$$\nu_R(x, y) \le \bigwedge_{z \in X}(\mu_R(x, z) \vee \mu_R(z, y))$$

If R satisfies (1) and (2), then R is called an IFSR; If R satisfies (1), (2), and (3), then R is called an intuitionistic fuzzy equivalent relation.

3 ISOMORPHISM AND HOMOMORPHISM OF IFS AND IFR

Definition 3.1 Let $f : A \to f(A)$ be a mapping from $IFS(X)$ to itself. If for $\forall x, y \in X$

1. $IFV_A(x) < IFV_A(y) \Leftrightarrow IFV_{f(A)}(x) < IFV_{f(A)}(y)$

$$IFV_A(x) = IFV_A(y) \Leftrightarrow IFV_{f(A)}(x) = IFV_{f(A)}(y)$$

Then f is called an isomorphism on $IFS(X)$, and we call A is isomorphic to $f(A)$, denoted by $A \cong f(A)$.

2. $IFV_A(x) \le IFV_A(y) \Rightarrow IFV_{f(A)}(x) \le IFV_{f(A)}(y)$

$$IFV_{f(A)}(x) < IFV_{f(A)}(y) \Rightarrow IFV_A(x) < IFV_A(y)$$

Then f is called a homormorphism on $IFS(X)$, and we call A is homomorphic to $f(A)$, denoted by $A \sim f(A)$.

Definition 3.2 Let $f : R \to Q$ be a mapping from $IFS[X \times X]$ to itself. If for $\forall (x,y), (u,v) \in X \times X$

1. $IFV_R(x,y) < IFV_R(u,v) \Leftrightarrow IFV_Q(x,y) < IFV_Q(u,v)$

$$IFV_R(x,y) = IFV_R(u,v) \Leftrightarrow IFV_Q(x,y) = IFV_Q(u,v)$$

Then f is called an isomorphism on $IFS[X \times X]$, and we R is isomorphic to Q, denoted by $R \cong Q$.

2. $IFV_R(x,y) \le IFV_R(u,v) \Rightarrow IFV_Q(x,y) \le IFV_Q(u,v)$

$$IFV_Q(x,y) < IFV_Q(u,v) \Rightarrow IFV_R(x,y) < IFV_R(u,v)$$

Then f is called a homormorphism on $IFS[X \times X]$, and we call R is homomorphic to Q, denoted by $R \sim Q$.

4 CLASSIFIED ISOMORPHISM BASED ON IFSR

Theorem 4.1 Let R be an IFER on X, $R_{\langle \lambda, \mu \rangle} = \{(x,y) \in X \times X | \mu_R(x,y) \ge \lambda, \nu_R(x,y) \le \mu \}$ is the $\langle \lambda, \mu \rangle$−cut set of R. A binary relation "$\sim_{R_{\langle \lambda, \mu \rangle}}$" is defined by $R_{\langle \lambda, \mu \rangle}$ as follows:

$x \sim_{R_{\langle \lambda, \mu \rangle}} y \Leftrightarrow$ there exist some $z_i \in X (i = 1, 2, \ldots, n)$, such that $(x, z_1), (z_1, z_2), \ldots, (z_n, y) \in R_{\langle \lambda, \mu \rangle}$.

Then "$\sim_{R_{\langle \lambda, \mu \rangle}}$" is an equivalent relation.

Proof: Since R is an IFSR, then:

(1) R is reflexive. Then for $\forall x \in X$, we have $\mu_R(x,x) = 1 \ge \lambda, \nu_R(x,x) = 0 \le \mu$, thus $(x,x) \in R_{\langle \lambda, \mu \rangle}$, therefore, $x \sim_{R_{\langle \lambda, \mu \rangle}} x$ holds.

(2) R is symmetric. Then for $\forall x, y \in X$, we have $\mu_R(x,y) = \mu_R(y,x), \nu_R(x,y) = \nu_R(y,x)$. If $x \sim_{R_{\langle \lambda, \mu \rangle}} y$, then there exist some $z_i \in X (i = 1, 2, \ldots, n)$, such that $(x, z_1), (z_1, z_2), \ldots, (z_n, y) \in R_{\langle \lambda, \mu \rangle}$. Since $(z_n, y) \in R_{\langle \lambda, \mu \rangle}$, by the symmetry of R, we have $\mu_R(y, z_n) = \mu_R(z_n, y) \ge \lambda$ and $\nu_R(y, z_n) = \nu_R(z_n, y) \le \mu$, therefore, $(y, z_n) \in R_{\langle \lambda, \mu \rangle}$ holds.

Similarly, we can get $(z_n, z_{n-1}), (z_{n-1}, z_{n-2}), \ldots, (z_2, z_1), (z_1, x) \in R_{\langle \lambda, \mu \rangle}$, then $y \sim_{R_{\langle \lambda, \mu \rangle}} x$.

Therefore, $x \sim_{R_{\langle \lambda, \mu \rangle}} y \Rightarrow y \sim_{R_{\langle \lambda, \mu \rangle}} x$ holds.

(3) If $x \sim_{R_{\langle \lambda, \mu \rangle}} y$, then there exist some $z_i \in X (i = 1, 2, \ldots, n)$, such that $(x, z_1), (z_1, z_2), \ldots, (z_n, y) \in R_{\langle \lambda, \mu \rangle}$. If $y \sim_{R_{\langle \lambda, \mu \rangle}} z$, then there exist some $z_i \in X (i = n+1, n+2, \ldots, n+m)$, such that $(y, z_{n+1}), (z_{n+1}, z_{n+2}), \ldots, (z_{n+m}, z) \in R_{\langle \lambda, \mu \rangle}$. So there exist $z_i \in X (i = 1, 2, \ldots, n+m)$, such that $(x, z_1), (z_1, z_2), \ldots, (z_n, y), (y, z_{n+1}), (z_{n+1}, z_{n+2}), \ldots, (z_{n+m}, z) \in R_{\langle \lambda, \mu \rangle}$, thus we get $x \sim_{R_{\langle \lambda, \mu \rangle}} z$.

Therefore, $x \sim_{R_{\langle \lambda, \mu \rangle}} y$ and $y \sim_{R_{\langle \lambda, \mu \rangle}} z \Rightarrow x \sim_{R_{\langle \lambda, \mu \rangle}} z$ holds.

By (1), (2), and (3), "$\sim_{R_{\langle \lambda, \mu \rangle}}$" is an equivalent relation on X.

Now "$\sim_{R_{\langle \lambda, \mu \rangle}}$" is called the classified relation determined by $R_{\langle \lambda, \mu \rangle}$, so we can get the classification of X which is called the $\langle \lambda, \mu \rangle$—classification of X and denoted by $R[\lambda, \mu]$, and $X(R) = \{R[\lambda, \mu] | \langle \lambda, \mu \rangle \in Im(R)\}$ is called the classified structure of X.

Definition 4.1 Let R and Q be any two IFSRs on X, $X(R)$ and $X(Q)$ are two classified structures of X. If $X(R) = X(Q)$, then R is classified isomorphic to Q.

Theorem 4.2 Let R be an IFSR on X. $\bar{R} = t(R)$ is the transitive closure of R. $R[\lambda, \mu]$ and $\bar{R}[\lambda, \mu]$ are two $\langle \lambda, \mu \rangle$—classifications of X which are determined by $R_{\langle \lambda, \mu \rangle}$ and $\bar{R}_{\langle \lambda, \mu \rangle}$ respectively. Then $R[\lambda, \mu] = \bar{R}[\lambda, \mu]$, i.e. the IFSR R is classified isomorphic to its transitive closure.

Proof: Since R is a similarity relation, so it is reflexive, i.e. $I \subseteq R$. And since X is finite, so there exists k such that $\bar{R} = R^k$. By the composition properties of relation, we have $I \subseteq R \subseteq R^2 \subseteq \cdots \subseteq R^k = \bar{R}$. So for $\forall \langle \lambda, \mu \rangle \in \langle I \rangle$, $R_{\langle \lambda, \mu \rangle} \subseteq \bar{R}_{\langle \lambda, \mu \rangle}$ holds. The $\langle \lambda, \mu \rangle$-classification containing x is determined by $R_{\langle \lambda, \mu \rangle}$ and $\bar{R}_{\langle \lambda, \mu \rangle}$, which is denoted by $[x]_{R_{\langle \lambda, \mu \rangle}}$ and $[x]_{\bar{R}_{\langle \lambda, \mu \rangle}}$ respectively, then we have

$$[x]_{R_{\langle \lambda, \mu \rangle}} = \{y | x \sim_{R_{\langle \lambda, \mu \rangle}} y\} \subseteq \{y | x \sim_{\bar{R}_{\langle \lambda, \mu \rangle}} y\} = [x]_{\bar{R}_{\langle \lambda, \mu \rangle}};$$

Contrarily, for $\forall z \in [x]_{\bar{R}_{\langle \lambda, \mu \rangle}}$, $(x, z) \in \bar{R}_{\langle \lambda, \mu \rangle}$ holds, then $R^k(x, z) \ge \langle \lambda, \mu \rangle$, and since

$$R^k(x, z) = R \circ R^{k-1}(x, z) = \bigvee_{u_1 \in X} (R(x, u_1) \wedge R^{k-1}(u_1, z)),$$

there exists $u_1 \in X$ certainly, such that $R(x, u_1) \ge \langle \lambda, \mu \rangle$, thus $(x, u_1) \in R_{\langle \lambda, \mu \rangle}$.

Similarly, R^{k-1} is decomposed by the composition of relation, then there exists $u_2 \in X$ certainly, such that $(u_1, u_2) \in R_{\langle \lambda, \mu \rangle}$. By the same method, we can find $u_1, u_2, \ldots, u_{k-1} \in X$, such that $(x, u_1), (u_1, u_2), \ldots, (u_{k-1}, z) \in R_{\langle \lambda, \mu \rangle}$. By the theorem 4.1, we can get $x \sim_{R_{\langle \lambda, \mu \rangle}} z$, then $z \in [x]_{R_{\langle \lambda, \mu \rangle}}$, thus $[x]_{\bar{R}_{\langle \lambda, \mu \rangle}} \subseteq [x]_{R_{\langle \lambda, \mu \rangle}}$.

Thereby, R is classified isomorphic to its transitive closure.

5 APPLICATION

Now a new clustering method will be designed.

1. Determine the order relation of samples based on some similarity, and determine an intuitionistic fuzzy relation matrix based on the order relation of samples. Generally, we may obtain an Intuitionistic Fuzzy Similarity Matrix (IFSM) according to the order from big to small in the elements of R, and only find x_j satisfying $r_{ij} \ge \langle \lambda, \mu \rangle$ from each row (or column) of R. Then all x_j

Chart 1. Characteristic information.

	C_1	C_2	C_3	C_4	C_5
G_1	[0.3, 0.5]	[0.6, 0.7]	[0.4, 0.6]	[0.2, 0.6]	[0.5, 0.8]
G_2	[0.6, 0.9]	[0.5, 0.8]	[0.8, 0.9]	[0.4, 0.9]	[0.3, 0.4]
G_3	[0.4, 0.7]	[0.6, 0.9]	[0.5, 0.9]	[0.9, 1]	[0.6, 0.7]
G_4	[0.8, 0.9]	[0.7, 0.9]	[0.6, 0.8]	[0.8, 0.9]	[0.7, 0.9]
G_5	[0.1, 0.4]	[0.3, 0.4]	[0.4, 0.5]	[0.2, 0.5]	[0.6, 0.8]
G_6	[0.5, 0.6]	[0.4, 0.7]	[0.3, 0.8]	[0.7, 0.9]	[0.5, 0.7]

are clustered into a class, and we can get the original clustering $\{A_1, A_2, \ldots, A_m\}$.

2. Classes with non empty intersections are united, i.e.,

$$\bar{A}_i = \cup\{A_l \,|\, A_l \cap A_i \neq \varnothing\}, i = 1, 2, \cdots, k.$$

(1) If $\bar{A}_i \cap \bar{A}_j = \varnothing (i \neq j)$, and $\sum\limits_{i=1}^{k} card(\bar{A}_i) = n$, then we can get the classification $R[\lambda, \mu] = \{\bar{A}_1, \bar{A}_2, \cdots, \bar{A}_k\}$ determined by $R_{\langle\lambda,\mu\rangle}$;

(2) If $\bar{A}_i \cap \bar{A}_j \neq \varnothing (i \neq j)$, then $\{\bar{A}_1, \bar{A}_2, \ldots, \bar{A}_k\}$ will be classified by step 3 again, until we get classification $R[\lambda, \mu] = \{A'_1, A'_2, \ldots, A'_p\}$ determined by $R_{\langle\lambda,\mu\rangle}$ satisfying

$$A'_i \cap A'_j = \varnothing \ (i \neq j) \text{ , and } \sum\limits_{i=1}^{p} card(A'_i) = n.$$

Example: A car market will classify five cars $C_j (j = 1, 2, \ldots, 5)$ and every car will be evaluated by six factors: fuel wastage (G_1), friction degree (G_2), price (G_3), comfortable degree (G_4), design (G_5), security (G_6). Now every car's characteristic information is expressed by an intuitionistic fuzzy value under the evaluation factors. See Chart 1.

Now we establish an IFSM $R = (r_{ij})_{5\times5}$ by the above information and equation of similarity measure :

$$r_{ij} = \left\langle 1 - \sqrt[p]{h^*(C_i, C_j)}, \sqrt[p]{h_*(C_i, C_j)} \right\rangle$$

where

$$h^*(C_i, C_j) = \max_k \{\alpha \left| t_{c_i}(G_k) - t_{c_j}(G_k) \right|^p + \beta \left| f_{c_i}(G_k) - f_{c_j}(G_k) \right|^p$$
$$+ \gamma \left| \pi_{c_i}(G_k) - \pi_{c_j}(G_k) \right|^p \}$$

$$h_*(C_i, C_j) = \min_k \{\alpha \left| t_{c_i}(G_k) - t_{c_j}(G_k) \right|^p + \beta \left| f_{c_i}(G_k) - f_{c_j}(G_k) \right|^p$$
$$+ \gamma \left| \pi_{c_i}(G_k) - \pi_{c_j}(G_k) \right|^p \}$$

(1) Suppose similar coefficient $p = 2, \alpha = \beta = \gamma = 1/3$, then we have

$$R = \begin{pmatrix} \langle 1,0 \rangle & \langle 0.78,0.08 \rangle & \langle 0.72,0.08 \rangle & \langle 0.64,0 \rangle & \langle 0.63,0.08 \rangle \\ \langle 0.78,0.08 \rangle & \langle 1,0 \rangle & \langle 0.78,0.08 \rangle & \langle 0.71,0.08 \rangle & \langle 0.71,0 \rangle \\ \langle 0.72,0.08 \rangle & \langle 0.78,0.08 \rangle & \langle 1,0 \rangle & \langle 0.69,0.14 \rangle & \langle 0.59,0.08 \rangle \\ \langle 0.64,0 \rangle & \langle 0.71,0.08 \rangle & \langle 0.69,0.14 \rangle & \langle 1,0 \rangle & \langle 0.63,0.08 \rangle \\ \langle 0.63,0.08 \rangle & \langle 0.71,0 \rangle & \langle 0.59,0.08 \rangle & \langle 0.63,0.08 \rangle & \langle 1,0 \rangle \end{pmatrix}$$

By computing, we can get the following $\langle\lambda, \mu\rangle$–classified structure:

Determined by $R_{\langle1,0\rangle}$: $\{(C_1), (C_2), (C_3), (C_4), (C_5)\}$
Determined by $R_{\langle0.78,0.08\rangle}$:

$$\{(C_1, C_2), (C_1, C_2, C_3), (C_2, C_3), (C_4), (C_5)\} \xrightarrow{\text{merger}} \text{Deter-}$$
$$\{(C_1, C_2, C_3), (C_4), (C_5)\}$$

mined by $R_{\langle0.71,0.08\rangle}$:

$$\{(C_1, C_2, C_3), (C_1, C_2, C_3, C_4), (C_2, C_4), (C_5)\} \xrightarrow{\text{merger}} \text{De-}$$
$$\{(C_1, C_2, C_3, C_4), (C_5)\}$$

termined by $R_{\langle0.69,0.14\rangle}$:

$$\{(C_1, C_2, C_3), (C_1, C_2, C_3, C_4, C_5), (C_1, C_2, C_3, C_4), (C_2, C_3, C_4),$$
$$(C_2, C_5)\} \xrightarrow{\text{merger}} \{(C_1, C_2, C_3, C_4, C_5)\}$$

where detailed process of $\langle\lambda, \mu\rangle-$, classified structure determined by $R_{\langle0.78,0.08\rangle}$ is as follows (others are similar):

$$\{A_1, A_2, A_3, A_4, A_5\} = \{(C_1, C_2), (C_1, C_2, C_3), (C_2, C_3), (C_4), (C_5)\}$$
$$\xrightarrow{\text{merger}} \{(C_1, C_2, C_3), (C_4), (C_5)\} = \{\bar{A}_1, \bar{A}_2, \bar{A}_3\} = R[0.78,0.08]$$

where

$$\bar{A}_1 = A_1 \cup A_2 \cup A_3 = \cup\{(C_1, C_2), (C_1, C_2, C_3), (C_2, C_3)\},$$
$$= (C_1, C_2, C_3)$$
$$\bar{A}_2 = A_4 = (C_4);$$
$$\bar{A}_3 = A_5 = (C_5).$$

(2) Suppose similar coefficient $p = 2, \alpha = 2/5, \beta = 2/5, \gamma = 1/5$, then we have

$$T = \begin{pmatrix} \langle 1,0 \rangle & \langle 0.77,0.08 \rangle & \langle 0.75,0.08 \rangle & \langle 0.62,0 \rangle & \langle 0.59,0.08 \rangle \\ \langle 0.77,0.08 \rangle & \langle 1,0 \rangle & \langle 0.78,0.08 \rangle & \langle 0.71,0.08 \rangle & \langle 0.68,0 \rangle \\ \langle 0.75,0.08 \rangle & \langle 0.78,0.08 \rangle & \langle 1,0 \rangle & \langle 0.71,0.15 \rangle & \langle 0.55,0.09 \rangle \\ \langle 0.62,0 \rangle & \langle 0.71,0.08 \rangle & \langle 0.71,0.15 \rangle & \langle 1,0 \rangle & \langle 0.63,0.08 \rangle \\ \langle 0.59,0.08 \rangle & \langle 0.68,0 \rangle & \langle 0.55,0.09 \rangle & \langle 0.63,0.08 \rangle & \langle 1,0 \rangle \end{pmatrix}$$

By computing, we can get the following $\langle\lambda, \mu\rangle$-classified structure:

Determined by $T_{\langle1,0\rangle}$: $\{(C_1), (C_2), (C_3), (C_4), (C_5)\}$
Determined by $T_{\langle0.78,0.08\rangle}$: $\{(C_1), (C_2, C_3), (C_4), (C_5)\}$
Determined by $T_{\langle0.77,0.08\rangle}$:

$$\{(C_1, C_2), (C_1, C_2, C_3), (C_2, C_3), (C_4), (C_5)\} \xrightarrow{\text{merger}} \text{Deter-}$$
$$\{(C_1, C_2, C_3), (C_4), (C_5)\}$$

mined by $T_{\langle0.71,0.08\rangle}$:

$$\{(C_1, C_2, C_3), (C_1, C_2, C_3, C_4), (C_2, C_4), (C_5)\} \xrightarrow{\text{merger}} \text{De-}$$
$$\{(C_1, C_2, C_3, C_4), (C_5)\}$$

termined by $T_{\langle0.63,0.08\rangle}$:

$$\{(C_1, C_2, C_3), (C_1, C_2, C_3, C_4, C_5), (C_2, C_4, C_5)\} \xrightarrow{\text{merger}}$$
$$\{(C_1, C_2, C_3, C_4, C_5)\}$$

The $\langle\lambda, \mu\rangle$-classified structure is determined by $T_{\langle0.59,0.08\rangle}$ and $T_{\langle0.55,0.09\rangle}$ which is the same as that by $T_{\langle0.63,0.08\rangle}$.

163

(2) Suppose similar coefficient $p = 4, \alpha = {}^2/_5, \beta = {}^2/_5, \gamma = {}^1/_5$, then we have

$$Q = \begin{pmatrix} \langle 1,0 \rangle & \langle 0.75,0.09 \rangle & \langle 0.72,0.09 \rangle & \langle 0.41,0 \rangle & \langle 0.57,0.09 \rangle \\ \langle 0.75,0.09 \rangle & \langle 1,0 \rangle & \langle 0.76,0.09 \rangle & \langle 0.67,0.09 \rangle & \langle 0.66,0 \rangle \\ \langle 0.72,0.09 \rangle & \langle 0.76,0.09 \rangle & \langle 1,0 \rangle & \langle 0.67,0.16 \rangle & \langle 0.53,0.09 \rangle \\ \langle 0.41,0 \rangle & \langle 0.67,0.09 \rangle & \langle 0.67,0.16 \rangle & \langle 1,0 \rangle & \langle 0.58,0.09 \rangle \\ \langle 0.57,0.09 \rangle & \langle 0.66,0 \rangle & \langle 0.53,0.09 \rangle & \langle 0.58,0.09 \rangle & \langle 1,0 \rangle \end{pmatrix}$$

By computing, we can get the following $\langle \lambda, \mu \rangle$-classified structure:

Determined by $Q_{\langle 1,0 \rangle}$: $\{(C_1), (C_2), (C_3), (C_4), (C_5)\}$
Determined by $Q_{\langle 0.76,0.09 \rangle}$: $\{(C_1), (C_2, C_3), (C_4), (C_5)\}$
Determined by $Q_{\langle 0.75,0.09 \rangle}$:

$$\{(C_1, C_2),(C_1, C_2, C_3),(C_2, C_3),(C_4),(C_5)\} \xrightarrow{\text{merger}} \{(C_1, C_2, C_3),(C_4),(C_5)\}$$ Deter-

mined by $Q_{\langle 0.67, 0.09 \rangle}$:

$$\{(C_1, C_2, C_3),(C_1, C_2, C_3, C_4),(C_2, C_4),(C_5)\} \xrightarrow{\text{merger}} \{(C_1, C_2, C_3, C_4),(C_5)\}$$

Determined by $Q_{\langle 0.58, 0.09 \rangle}$:

$$\{(C_1, C_2, C_3),(C_1, C_2, C_3, C_4, C_5),(C_2, C_4, C_5)\} \xrightarrow{\text{merger}} \{(C_1, C_2, C_3, C_4, C_5)\}$$

The $\langle \lambda, \mu \rangle$-classified structure determined by $Q_{\langle 0.57,0.09 \rangle}$ and $Q_{\langle 0.53,0.09 \rangle}$ is same to that by $Q_{\langle 0.58,0.09 \rangle}$. Comparing the above results, we know that:

(1) In the process of choosing $\langle \lambda, \mu \rangle$, if $\langle \lambda, \mu \rangle = \langle \lambda_i, \mu_i \rangle$ and the corresponding clustering is only one that is composed of all samples, then $\langle \lambda, \mu \rangle$-classified structure determined by $R_{\langle \lambda_k, \mu_k \rangle}$ is same to that by $R_{\langle \lambda_i, \mu_i \rangle}$, where$\langle \lambda_k, \mu_k \rangle \leq \langle \lambda_i, \mu_i \rangle$, $\langle \lambda_k, \mu_k \rangle \in Im(R)$.
(2) Although IFSMs R, T and Q are different and based on different rules, T is isomorphic to Q, R is homomorphic to Q, and T is isomorphic to Q We also find that classified structures are isomorphic which are determined by T and Q respectively. And classified structures are not isomorphic which are determined by R and Q respectively, but the one must be a refinement of another.

The results show that: if the two IFSRs are isomorphic, then the order relation among elements is completely consistent, and they are certain classified isomorphism. However, if the two IFSRs are homomorphic, then the order relation among elements is consistent by and large, and they are not always classified isomorphism, but the one must be a refinement of the other.

6 CONCLUSION

The isomorphism and homomorphism theories have explained that the essence of IFSs (IFRs) is the order relation among elements. In other words, an IFS (or IFR) is determined by the order relation between elements on X. In practice, although different IFS (or IFR) functions are given because of users' subjective factors, as long as the users' essential comprehension of the object is the same or similar, conclusions are certainly the same or similar, and then they would get the same results. Because their judgments do not always rely on computing, but also through comparing, then understanding this thought will help to solve problems and reduce the process of solving problems greatly.

ACKNOWLEDGMENT

This work is supported by NSF of China (70940007), Hainan Natural Science Fund 110008 and 612136.

REFERENCES

C.-y. Zhang, B.-c. Wei, H.-y. Zhou, On consistency of fuzzy clustering analysis, IEEE International Conference on Granular Computing, 2007, pp. 110–113.
H.-m. Zhang, Z.-s. Xu, Q. Chen, On clustering approach to intuitionistic fuzzy sets, Control and Decision, 2007, 22 (8): 882–888.
K. Atanassov, Intuitionistic fuzzy sets. Fuzzy Sets and Systems, 1986, 20(1): 87–96.
K. Atanassov, Intuitionistic fuzzy relations. First Scientific Session of the Mathematical Foundation Artificial Intelligence, Sofia IM-MFAIS, 1989, pp. 87–96.
L.A. Zadeh, Fuzzy sets, Information and Control, 1965, 8(3):338–356.
T.T. Buhaescu, Some observations on intuitionistic fuzzy relations, Itimerat Seminar on Functional Equations, pp. 111–118.
Z. Le, Fuzzy relation compositions and pattern recognition, Inf.Sci., 1996, 89: 107–1306.
Z. Wu, R. Leathy, An optimal graph theoretic approach to data clustering: theory and its application to image segmentation, IEEEPAMI, 1993, 15 (11): 1101–1137.
Z.-p. Fan, C.-h. Yu, T.-h. You, An FCM clustering algorithm for multiple attribute information with triangular fuzzy numbers[J], Control and Decision, 2004, 19 (12): 1407–1411.
Zhang Ling, Zhang Bo, Theory of Fuzzy Quotient Space (Methods of Fuzzy Granular Computing), Journal of Software, 2003, 14(4): 770–776. (in Chinese)

Electronics, Information Technology and Intellectualization – Song & Kwak (Eds)
© 2015 Taylor & Francis Group, London, ISBN 978-1-138-02741-1

Designing human-computer interaction software

Yang Yang
Faculty of Arts and Media, Kunming University of Science & Technology, Kunming, China

ABSTRACT: With the proliferation of newer types of digital interfaces—Facebook and MySpace, blogs and wikis, cell phones and iPods—come new opportunities, new frustrations, and new rhetorical choices for writers and designers. For every time we have received a well-crafted email message from a colleague or had a search engine give us a relevant answer on the first try, we have received dozens of spam emails and clicked hundreds of irrelevant links. This article describes the importance and the main detail of designing interactive software interface. Some designing principles are clearly pointed out when the interactive interface is designed. Much useful technology is also put forward for the software interface design.

Keywords: Human-computer interaction; designing principles; technology

1 INTRODUCTION

Generally speaking, the human-computer interaction interface has two kinds of interface, a hardware interface and a software interface. For computer software, the human-computer interaction interface is mainly composed of software interface. In the past ten years, I have designed and developed a number of different enterprises successively for computer information systems, and engaged in the research on the development of the ERP (enterprise resource planning) system and implementation of the work. In addition, I have studied in depth and in detail some large foreign ERP software projects about the human-computer interaction interface. The good or bad aspects of the interactive software interface design are directly related to the promotion and use of the software, related to the whole system function of the play, even other perfect relationship to the software part of the performance. So this article discusses the human-computer interaction which is conducted in the process of software interface design and input/output interface design. In addition, some designing principles are also included.

2 INPUT AND OUTPUT DESIGN OF SOFTWARE INTERFACE OF HUMAN-COMPUTER INTERACTION

The successful design of software interface of human-computer interaction is indispensable to the excellent design input and output interface in the process of input and output design in human-computer interaction., Here, the development practices over the years as well as the ERP software implementation experience

abroad, and some of the design technology and principles, especially the real difficulty of a software system for the users from the view of man-machine engineering, will be discussed and analysed. Next, there is an assessment on common mistakes which are easy to make in the design process. For example, whether too much jargon and too many acronyms have been used or not, whether the needs of users have been understood or not, and whether users can or cannot understand the method to solve any problems they might encounter during operation, if there is no consistency and design style, etc.

2.1 *The principle of consistency*

The consistency of the human-computer interaction interface is mainly reflected in the input and output. The consistency of the interaction effect on input and output interface specifically refers to software systems with a similar interface appearance, layout, similar way of human-computer interaction, similar information display format, and so on. General consistency can even be extended to a certain consistency among the application software platform, for example, Illustrator, Photoshop, PageMaker and some other software which run under the Macintosh system platform. Their interface design maintains a high degree of consistency. Only in this way will the user not spend too much time on learning to master software after software. The Windows platform and a great many other application software platforms in Linnus have such characteristics. Obviously, the consistency principle is helpful for users to become familiar with the use of the software as soon as possible and to reduce the error and the amount of memory in the process of using the software. However, the principle of consistency is often overlooked. In fact, as long as clear thought and perseverance is kept throughout the design process, can the engineers avoid such mistakes? In particular, organizers who are responsible for the overall design of the project must seriously apply consistency to carrying out every design in the link. In the process of interactive interface design, they must be strict with each group of the design team to remind them of the consistency of the concept model, the semantics, the command language grammar and display formats, and so on, to maintain a certain standard. In general, if there is a similar interaction situation, general requirements remain the same, that is, a strict consistent sequence of operations and prompt style, namely in tips, menus, and helps using the same terms and using consistent style from beginning to end.

2.2   *Window design*

Screen window design is one of the key contents in the process of human-computer interaction interface design which must be considered. It must have a reasonable design with a scientific arrangement on the regional distribution according to the importance of information, and a clear degree on the screen window so as to make the window interface look fresh. As far as this is concerned, there are some experiences which greatly enhance the implementation of the foreign ERP large-scale software so that the eye does not tire easily, or important interaction information is not arranged in an easily overlooked area. Helpful information or reminders are often arranged at the bottom of the screen. As a result, it forms a concise, clear and reasonable layout on the space arrangement of the window. In addition, some other individual parts of the world have inserted by the blank space which is used to highlight some of the display elements. Developers like to use the common tools provided by many common controls when designing software, such as using a single text box, multi-line text boxes, drop-down list boxes, radio buttons, check boxes, command buttons, list boxes, and so on. In summary, it is more necessary to have a reasonable arrangement of the white space around the forms between controls and controls.

2.3   *Interface effect*

Interface effect is the concrete embodiment of the effect of the human-computer interaction interface in the end. Text and black and white colour causes the users to become fatigued quickly. However, colour and image media interface can increase the visual appeal so as to reduce fatigue; graphics and images are more intuitive because of the advantages of large amounts of information. Therefore, the use of multimedia to show some entities or operations, can make the user's operation feel more immediate, visible, and clear, as well as enhance the understanding of the software system and its ease of use. Of course, each user has quite different attitudes towards analogy between the figure and image, as well as towards the love of colour and media tastes. Improper collocation of media or

colour can cause strong emotional ups and downs. It is also likely to be too fancy, which produces the opposite effect. Therefore, multimedia cannot be used with abuse, unless there is a certain artistic designer who understands the scientific and reasonable use of these media, even the combination effects of static and dynamic can be dangerous (such as special display attributes, font and so on.). For general users, therefore, it is best to adopt relatively mild media, such as softer, more neutral colour. Sometimes, individual colours can be used to effectively render and highlight the important information to attract the attention of the user to important areas. Of course, some principles as referred to above must be adhered to, such as trying to keep the consistency of the colour and types. Here, we must underline the use of colour in the design of interface effect again: if colour is applied properly it can produce extraordinary charm and different feelings and impressions. Therefore, we should maximize colour combination of visual interfaces to create the ideal effect.

3   DESIGNING PRINCIPLES

3.1   *The principle of users*

The first principle of good design is to be aware of the differences between users of a proposed computer interface, be it a piece of software, a website, or an electronic gadget. That shows you know your user. Knowing your users includes gathering basic demographic information about them, but also can involve dividing the user groups into their expertise level, categorizing them based on the kinds of tasks they want to complete by using the system, and categorizing them by their preferred method of interacting with the system design of the human-computer interface. Therefore, we must first establish the type of user. Types can be determined from different angles, depending on the actual situation. Predictions of their response to the different interface according to its characteristics, need to be conducted after making sure of the type of user. This analysis can be designed from several aspects.

3.2   *The principle of minimum information*

The design of the human-computer interface must minimize the burden on the memory of the user. At the same time, a scheme which helps the user improve their memory can be adopted. Understanding how humans perceive their environment—the senses involved, the mental maps and metaphors that they construct, and the decomposition of complicated layouts into their component parts—is critical to good design.

3.3   *The principle of helps and tips*

The human-computer interface must be designed to respond to the user operation command as well as to help the user deal with problems. The system must be designed with the ability to recover error sites. Moreover, there should be prompts in the system's internal processing work so as to provide the initiative to the user as far as possible. In addition, by relying on perceived affordance as a way to figure out how an object will behave, we need to think how users also rely heavily on conventions and constraints to guide them in how to use new objects within a system. Consider the example of a scroll bar: the bar has no inherent meaning as a device to humans (it is a made-up, digital construct), but our reaction to the scroll bar (we know it represents "more text than can fit on the screen," and we know that we can click it, pull it down, and expect to move along in the text) is learned and is culturally dependent. We have not encountered scroll bars outside of digital interfaces, and if we encountered a scroll bar that suddenly acted differently, we would be surprised and confused.

3.4 *The principle of the best combination of media*

The success of the multimedia interface lies not only in rich media being only available to the user, but also by paying good attention to detail when considering the relationships between various media, even a proper selection under the guidance of the relevant theory.

3.5 *The principle of perceived affordance*

The term "perceived affordance" refers to the set of potential actions held by a physical object. We can explain this in terms of simple objects, for example, when a human sees a handle, she wants to grab it, and when she sees a ball, she wants to roll it. The perceived affordance of an interface, therefore, refers to the way a user can tell how to interact with the system upon first experiencing it. The sensory qualities that the user perceives in a system imply how to use it. Perceived affordances of an object, therefore, are subject to each user's ability to sense something, as well as to their experiences, their backgrounds, their memories, etc. This is an important distinction; it is not solely the inherent qualities of the object itself that imply its use. These inherent qualities will always be complemented (and complicated) by the very powerful knowledge that exists in the user's own mind.

4 CONCLUSION

In both composition and computer science, there is a strong tradition of "Know Your User," and both disciplines expect that the designer will take into account the diversity, experience, or expectations of the user or audience. However, we found that in both our writing and computer science classes, students sometimes made design decisions based on only a limited understanding of their audience or user. As more and more of the information available to readers is mediated through digital interfaces, it becomes increasingly necessary for writing students to understand how an "audience" becomes a "user." How does the audience interact with the interfaces the writer creates? How is it that these "little machines" that translate human-readable programming code into zeroes and ones can also serve up faithful visual and auditory representations of a writer's ideas? How will our audience react to the interface that we design to express our ideas, and how will we know if our choices were successful? Designing digital interfaces requires writers to make technical choices that are also rhetorical in nature and to engage in activities that parallel the design, implementation, and evaluation cycles typical in software development.

REFERENCES

Lean UX: Applying Lean Principles to Improve User Experience, [M]. Jeff Gothelf, Josh Seiden, translated by Dai Zhang. Beijing: Publishing House of electronics industry, 2013.

Microinteractions: Designing with Details. [M]. Dan Saffer, Songfeng Liu, translation. Beijing: Publishing House of electronics industry, 2013.

Quantifying the User Experience – Practical Statistics for User Research [M]. Jeff Sauro, James R. Lewis, translated by Wenqian Qing. Beijing: Publishing House of Mechanical industry, 2014.

The Elements of User Experience – User-centered Design for the Web and Beyond Second Edition. [M]. Jesse James Garrett, Translated by Xiaoyan Fang. Beijing: Publishing House of Mechanical industry, 2011.

User Experience and Product Innovation Design. Shiqian Luo, Shangshang Zhu. Beijing: Publishing House of Mechanical industry, 2010.

Electronics, Information Technology and Intellectualization – Song & Kwak (Eds)
© 2015 Taylor & Francis Group, London, ISBN 978-1-138-02741-1

Analysis and simulation of a flue gas flow model with a 90° rectangular large-diameter bend

Lianhe Li, Ping Wang, Jin Zhou, Xu Liu & Qian Song
Tianjin Polytechnic University, Tianjin, China

ABSTRACT: To utilize FLUENT 6.3 software to make CFD simulation of the air flow in a rectangular bend with a large-diameter of 90°, and through adopting the standard k-ε model to achieve the quantitative distribution of speed and pressure from all the positions of the pipeline. In order to establish a more accurate mathematical model of the flow field, we explore the quantitative description method of change rule of a non-fully developed flow filed pipe. Nowadays, it is difficult to make flow tests of large-diameter non-fully flow pipes, and there is no such a test in domestic country.

1 INTRODUCTION

The chimney of a coal-fired plant is an inevitable waste gas discharge point. At present there are no plants producing relative instruments to calculate the flow of their waste gas. The reason is that it is too difficult to master the distribution characteristics of gas flow in large-diameter rectangular bends, and furthermore, it restricts the development of measurement devices. With the development of the individual computer, many domestic and overseas scholars make use of CFD to research the numerical simulation of the distribution of parts of small-diameter, non-fully developed pipe flow fields, and the measurement of flow. The scholars in our nation have achieved some achievements regarding the aspect of numerical simulation of the distribution of large-diameter short air flow fields, which stress the aspect of measurement of the wind tunnel.

The numerical simulation is performed and a model is conducted in the environment of flue gases, but these things are less used in practical research. All workers who are engaged in the research of internal-flow pay attention to the issue of the flow of liquid in rectangular ducts. With the development of computational fluid mechanics, computer technology and the numerical simulation method, the calculation of computation fluid has become the important method of simulation and research of turbulent flow. This paper adopts the k-ε model to make numerical simulation of a rectangular bend with a 90° diameter. Though some scholars research small-diameter flow pipes, recently, the research and materials which are about the measurement of small-diameter non-fully developed pipe flow fields are lacking.

2 GOVERNING EQUATION

The used fluid is air, the temperature is 20°C, density-ρ is 1.225 kg/m³, Kinetic viscosity-μ is 1.7894 × 10^{-5} kg/(m.s^{-1}). Suppose air flows steadily through the rectangular pipe at 15 m/s flow rate. Because of not large flow rate, fluid can be regarded as incompressible fluid approximately, through the continuity equation and N-S equation homogenizing treatment, then getting steady of Cartesian coordinate, adiabatic, viscosity, incompressible fluid flow equation.

Continuity equation:

$$\frac{\partial \bar{u}_i}{\partial x_i} = 0 \tag{1}$$

Motion equation:

$$\frac{\partial}{\partial t}(\rho \bar{u}_i) + \frac{\partial}{\partial x_i}(\rho \bar{u}_i \bar{u}_i) = -\frac{\partial p}{\partial t} + \frac{\partial}{\partial x_i}\left[\frac{\partial \bar{u}_i}{\partial x_i} - \rho \overline{u_i u_i}\right] + S_i \tag{2}$$

$\bar{u}_i$–average speed of air, p–average pressure, S_i–source term, $\rho u_i u_i$–Reynolds stress, it can be calculated by Boussinesque, such as an eddy viscosity model. Transport equation of turbulent kinetic energy–k:

$$\frac{\partial}{\partial t}(\rho k) + \frac{\partial}{\partial x_i}\left[\rho u_i k - (\mu + \frac{u_i}{\delta_k})\left(\frac{\partial k}{\partial x_i}\right)\right] = G - \rho \varepsilon \tag{3}$$

Transport equation of turbulent dissipation rate–ε:

$$\frac{\partial}{\partial t}(\rho \varepsilon) + \frac{\partial}{\partial x_i}\left[\rho u_i \varepsilon - (\mu + \frac{u_i}{\delta_\varepsilon})\left(\frac{\partial \varepsilon}{\partial x_i}\right)\right] = C_{\varepsilon} G \frac{\varepsilon}{k} - C_{\varepsilon 2} \rho \frac{\varepsilon^2}{k} \tag{4}$$

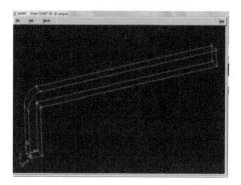

Figure 1. The sketch map of the 90° rectangular pipe.

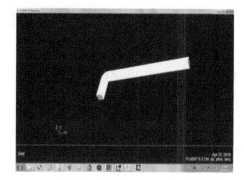

Figure 2. Grid map bending pipe.

Int:

$$\mu_t = C_\mu \rho \frac{k^2}{\varepsilon}, \qquad G = \mu_t \left(\frac{\partial u_i}{\partial x_j p} + \frac{\partial u_j}{\partial x_i} \right) \frac{\partial u_i}{\partial x_j}.$$

The constants of the above equations are achieved from chart: $C_\mu = 0.09$, $\delta_k = 1.0$, $\delta_\varepsilon = 1.3$, $C_{\varepsilon 1} = 1.44$, $C_{\varepsilon 2} = 1.92$.

The standard k–ε model is one of the most widely used turbulence models. A large amount of project application practices show that the model can calculate comparative complicated turbulence and better forecast wall boundary layer flow, pipe flow, channel flow, and so on. The model has not shown more obvious advantages like the algebraic model. With the development of cavitation flow theory and calculating measurement, numerical calculation has become the powerful tool of cavitation phenomenon. Turbulence flow as a complicated turbulent flow is simulated by a turbulent flow model which is an important aspect.

3 GEOMETRIC CONSTRUCTION AND GRID GENERATION

Firstly, a 90° rectangular pipe as a physical model is established by Gambit software and its size is shown in Figure 1. In order to analyse it, the pipeline is divided into three parts: the upstream straight pipe, the bending pipe and the downstream straight pipe. The section is rectangular, 1 m × 1 m. The radius of the bending pipe's internal surface is 0.092 m, the radius of the external surface is 1.092 m. The length of the upstream straight pipe is 3 m, the length of the downstream exit pipe is 15 m; origin—o is located in the rotation centre of the curvature of the bending pipe; θ in the section of the entrance of main flow is defined 0°, θ in the exit of the section of the bending pipe is defined as 90°. The boundary layer near the wall where the use of grid, the initial width of the wall boundary layer grid is 0.05m, the growth factor is 1.2, the total length is 0.268, the number of layer is four, the form of grid is Quad, the type is "map", the interval size is 0.03. Because the structural shape is not complicated, we chose structural mesh. All of the section uses "map",

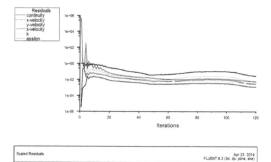

Figure 3. Sectional residuals plot.

a tetrahedral structural grid unit which is shown in Figure 2.

4 BOUNDARY CONDITION

The condition of simulation equipment gives the entrance velocity which is 15 m/s at the entrance and the direction is vertical entrance. The wall adopts a non-slip wall, the outlet adopts outlet—pressure. The outlet pressure gauge is 0 Pa. The section is non-circular, so the circular sectional diameter is replaced by an equivalent diameter—D = 2A/P, A is the area of the rectangular section, P is the perimeter of the section. The Reynolds number of the model is Re = ρVD/μ. Then the calculated result is a high Reynolds number, the Hydraulic Diameter is replaced by an equivalent diameter.

5 PHOTOGRAPHS AND FIGURES

5.1 The analysis of the sectional residuals plot

From Figure 3, we can see that 120 time's iteration get convergence, and it shows that the convergent speed of the tetrahedral structural grid is comparatively slow, the calculation time is comparatively short, and the numerical diffusion is reduced under the high Reynolds number. It is easy to adopt a hexahedral grid to speed up convergent velocity for a low Reynolds

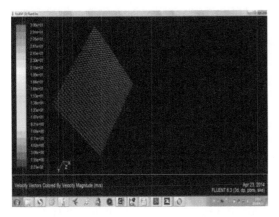

Figure 4. The diagram of the distribution of the speed arrow.

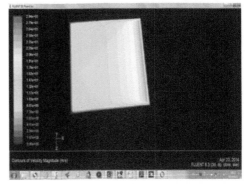

Figure 5. The diagram of speed distribution of section 1.

Chart 1. The data of iteration's first and last 5 times .

time	x(m/s)	y(m/s)	z(m/s)	time	x(m/s)	y(m/s)	z(m/s)
1	3.2447 e-06	6.8602 e-05	3.8929 e-06	116	3.1787 e-03	6.4948 e-03	1.3312 e-03
2	1.9730 e-01	7.1814 e-01	7.5398 e-04	117	3.0300 e-03	6.1600 e-03	1.2997 e-03
3	2.2373 e-01	1.1422 e-01	1.2791 e-03	118	2.9671 e-03	5.7927 e-03	1.2488 e-03
4	1.8008 e-02	3.8769 e-02	2.1985 e-03	119	2.8325 e-03	5.5355 e-03	1.2255 e-03
5	5.5143 e-02	2.2630 e-01	2.3863 e-02	120	2.8193 e-03	5.2788 e-03	1.1876 e-03

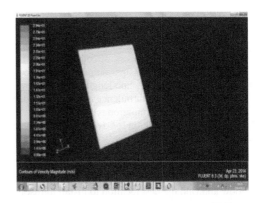

Figure 6. The diagram of speed distribution of section 2.

number and a comparative simple structure. The number of iterations is less than 60 for the 90° bending pipe with a circular section and the convergent result is better. Because of comparatively more iterations, only the first five times and the last five times of iteration of the rectangular pipe are shown in Chart 1, and the diagram of the distribution of speed arrow is shown in Figure 4.

5.2 The analysis of the diagram of speed distribution

Figure 5 is formed to be the velocity of section 1 by the three points, $(0, 3, 0)$ $(1, 3, 0)$ $(1, 3, 1)$, of Cartesian coordinates rule, Figure 6 is formed to be velocity of section 2 by the three points, $(1.092, 3.092, 0)$ $(1.092, 4.092, 0)$ $(1.092, 3.092, 1)$, of coordinates rule.

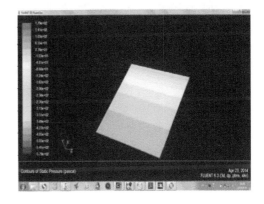

Figure 7. The diagram of pressure of section 1.

5.3 The analysis of the diagram of pressure

Figure 7 is the diagram of pressure of section 1. Figure 8 is the diagram of pressure of section 2. From the above diagrams, we can see that when fluid flows from a straight pipe into a bending pipe and it is restricted by curvature and external face, the fluid produces the effect of a centrifugal force and the pressure of the external face increases and the pressure of the internal face decreases. As a result of this, the tangential velocity of the external face decreases and the tangential velocity of the internal face increases. The pressure growth of section 1 is greater than that of section 2. The reason is that the state of flow of fluid exists a secondary flow and meanwhile, the secondary flow leads the exchange of kinetic energy with energy of the bending pipe. The pressure of fluid flow from section 1 to section 2 gradually tends to uniform distribution.

171

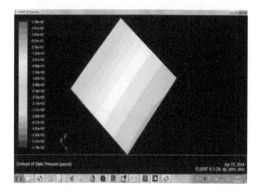

Figure 8. The diagram of pressure of section 2.

6 CONCLUSIONS

1) The astringent of standard k-ε model which is adopted during the process of calculation is very poor and a standard dissipation rate equation cannot reach the suitable scale length case. However, compared with other turbulence models, the model can achieve better simulation results.

2) Through the simulation of rectangular bending pipes with large diameters, we can gather experience of simulation of fluid flow in complicated bending pipes. The simulation can provides the bases for ultrasonic flowmeters to measure outlet flow and find the best location to install ultrasonic sensors.

3) FLUENT6.3 is one of the most fashionable software programs. Compared with other software such as ICEM, FLUENT6.3 can be more widely used and it has better stability and more advantageous simulation.

REFERENCES

Gui Keting, Wang Jun, Wang Qiuying. 2011. Engineering Fluid Mechanics. Beijing. Science Press.

Jiang Shan, Zhang Jingwei, Wu Chongjian. 2008. Based on FLUENT 90° elbow internal circular flow field. China Ship Research. 3(1):37–41.

Yao Chaohui, Zhou Qiang. 2012. CFD entry. Beijing: Qinghua University Press.

Zhang Deliang. 2010. CFD Tutorial. Beijing: Higher Education Press.

Zhou Junbo, Liu yang. 2012. FLUENT6.3 flow field analysis from entry to the master. Beijing: Machinery Industry Press.

Electronics, Information Technology and Intellectualization – Song & Kwak (Eds)
© 2015 Taylor & Francis Group, London, ISBN 978-1-138-02741-1

A new fast gas recognition algorithm based on the ion mobility spectrum

Bo Liu, Yanwei Jiang, Jiang Zhao, Genwei Zhang & Jie Yang
Research Institute of Chemical Defense, Beijing, China

ABSTRACT: Ion mobility spectrometry is one of the most widely used technologies for trace chemicals detection. After briefly introducing the principle of the ion mobility spectrum technology, this paper presents a new fast gas recognition algorithm according to induce the resilient BP neural network method. The experiments are completed. The results show that under the condition of 0.001 training error and 0.20 training speed, the resilient BP neural network generated by the proposed algorithm has a higher identification rate in different threshold conditions. When the training speed is 0.30, the iteration number and training time to achieve the expected training error are less.

1 INTRODUCTION

Ion Mobility Spectrometry (IMS) is one of the most widely used technologies for trace chemicals detection. With the advantage of wide detection species, high analytical sensitivity, short response time and other prominent advantages, IMS is widely used in the field of drugs, explosives and chemical agent detection (Eiceman 1994, Eiceman 2004). Meanwhile, IMS has great application prospect in many fields, such as expert testimony, environmental monitoring, determination of the freshness of meat food, clinical medical diagnosis, analysis of bacterium and macromolecule (Krebs 2005, Snyder et al 1991).

Commonly the opening time of an ion gate in a classic IMS detector is 0.2 ms. Compared with the analytical period 20~40 ms, the coefficient of ion utilization is below 1%. Although it will improve the coefficient of ion utilization by means of increasing the opening time of the ion gate, the detecting resolution may decrease disastrously. The sample quantity to be detected is between pg and ng. The amplitude of current output by drift tube is approximately 0.01~0.4nA. In this condition, the useful signals are rather weaker, and are easy to submerge by lots of the noises. Therefore, the signal process is indispensable to IMS detectors.

IMS signal procedure is usually composed of two steps. The first is signal modulate by the IMS hardware. The weak current signals are amplified 1,012 times. For the process of ion mobility is repeatable, the signal noise ratio could be significantly improved according to multiply acquisition and averaging. The second step is signal process by the IMS software. Using signal process algorithms, fast recognization of the chemical warfare and industrial toxic and harmful gas can be realized.

The IMS signal process is quite complex in common conditions, which consists of a data smoothing filter, evaluating the noise level, adjusting the baseline, estimating the peak value and agent type identification. Due to the reduced ion mobility instability caused by the instrument drift, zero shift, temperature and pressure sensor errors, environmental disturbance, etc., false alarming and missing alarming of the IMS detectors occur. In this paper, the resilient BP neural network algorithms are introduced for rapid identification of chemical warfare agents, and a new method for rapid identification of a chemical warfare agent algorithm is proposed.

2 PRINCIPLE OF IMS

In a classic IMS detector, as Figure 1 shows, a drift tube is composed of the ionization zone, the reaction zone, the ion gate, the drift zone and detection electrode. The operational principle of IMS detectors is as follows: The measured sample gas driven by the carrier gas gets into the ionization reaction zone of the drift tube. The ionization gas builds up in the front of the ion gate. When the ion gate opens, the ions enter the drift zone and drift under the action of an electric field force. Supposing the length of the drift zone is L(cm), then the electric field intensity is E(V/cm). Under the action of the electric field force, the ions with the characteristic of ion mobility ratio K reach the Faraday pam after a time span t_d(ms). Therefore, a weak ion current signal emerges. The IMS data is stored into SDRAM after amplifying, filtering and acquisition. By means of the signal smoothing filter, the peak extraction and the recognition algorithms, the type of samples are identified fast and accurately, and the concentration of the sample is estimated approximately.

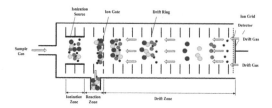

Figure 1.   Principle of IMS drifts tube.

The brief calculation process of the reduced ion mobility ratio is:

$$V_d = L / t_d \tag{1}$$
$$K = V_d / E \tag{2}$$
$$E = V_h / L \tag{3}$$
$$K_0 = K(273/T)(p/760) \tag{4}$$

In the equations, T is the absolute temperature (K), p is the pressure of the gas passed by the ion group (torr or mmHg), V_d is the speed of ion mobility (cm/s), V_h is the load voltage of the drift zone (V). The temperature of the drift tube T is measured detection circuit. The pressure p is gained by the drift tube pressure monitoring circuit.

3   PREPROCESSING OF IMS SIGNALS

3.1   Filtering algorithm

An IMS signal filtering algorithm can be divided into spatial filtering and time filtering algorithms. To reduce the difference of the reactive ion peak positions in different IMS spectrum periods, the spatial filtering algorithms fulfil multiple IMS spectrum acquisition as the repeatability of the process of the ion mobility, and deal with the acquisition points in the different periods but with the same drift time. Time smoothing algorithms are aiming at the adjacent acquisition points in the same period, and can efficiently decrease the influence of the zero mean noises and the periodic interference signals. A time filtering algorithm can be regarded as the procedure, when a fixed length L window slides along the time axis, the system outputs a mean value in each slide cycle. The principle of the algorithm is that firstly, a length of L time window is determinate, where L is odd. Secondly, sort the data in the window and remove the maximum and minimum of two data. Average the remaining data in the window, and replace the original data points in the centre of the window with the average value. Then, move the window right by one data point, and repeat the above average process until the traversal of all data points.

3.2   Estimation of noise level

The approach of estimation of noise level is that, (1) select N points in equal distance from the spectrum subtracted by the background noises, whether the points are in the peak or not; (2) calculate the standard average μ_1 and the average standard windage σ_1; (3) remove the points whose intensity is bigger than $\mu_1 + 3\sigma_1$ from the N points group, and the remaining N_1 points; (4) calculate the average μ_2 and the average standard windage σ_2 of the remaining points; (5) subtract the points whose intensity is bigger than $\mu_2 + 3\sigma_2$, and the remaining N_2 points; (6) Compare the former and the latter μ, σ. If the difference is remarkably large, return to step (2) and circularly calculate the data. After several times computing, when the former, μ, σ is nearly equal with the latter, the final, μ, σ is achieved. Select a proper noise level, such as $3\sigma_n, 5\sigma_n$.

3.3   Baseline adjustment

The approach towards eliminating the baseline excursion is commonly composed of a clinical correction method, a parabolic fitting method and a median filtering method based on a moving window. The clinical correction method is effective in that the baseline drift is light. The parabolic fitting method is complex and its real-time is poor. In the example below, estimating and extracting the baseline drift by means of the median filtering method are briefly introduced.

Supposing the original IMS signal to be dealt with is $sig1$ and the length of the signals is L. The realization of the algorithm is as follows:

(1)   Select the appropriate window width of W, in order to facilitate the processing, general W odd.
(2)   To avoid appearing edge effect, the ends of the original signal $sig1$ are continued, and the signal $sig2$ generates. The length of the signal $sig2$ is $L + W - 1$.
(3)   Setup the window for the signal sig2: the signal in the window is filtered by the median filter. That is, to sort the signal in the window, and then use the median instead of the point in the centre of the window. Move the window, traverse the signal sig2, and fit the baseline drift BL.
(4)   The original signal $sig1$ is subtracted from BL, and we get rid of the signal $sig3$.

3.4   Peak evaluation

From the equation (4), it is known that the reduced ion mobility ratio is mainly related to the temperature of the drift, the pressure of the drift, the electric field intensity and the drift time. The three parameters of the drift tube temperature, the drift tube pressure and the electric field intensity, are directly associated with the working states of the drift tube. Its value is relatively simple, and does not change with the interfered noise. Therefore, the reduced ion mobility ratio is only decided by the drift time. Accurately measuring its value becomes very important, which is commonly referred to as the "peak".

In order to accurately measure the peak of the ion mobility spectrum, it is necessary that the original signal will be dealt with average smoothing and filtering. The dealt IMS signals nearly include the sharp pulse noise and high frequency noise, which lay the foundation for the detection of the peak. A common method to seek peak position is the derivative method. We know that the curve derivative of extreme points are zero, the derivative of the inflexion points are extremum (maximum or minimum). After derivation, the zero points are changed into the extreme points, and the extreme points by the inflection points are changed into the zero points. According to the theorem, the ion peak position can then be accurately evaluated. The first derivatives are zero points, determined as the "peaks" or "valleys", and the minimal two derivative values are determined as the" summit".

4 RECOGNITION ALGORITHM

4.1 Deficiency of BP neural network

Though the BP neural network algorithm is widely used, it has limitations and shortcomings. The shortages mainly consist of the slow convergence speed, requiring longer training time; it is easy to fall into local minima. Aiming at the problems of the BP algorithm, the scholars at home and abroad have done in-depth research, and proposed many improved algorithms. The improvements of the BP algorithm have two options. One is to use heuristic learning methods, such as the additional momentum method and the adaptive learning rate method; the other is to use numerical optimization methods, such as the conjugate gradient method, Newton method, Levenberg-Marquardt method and the resilient BP method (Sussmann 1992, Hornik 1991, Shen 2006).

The resilient BP method is to eliminate the adverse effect of the gradient magnitude. Therefore, it only uses the partial derivatives of the symbol in the revision of the weight, but the amplitude, which depends on the correction independent of the amplitude, does not affect the correction weights.

4.2 Structure design of resilient BP neural network

In the detection process of ion mobility spectrometry, besides the relatively fixed factors of the design of the sensor detection system hardware module performances, the main factors affecting the sample type identification includes the drift tube temperature, the pressure, the voltage of the tube drift region, the gas flow ratio and the ambient humidity; the above factors indirectly have an effect on the sample type recognition function according to the output signal of ion mobility spectrometry. Therefore, the output signal of ion mobility spectrometer must be selected in order to build the input layer of the neural network. The gate cycle of the ion mobility spectrometer signal processing module is 30ms, meaning each 30 ms, the module

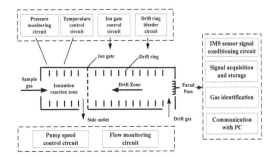

Figure 2. Overall block diagram of an ion mobility spectrometer.

outputs a set of independent and complete mobility spectrum signals. The acquisition chip, gets 1,024 independent discrete data each 30 ms, so the neurons number of the neural network input layer is 1,024. The output layer of the neural network is the sample type, the number of neurons is 1.

Gas type identification is divided into two processes. The first step is resilient BP neural network training. It is necessary to acquire a large number of experimental samples of different gas input vectors and output vectors under known detection conditions. Train the resilient BP neural network. Get the nonlinear mapping relationship between the input vector and output vector according to modify the neural network weights. The second step is the recognition of unknown types of gas. According to detect an unknown gas, we gain multiple groups of ion mobility spectrometry signals. Using the first step of the trained neural network, it is possible to identify the gas type.

5 EXPERIMENT AND RESULT

Figure 2 is the overall block diagram of an ion mobility spectrometer. The hardware mainly includes the pressure monitoring circuit, the temperature control circuit of the drift tube, the gate control circuit, the high voltage power supply circuit of the drift ring, the pump speed control circuit, the flow monitoring circuit, the IMS sensor signal modulated circuit, the signal acquisition and data storage circuit, the main control circuit, the display circuit, the key scanning circuit and the communication circuit. The software includes the ARM main control board memory allocation and management algorithm, the signal acquisition and storage algorithm, the signal processing algorithm, the temperature and pressure flow sensor control algorithm, the display and key control algorithm and the communication algorithm. Signal processing algorithms include data filtering, noise level estimation, baseline correction and peak extraction algorithms.

Using an ion mobility spectrum instrument shown in Figure 2, the experiments on toluene, acetone and air are carried out. We achieve 100 groups of ion mobility spectrum signals. Based on these signals, 100 experimental swatches are constructed. 70% of data samples

Table 1. Gas recognition of the resilient BP neural network.

Training speed	Iteration number	Training time (s)	Recognition rate (%) <0.10	<0.15	<0.20
0.05	452	21.4	73.3	90.0	100
0.10	413	19.8	76.7	93.3	100
0.15	695	33.3	80	86.7	100
0.20	568	27.1	96.7	100	100
0.25	760	36.5	76.7	86.7	100
0.30	399	18.9	86.7	100	100
0.35	725	34.6	90.0	100	100
0.40	552	26.4	53.3	90.0	100

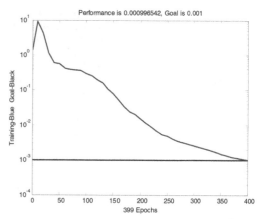

Figure 4. Training procedure of the resilient BP neural network.

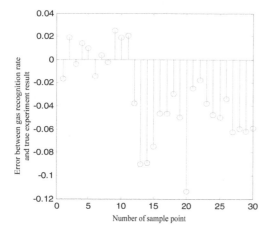

Figure 3. Gas recognition result of the resilient BP neural network.

are used to train the resilient BP neural network, the remainder are used to verify the network. In the experiment the numbers of the input layer, connotative layer and output layer are respectively 1,024, and 10, 1. The values that toluene, acetone and unknown gas correspond to in the output layer are respectively 1, 2, and 0. Table 1 is the iteration number of the resilient BP neural network, the training time and the sample recognition ratio.

By analysing Table 1, it is confirmed that under the condition of 0.001 training error and 0.20 training speed, the resilient BP neural network generated by the proposed algorithm has a higher identification rate in different threshold conditions, as shown in Figure 3. When the training speed is 0.30, the iteration number and training time to achieve the expected training error are less, as shown in Figure 4.

completed. The results show that under the condition of 0.001 training error and 0.20 training speed, the resilient BP neural network generated by the proposed algorithm has a higher identification rate in different threshold conditions. When the training speed is 0.30, the iteration number and training time required to achieve the expected training error, are less.

REFERENCES

Eiceman, G. A. & Karpas, Z. (1994). Ion Mobility Spectrometry: 5–21. Boca Raton, CRC Press.
Eiceman, G.A. & Stone, J.A. (2004). Ion mobility spectrometers in national defence 76(21): 390–397. American Chemical Society, Analytic Chemistry.
Krebs, M. D., Zapata, A. M., Nazarov, E.G et al. (2005). Detection of biological and chemical agents using differential mobility speetrometry (DMS) technology 5(4): 696–703. IEEE Sensors Council, IEEE Sensors Journal.
Snyder, A.P., Shoff, D.B., Eiceman, G.A et al. (1991). Detection of bacteria by ion mobility spectrometry 63(5): 526–529. American Chemical Society, Analytic Chemistry.
Sussmann, H. J. (1992). Uniqueness of the Weights for Minimal Feed Forward Nets with a Given Input-Output Map 5(3): 589–593. Neural Networks.
Hornik. K. (1991). Approximation Capabilities of Multilayer Feed Forward Networks 4(2): 251–257. Neural Networks.
Shen, Z. H. & Liu, F. (2006). Applying Improved BP Neural Network in Underwater Targets Recognition: 2588–2592, July. Vancouver, BC, Canada, International Joint Conference on Neural Networks.

6 CONCLUSION

After briefly introducing the principle of ion mobility spectrum technology, this paper presents a new fast gas recognition algorithm in order to induce the resilient BP neural network method. Under the experiments are

Electronics, Information Technology and Intellectualization – Song & Kwak (Eds)
© 2015 Taylor & Francis Group, London, ISBN 978-1-138-02741-1

An improved medical imaging denoising method based on independent component analysis

Hao Wang & Zhiqiang Li
Department of Interventional Radiology, Affiliated Hospital of Hebei University, Hebei, China

Xiaodong Kang & Wenyong Yu
College of Medical Imaging, Tianjin Medical University, Tianjin, China

ABSTRACT: This paper proposes an improved medical imaging denoising method based on independent component analysis. When the image is different from the same noise variance by two different sources of pollution, the ICA in the separation process can obtain two new hybrid images. Then the separation matrix, an original multiplication image and a noise image are separated from the mixed image. Experiments show that this algorithm is more effective compared with other methods obtained by the PSNR and RMSE values.

1 INTRODUCTION

Medical images may become distorted due to the noise in the reconstruction or the transmission and storage process, resulting in degradation of the image quality, thereby influencing the clinical diagnosis and treatment. One purpose of medical image restoration is to try to reduce or eliminate the noise in the image distortion and to get a high quality image, in order to do further processing. Currently, Independent Component Analysis (ICA) as a new signal processing method, has very important applications in image denoising and image feature extraction etc.

Independent Component Analysis (ICA) is an important branch of blind signal processing (BSP), which originated in the famous "Cocktail Party Problem".

According to the differences in the assumptions and approaches, BSP can be divided into Independent Component Analysis (ICA), Factor Analysis (FA), Independent Factor Analysis (IFA), Principal Component Analysis (PCA), and so on.

An ICA algorithm was first proposed by the French, J. Herault and C.Jutten in the mid-80s, called the "H-J algorithm".

In 1989, the 1st International Conference on Higher Order Spectral Analysis, by J.F.Cardoso and P. Comon (ICA's early papers) was published. J.F.Cardoso proposed a based cumulate algebraic method that eventually formed the JADE algorithm. P. Comon then systematically expounded ICA issues and put forward the concept of independent component analysis based on minimum mutual information of the objective function. During the supremacy of the mid-1990s, A. J. Bell and T. J. Sejnowski proposed based on the principle of maximizing the information InfoMax method; from ICA has been widespread concern. In applications, E. Oja and A. Hyvarinen and others, proposed an ICA based on image feature extraction and noise reduction methods, whilst L.K.Hansen made a blind separation method of noisy images.

The traditional method of image denoising is where the image data and noise data are treated equally, that image and noise in a transform domain distributed in the interval range of different noise filtering, the interval set information to remove noise.

Although these methods have achieved certain results, because the image and noise are usually not of a completely different range in a domain, but do overlap together, this method has two aspects: one is to filter the noise which cannot be set outside the interval; two, is to simultaneously erase the image information of the interval within the. Denoising method based on the ICA image in two ways. One way is to start from the traditional denoising methods; the noise image to the ICA domain, for each independent component is a method for removing noise. Another method is to proceed from the angle of separation, and believes that the noise and the image data are independent of each other. The contaminated image as source image and the noise and mixed noise, trying to from contaminated to realize image denoising is separated in the image, and the difficulty and the key to the second method is how to select the virtual observation map as input for the blind separation algorithm.

This paper presents an adaptive image denoising method based on ICA. This method can obtain a very high peak signal-to-noise ratio, noise pollution is serious in the image and the reference image noise can be very good in order to restore the original image.

2 ICA MODEL

2.1 *ICA basic model*

From a mathematical point of view, ICA is a kind of multivariate data of non-orthogonal linear transformation method; its main purpose is to determine a linear switching matrix, so that the output component is transformed as statistically independent. ICA can also be seen as an extension of PCA. PCA only requires that the decomposition of the components are orthogonal to each other, i.e. uncorrelated, it only takes into account the two order statistics signals; each component and ICA decomposition not only each other, and are statistically independent, they also consider the high order statistical property of the signal. So ICA has a better separation effect.

The instantaneous linear mixing process, provided that there are N independent source signals, is expressed in vector form as: $s(t) = [s_1(t), s_2(t), \ldots, s_N(t)]^T$. T is the transpose of the vector, $t = 0, 1, 2, \ldots$, and M observed signals $x(t) = [x_1(t), x_2(t), \ldots, x_M(t)]^T$. M signal is composed of N source signal linear instantaneous mixed, where each time t has:

$$x_i(t) = \sum_{j=1}^{N} a_{ij} N_j(t), \quad i = 1, 2, \ldots, M, \tag{1}$$

In vector matrix form:

$$x(t) = As(t), \tag{2}$$

where A is the mixing matrix composed of mixing coefficient $\{a_{ij}\}$, $s(t)$ of source signals and mixing matrix A is unknown, only the mixed signal $x(t)$ can be observed. The goal of ICA is to obtain a separation matrix W, and through the W from the observed signal $x(t)$ to restore the source signal $s(t)$.

In practical application, it was always observed that signals contains some noise, where a noise and the source signals are independent of each other, and the noise is Gauss white noise. The ICA noise model is:

$$x(t) = As(t) + n \tag{3}$$

where $n = (n_1, n_2, \ldots, n_M)^T$ is the noise signal vector, the covariance matrix is $\delta^2 I$, δ is noise variance, I is the unit matrix.

ICA is actually an optimization problem, usually through whitening pre-processing; establishing the objective function and the optimization objective function is accomplished. The whitening pre-processing often removes correlation by PCA method. The objective function or to separate the matrix W is due to the objective function of $L(W)$ variables, it reflects the independence of the output random variable y between each component of the target function, commonly used with high order cumulant, maximum entropy, mutual information minimization and the maximum likelihood estimation, and the optimization objective function can be realized with the help of the natural gradient algorithm and the fixed-point algorithm.

The goal of ICA is through the $y = wx$ transform; by solving the source signals from the observed signals, we need to set a target function of $L(w)$, where the $L(w)$ reached maximum, $\bar{w}$ is the solution, so that the y and the s correspond. In practical applications a negative entropy is often used based on information theory as a non-Gaussian target function to describe the signal. The negative entropy of random variable x is defined in $J(x)$:

$$J(x) = H_c(x) = \int p(x) \log \frac{p(x)}{p_c(x)} dx \tag{4}$$

where $p_c(x)$, with the distribution function of the random variable x, has the same mean and variance of the Gaussian random variable. When x is a Gaussian distribution, the negative entropy is 0; when x is another non-Gaussian distribution, the negative entropy is greater than 0. In addition, all reversible linear transformations, the negative entropy values remain constant.

It is based on the approximate calculation of the maximum entropy principle of negative entropy:

$$J(x) \approx [E\{G(x)\} - E\{G(y)\}]^2 \tag{5}$$

where y is a standard Gaussian random variable, G is a form of non-quadratic function, the general election even symmetric functions with convex.

2.2 *Fast ICA model*

In order to solve the optimization problems of the slow convergence speed of the target function, A.Hyvărinen et al.proposed a fast fixed-point algorithm based on the kurtosis of the objective function, called" Fast ICA".

To solve $y = wx$ projection direction, which is the;

$$J_c(w) = [E\{G(w^T x)\} - E\{G(y)\}]^2 \tag{6}$$

where w is the m-dimensional variable.

Since y and x are Gaussian variables with the same mean and covariance matrix, the maximization problem can be transformed into $E\{G(w^T x)\}$ optimization problems. By Kuhn-Tucker conditions, $E\{G(w^T x)\}$ under conditions of $E\{G(w^T x)\} = ||w|| = 1$ were obtained by solving the following equation:

$$E\{xg(w^T x)\} - \beta w = 0 \tag{7}$$

where β is constant, w_0 initial value can be obtained: $\beta = E\{w_0^T xg(w_0^T x)\}$. $g(\cdot)$ is $G(\cdot)$ derivative.

To solve Equation (7) w, the objective function $F(w) = E\{xg(w^T x)\} - \beta w$, using the Newton iterative method for the optimal iterative:

$$w_{k+1} = w_k - \frac{E\{xg(\mathbf{w}_k^T x)\} - \beta \mathbf{w}_k}{E\{g'(\mathbf{w}_k^T x)\} - \beta} \tag{8}$$

178

where $\beta = E\{w_k^T xg(w^T x)\}$. After each iteration, w_{k+1} is normalized, $w_{k+1} = \frac{w_{k+1}}{||w_{k+1}||}$.

By multiplying both sides for $\beta - E\{g, (w^T x)\}$ on equation (8), the algorithm obtained is an iterative equation:

$$w_{k+1} = E\{xg(w_k^T x)\} - E\{g'(w_k^T x)\}w_k \qquad (9)$$

Because of fast convergence Fast ICA, which is the mainstream of the ICA, an algorithm is used.

3 IMPROVED IMAGE DENOISING ALGORITHM BASED ON ICA

Consider that a separate component of the objective function does not require a priori knowledge, and is easy to implement a universal numerical calculation, which is expressed as:

$$J(y) = |E_y[G(y)] - E_v[G(v)]|^p \qquad (10)$$

where G is a non-quadratic function, and is sufficiently smooth, v is the standard Gaussian random variable, $p = \log 2$. Non-quadratic function Gs often take the form:

$$G_1(u) = logcosha_1u \quad \text{or}$$
$$G_2(u) = \exp(-\frac{a_2 u^2}{2}) \qquad (11)$$

where for sup-Gaussian variables, take G_1; for sub-Gaussian variables take G_2.

Observational data were pre-processed once Fast ICA iteration, iterative process of:

$$W(k) = E\{xg[w(k-1)^T x]\} - \\ E\{g'[w(k-1)^T x]\}w(k-1) \qquad (12)$$

In each iteration, the new w will be normalized, $w(k) = \frac{w(k)}{||w(k)||}$. Then, the function g is the derivative of the objective function Equation (11) generic non-quadratic function.

4 ALGORITHM DESIGN

By setting an image by two different pollutions of the same noise source, when the noise variance is not at the same time, if the original image is S, the reference noise image as the variance is equal to 1, the Gauss white noise image of n, two mixed noise variances are δ_1 and δ_2, then the mixing process can be expressed as:

$$\begin{cases} x_1 = \text{round}(S + \delta_1 \mathbf{n}) \\ x_2 = \text{round}(S + \delta_2 \mathbf{n}) \end{cases} \qquad (13)$$

where round $(\cdot)$ is rounded off symbol, while the conversion of less than 0 is 0, the conversion of is greater than 255 to 255.

The image noise is mixed by x_1, and x_2 is converted to a column vector of the row vector $x_1 = [x_1(1), x_1(2), \ldots, x_1(N)]^T$ and $x_2 = [x_2(1), x_2(2), \ldots, x_2(N)]^T$, where N is the number of pixels in the image, in the gray level of the pixel, 255 is equal to the position of a process:

$$\begin{cases} p(i) = 1 & x_1(i) = 255 \text{ or } x_2(i) = 255 \\ p(i) = 0 & \text{otherwise} \end{cases} i = 1, 2, \cdots, N \qquad (14)$$

x_1 and x_2 of the vector to make the filtering process, with $p(i) = 0$ retention element corresponding to obtain a new vector x_1' and x_2'. Assuming $\sum_{i=1}^{N} p(i) = M$ then the column vector x_1' and dimension x_2' of $(N - M)$ dimension.

The Fast ICA algorithm was applied to x_1' and x_2' separation to obtain independent components y_1', y_2' and separation matrix W. That is, the source signal can be estimated to be:

$$[z_1, \ z_2]^T = W[x_1, \ x_2]^T \qquad (15)$$

Because z_1 and z_2 in order and in the amplitude is uncertain, using Equation (13) to calculate $norm(z_1)$, $norm(-z_1)$, $norm(z_2)$ and $norm(-z_2)$, the correlation coefficient between the noise image containing noise variance is smaller. Here, $norm(\cdot)$ represents the vector in 2D gray images, and the gray level normalization for $0 \sim 255$. Finally, the maximum correlation coefficient is estimated for image noise images.

5 EXPERIMENTAL RESULTS AND CONCLUSION

To verify the effectiveness of the method respectively, an 8 bit, 256×256 grayscale 'Lena' image was used with a size of 512×512 and 256 gray standard medical imaging of head CT images as test samples, and experiments were conducted comparing the algorithm

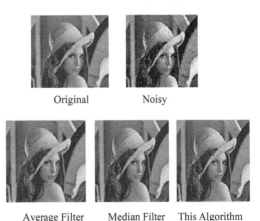

Original   Noisy

Average Filter   Median Filter   This Algorithm

Figure 1. Lena image denoising processing by different algorithms.

179

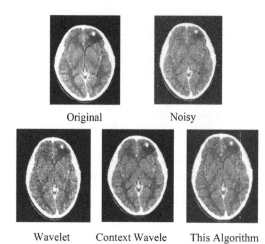

Original   Noisy

Wavelet   Context Wavele   This Algorithm

Figure 2.   CT denoising processing by a different algorithm.

Table 1.   Comparison of Lena images with a different denoising algorithm of RMSE & PSNR.

Model	RMSE	PSNR dB
Average Filter	7.25	30.92
Median Filter	8.50	29.55
This Algorithm	6.50	31.88

Table 2.   Comparison of CT images with a different denoising algorithm of RMSE & PSNR

Model	RMSE	PSNR dB
Wavelet	11.36	27.02
Context Wavelet	13.94	25.24
This Algorithm	8.73	29.31

and application experiments. The experiments were carried out using Dell E520, 2GB RAM, Windows XP, MATLAB 7.4 environment.

In the original Lena image and CT head images respectively, Gauss white noise was added as disturbance, with a different algorithm and the algorithm for the denoising processing. The evaluation index by using RMSE and PSNR, shows the denoising effects in Figures 1 and 2; the experimental results are shown in Tables 1 and 2.

From the experimental results it can be seen that the proposed algorithm on the RMSE and PSNR values is a very good optimization. From the subjective visual point of view, this algorithm to save the image in detail edges have a greater of improvement.

REFERENCES

A. Hyvärinen, P.O. Hoyer, E. Oja. (2002). Image denoising by sparse code shrinkage. *C. Intelligent Signal Processing.* 259–268
C. Jutten, J. Herault. (1991). Blind separation of sources, part I: An adaptive algorithm based on neuromimetic architecture. *J. Signal Processing. 24(1)*, 1–10
L.K. Hansen. (2000). Blind separation of noisy image mixtures. *C. In Advances in Independent Component Analysis,* 161–181
P. Gruber, K. Stadltharner, A.M. Tome, *et al.* (2004). Denoising using local ICA and a generalized eigendecomposition with time-delayed signals. *J. Independent Component Analysis and Blind Signal Separation. 3195*, 993–1000
H. Park, S. Oh, S. Lee. (2002). Adaptive noise canceling based on independent component analysis. *J. Electronics Letters. 38(15)*, 832–833
M.X. Wang. (2005). Independent Component Analysis and Applied Research in Image Processing. *D. Shanghai University*
W. Guo, P. Zhang, R.S. Wang (2008). Independent Component Analysis and its Application in Image Processing. *J. Computer engineering and Applications. 44(23)*, 172–177
W. Guo, C.R. Zhu (2008). An Improved Fast ICA Algorithm and Application. *J. Computer Applications. 28(4)*, 960–962

Electronics, Information Technology and Intellectualization – Song & Kwak (Eds)
© 2015 Taylor & Francis Group, London, ISBN 978-1-138-02741-1

Design of a relay protection algorithm based on Matlab/Simulink and FPGA

Zhijuan Qu, Yuxiang Yuan, Pengfei Hu & Huihui Lu
State Grid Smart Grid Research Institute, Beijing, China

ABSTRACT: Currently in relay protection devices, FPGA is widely used to improve the security and stability of the power system. However, there are difficulties in voltage and current signals' generation and real-time monitoring during the FPGA development. According to the analysis of the common relay protection device's architecture, this paper proposes a new method to design a relay protection algorithm based on Matlab/Simulink tools and FPGA platforms. This method will have greater flexibility and shorten the relay protection device's development cycle.

Keywords: Relay protection algorithm, FPGA, Matlab/Simulink, real-time simulation

1 INTRODUCTION

Due to the development of the smart grid, the power system is becoming more and more complicated, so that relay protection devices have been expected to improve performance to ensure the power system's security and stability[1]. However, most relay protection devices are based on a single CPU, using plug-in architecture, which brings difficulties to the debugging and little room for performance improvement. In this case, mainstream manufacturers and research institutes have launched the research of relay protection, mainly based on Power PC+DSP, ARM+DSP, ARM+FPGA system architectures and so on.

In view of the flexible programming of FPGA and the high cost-effectiveness of ARM, the ARM+FPGA system architecture is adopted. However, there are difficulties in voltage and current signals' generation and real-time monitoring, which results in FPGA-based relay protection algorithms that are not easy to design and verify[1].

In this paper, we propose a new method to design relay protection algorithms, which can simplify the voltage and current signal's incentive mechanism and monitor the system in real-time, by combining Matlab with FPGA. This method uses a power system simulation model constructed by the Matlab/Simulink to calculate voltage signals and current signals, and then output them to FPGA in real-time through the Ethernet port. After receiving these signals, the relay protection module on FPGA processes the data, and then decides whether to activate the relay. This approach enhances the reliability, flexibility and portability of

the relay protection device, and greatly shortens the product development cycle.

2 ARCHITECTURE OF FPGA-BASED RELAY PROTECTION

In general, relay protection devices are composed of the data acquisition system, the MCU main system, the switch input and output system, the human-computer interaction system and the communication system[2,3]. Among them, the MCU main system includes a microprocessor, a program memory, data memory, timers, etc., that can complete the operating system, the relay protection algorithms and scheduling control. Normally, the relay protection algorithm is realized by software programming in C language. This system's architecture is shown in Figure 1.

In FPGA-based relay protection devices, the MCU, using an embedded operating system, could realize scheduling control and make GUI development easier. The FPGA carries out the functions of data

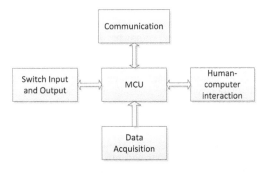

Figure 1. Common protection device architecture.

[1] Project supported by science and technology project of State Grid (SGRI-WD-71-13-010, SGRI-WD-71-13-014, SGRI-WD-71-13-008, SGRI-WD-71-13-011)

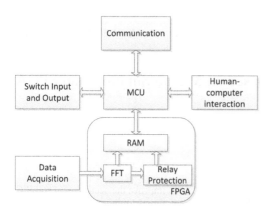

Figure 2. Relay protection device architecture based on FPGA.

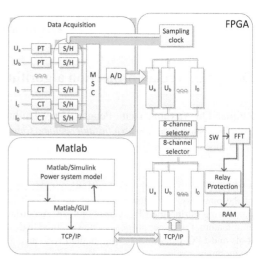

Figure 3. The architecture of the relay protection algorithm based on Matlab and FPGA.

sampling control, FFT and hardware protection algorithms. Between the MCU and FPGA, we define a standard interface that makes the FPGA to be an independent relay protection IC. The modular architecture allows hardware, functional, low coupling and easy to extend debug. This system architecture is shown in Figure 2.

3 COMMUNICATION BETWEEN MATLAB AND FPGA

The communication between Matlab and FPGA can be realized by joining Matlab API, Matlab/Simulink and Matlab/GUI.

Matlab is a kind of software for algorithm design, data visualization and data analysis, which has a high-tech programming language and an interactive programming environment. Matlab/Simulink is a platform for designing and simulating dynamic system models. It also provides an interactive graphical environment and a custom module library, as well as application specific extensions[4].

From the Matlab/Simulink 4.1 version, the Matlab/Simulink contains the Power System Blockset. It is mainly developed by the Canadian Hydro Quebec and TECSIM International Company. In Matlab/Simulink, we can use this library to build transmission lines and a power equipment simulation model to analyse the principle and performance of the system[5,6,7].

Matlab/GUI can complete the data exchange between Matlab/Simulink and Matlab, which is a graphical interface development module, built in Matlab. We can set the input and output of the Simulink simulation model by programming in Matlab/GUI. Therefore, the output of the power system simulation model based on Matlab/Simulink can be passed to Matlab/GUI.

Matlab Application Program Interface (API) has a lot of interface functions, and provides a series of functions for the TCP/IP port, e.g. to open, to close, to set parameters, and so on.

Therefore, Matlab can be programmed to send data through the Ethernet port to FPGA, and FPGA can receive data via the Ethernet port. Then FPGA processes signals and runs relay protection-related modules.

4 SYSTEM DESIGN

Through research of Matlab and FPGA communication, we propose a new relay protection algorithm design system. It uses Matlab/Simulink to construct a simulation model of the power system and calculate the signals in the model. Then Matlab sends the signals to FPGA in real-time through the Ethernet port.port. Finally, the relay protection module on FPGA processes the signals and decides whether to activate the relay. The architecture is shown in Figure 3.

In this paper, we use the distribution network transmission line as an example to do modelling, simulating, and verifying.

4.1 Fault simulation model

In Matlab/Simulink, the SimPowerSystem library offers a variety of power system components, including all the elements of the transmission line model[8]. For example, the transmission line's A phase grounding fault, as shown in Figure 4, is realized by existing elements in the SimPowerSystem library.

In this model, the voltage is 10 kV, the frequency is 50 Hz, the length of the A segment of the transmission line is 20 km, and the length of the B segment of the transmission line is 100 km, using a breaker to achieve the A phase grounding fault. A phase grounding fault occurs at 0.1 seconds after the start of the simulation, and a phase to ground fault eliminates at 0.15 seconds after the start of the simulation.

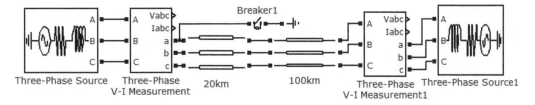

Figure 4.   The ground fault model of the transmission line.

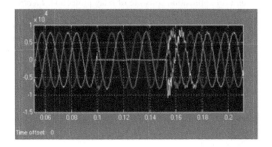

Figure 5.   Three phase voltage waveform.

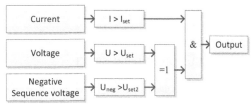

Figure 6.   Complex voltage overcurrent.

4.2   *Matlab/GUI real-time control Matlab/Simulink model*

There are several methods to realize the Matlab/GUI control and the Matlab/Simulink model in real-time: a) using the set_param() function to set the parameters; b) using the simset() function provided by Simulink to set the workspace of the Simulink model as the current workspace of Matlab.

In this paper, the set_param() function method is used to control the input and output of the Simulink model. The simulation results can be displayed in real-time to the GUI. After getting data from the transmission line model shown in Figure 4, GUI can display the voltage signal waveform. It is shown in Figure 5, wherein, A phase is yellow, B phase is purple, and C phase is cyan. As shown in Figure 5, at 0.1 seconds department, A phase occurs a ground fault, and A phase voltage declines to 0 volts quickly. At the same time, the B phase and C phase voltage is not affected. At 0.15 seconds department, the A phase ground fault is eliminated, and A phase, B phase and C phase voltage return to normal after the transient oscillations.

4.3   *Design communication of Matlab and FPGA*

Usually, network interfaces and serial ports are used to connect Matlab with FPGA. Considering that the highest rate for the serial port is 115200 bit/s, the speed is slower than the rate of the A/D module. Therefore, we choose to complete Matlab and FPGA communication through the network interface.

Matlab has a large number of APIs, providing the network interface to open and close, and a series of functions, such as network interface parameter settings operation. In Matlab, the relevant functions are used to create a TCP/IP module for receiving and sending data.

On the FPGA, TCP/IP protocol adopts a layer-based structure, being divided into the application layer, the transport layer, the network layer interconnect and the network interface layer. The system is divided into two parts. The first part is the physical layer and the MAC layer. The second part is the transport layer and the network layer, which is mainly realized by the software code.

Based on the above steps, the fault voltage signals and current signals are generated by Matlab, and then they are received by FPGA.

4.4   *Relay protection algorithm based on FPGA*

After receiving data from Matlab, the FFT module is run first to compute the amplitude of the input signal's fundamental wave and harmonics[9]. Usually, the fundamental wave is used for relay protection algorithms and electric energy metering; harmonics are used for power quality analysis. FFT is realized by Altera's FFT IP core.

In the hardware relay protection algorithm module, the results of FFT and the setting value are compared. According to the results of comparison, the action values are output in RAM. By scheduling control, MCU will control whether or not to send out a tripping signal.

For a large capacity of FPGA, the hardware relay protection algorithm module can contain a variety of protection algorithms. It can consist of complex voltage overcurrent protection, inverse-time overcurrent protection, low cycle load shedding protection, zero sequence overcurrent protection, zero-sequence over-voltage protection, etc. For example, the flow diagram of the complex voltage overcurrent protection algorithm based on FPGA is shown in Figure 6. After comparing the current, voltage and negative sequence voltage with the settings, the result is output.

4.5 *Verification*

To verify the validity of relay protection algorithms, this FPGA platform is Altera's DE2-70 development board, which is equipped with a DM9000 Ethernet chip. The vision of Matlab is 7.10.0. Through constructing the combination programming platform of Matlab and FPGA, the results of FPGA can be outputted. By observing the output waveform, we can test the effectiveness of the relay protection algorithm module. For example, when a grounding fault occurs, the output of hardware relay protection can be monitored.

5 CONCLUSION AND PROSPECT

This paper presents a method to design and verify relay protection algorithms by programming with Matlab and FPGA. This method makes full use of the Matlab/Simulink fault simulation and Matlab/GUI interactive programming to quickly implement a fully functional protection algorithm design system. It can improve the efficiency of FPGA development and shorten the product development cycle.

According to the research of relay protection algorithms based on FPGA, we can carry out the study of relay protection chips, which can be used in relay protection devices, to help reduce the volume of protection devices and improve devices' performance.

REFERENCES

Bo, Zhiqian, Zhang, Baohui, Dong, Xinzhou, He, Jinghan, Lin, Xiangning, Zeng, Xiangjun & Li, Bin. 2013. The development of protection intellectualization and smart relay network. *Power System Protection and Control*, 41(2), 1–12.

He, Jiali, Le, Yongli, Dong, Xinzhou & Le, Bin. 2010. Relay protection of power system (Fourth Edition), Beijing: China Power Press.

Zhang, Baohui & Yi, Xianggen. 2009. Power System Protective Relaying(Second Edition), Beijing: China Power Press.

Liu, Hao & Han, Jing. 2013. MATLAB R2012a self-taught one pass, Beijing: Publishing House of electronics industry.

Zhu, Ning, Wu, Chonghao, Li, Zhijian, Huang, Jibo & Li, Wei. 2012. Design of simulation and verification system for application software of relay protection. *Electric Power Automation Equipment*, 32(6)140–144.

Shi, Hongjie, Le, Xiufan & Xu, Dongli. 2009. Research of programmable logic microprocessor-based protection based on Matlab. *Power System Protection and Control*, 37(11): 82–85, 90.

Li, Yiqun, Wu, Guoyang & Zhang, Tao. 2002. The new software design method of module-based programmable digital relay. Automation of Electric Power Systems, 26(15): 66–69.

Yu, Qun & Cao, Na. 2011. MATLAB/Simulink modeling and Simulation of power system. Beijing, Machinery Industry Press, 2011.

Chen, Zhimin & Cao, Jian. 2007. Research and design of FPGA for digital protection in power system. *Central China electric power*, 20(3): 11–14, 18.

Electronics, Information Technology and Intellectualization – Song & Kwak (Eds)
© 2015 Taylor & Francis Group, London, ISBN 978-1-138-02741-1

Study on the development trend of fire detection alarm systems

W.G. Li, W.D. Zhao, K.K. Han, H.J. Chen & H.Y. Xiong
School of Civil Engineering, Central South University, Changsha, China

ABSTRACT: At the multi-stage, the centralized control mode is the mode most used in fire detection alarm systems, single source and fixed threshold techniques are the main methods used to fire detectors, and only two alarm signals 'fire' and 'non-fire' are used in the system because of their simple alarm logic; the RS-485 Bus technology is the main communication means of the system. The fact that the control area of the alarm controller is limited, the fundamental contradiction between sensitivity and accuracy of the system is unable to be resolved, the environmental interference signal which is similar to fire characteristics, may render the system vulnerable to omission and false alarm etc, are the existing main drawbacks of fire detection alarm systems. The paper summarizes the major trend of fire detection alarm systems through the analysis of the research status of fire detection alarm systems at home and abroad, and the application situation of advanced fire detection alarm technology. Sensitivity of the response speed, information fusion of the multi-sensor, complications of logical structure, intelligent detection and alarm algorithms, miniaturization of volume, classification of alarm signals, and the network of the system, are the development trends of fire detection alarm systems.

Keywords: Fire detection alarm system; intelligent detection; Fuzzy neural network; Information fusion

1 INTRODUCTION

Fire occurrences are burst and frequent, which poses a threat to life and property. According to related statistics, the loss caused by fire is about five times as much as an earthquake, only after the drought and flood damage. Early detection of fire has a great significance in controlling the scope and extent of the fire disaster. On the basis of detecting a fire early, locating the fire accurately, mastering the details of the fire, no false alarm etc., a fire detection alarm system makes use of advanced sensor technology, information processing technology, and rigorous logic relations to monitor the fire and deal with the alarm. So the study of fire detection alarm systems is necessary. Fire detection alarm systems have seen two centuries of development since the 1840s, and the development has never stopped, for example, the development of the detection signal from a single detection to compound detections, the detection algorithm of a simple switch quantity changed to a multi-threshold analog, the data processing structure changed from centralized control to distributed intelligent control, the development of the communication methods from multiple wires to the Bus, and again to the subsequent application of wireless technology. With the development of micro-processing technology and fuzzy neural network technology, sensitivity of the response speed, information fusion of the multi-sensor, complications of logical structure, intelligent detection and alarm algorithm, miniaturization of volume, classification of alarm signals, and the network of the system, are the development trends of fire detection alarm systems. As the core system of the fire control and protection system, a fire detection alarm system is used as a human control centre, just like the human brain, which is the linkage control centre of the fire control and protection system, and which plays an important safety role in the protection of natural resources and people's lives and property. This article will analyse the present situation and the study of development changes of the detection alarm system from the point of view of the system.

2 FIRE DETECTION ALARM SYSTEM STATUS ANALYSIS

Through the investigation and study of the fire detection alarm system in the practical application in engineering at the present stage, including the status research on the principles of fire signal detection, fire signal processing algorithms, fire detection alarm system signal processing mechanisms, and the development of fire detection alarm system communication, at the present stage, the main features and existing problems for fire detection alarm systems are as follows:

(1) At the multi-stage, the centralized control mode is the mode most used in fire detection alarm systems, that is, all devices (detection equipment, modules, response equipment, etc.) are centrally controlled by the controller. The detector does not have a signal so

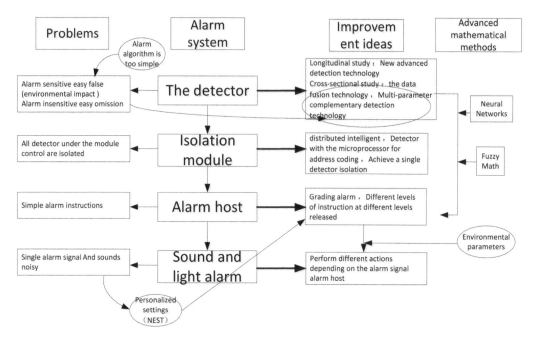

| Problems | Alarm system | Improvement ideas | Advanced mathematical methods |

Figure 1. Fire detection alarm system status analysis chart.

the processing functions will give the collected signals to the controller, and the controller respectively will deal with signals from the different address codes. The data exchange between the detector system and the controller is important in this mode; there are high demands on data communication, which is easily disturbed by environmental signals, and the system response is slow. The system cannot deliver a 100% secure job once the controller fails, and the entire fire detection alarm system will be paralysed and unable to work.

(2) At the multi-stage, single source and fixed threshold techniques are the methods most used by fire detectors. As different places have different fire characteristics due to the fire sort is more and difference is very big, fire detectors by using a single signal selection based on the need for fire detectors in different places, and many sites will need to install a number of different types of fire detectors, sending inconvenience to fire detection alarm system's design and construction. And the system using a single signal detector is likely to cause false positives, false negatives, which is a single signal with a single fixed threshold techniques insurmountable obstacle, if the threshold is raised, not false positives, but it is easy to underreporting, on the contrary, it is easy false positives. In short, a single signal detection technology cannot solve the contradiction between the detector sensitivity and accuracy.

(3) At this stage, fire alarm logic is simple, there are only two alarm signals 'fire' and 'non-fire, and fire detection alarm system in such an alarm logic will exist a contradiction between the sensitivity and accuracy requirements. In many cases, the interference signals

cannot be handled correctly, especially if there is a great impact on the system when the interfering signals are near the alarm threshold. And generally detect workplace environment parameters with the seasons, day and night, and the usage to be change, simple alarm algorithm can't fault-tolerant complex environment changes.

With the advent of intelligent fire detection systems, fire detection alarm system exists in a serious omission, misstatement has been significantly improved, and the reliability of the fire detection alarm system has been improved. However, due to the fire detection alarm system's low level of intelligence, the new fire sensing technology is still in the laboratory stage, the lack of integration with the existing fire detection alarm systems, fire detection alarm system misstatement omission phenomenon in practical project application is still plagued by the user. The fire detection alarm system analysis is shown in Figure 1 in the NEST's protect detection alarm inspired.

3 THE DEVELOPMENT DIRECTION OF FIRE DETECTION ALARM SYSTEMS

At this stage, the logical structure of fire detection alarm system mainly uses "1" and "0" as the two signals of fire, which conflicts with the requirements of a fire detection alarm system in sensitivity and accuracy. Moreover, fire data used in fire detection research is mostly obtained under laboratory conditions, besides some data of actual fires in monitored places. This results in insufficient fire data to determine the threshold of the fire alarm, which has greatly

186

limited the development of the fire detection and alarm technology. Therefore, future research in fire detection alarm systems' technology will be conducted with the following aspects in mind:

(1) Network Architecture

The key advantage of analogue fire detection systems is the ability to use a variety of signal processing algorithms to more accurately detect fires and reduce false alarms. The implementation of a fire detection signal processing algorithm can directly affect whether it gives full play to the role of signal processing algorithms, response speed, reliability, compatibility, costs, and a series of performance indicators of the whole system. At present, the vast majority of signal processing algorithms are implemented in software, wherein intelligent distributed signal processing architecture, as a development direction of implementing fire detection signal processing algorithms, has many advantages, such as flexible signal processing and high reliability.

Fire alarm technologies which mostly adopt integrated control models have some shortcomings. For example, once the controller fails, the system will result in paralysis, and the centralized control mode cannot be used in distributed architecture. Network fire alarm control technology uses a multiple area alarm controller to form a local area network based on a network communication. In this network, each area controller and their control equipment form a relatively independent alarm zone. Data exchange between controllers via a local area network, completes the interactive display and control functions. This not only prevents the entire system coming to a major standstill if the main controller fails, but also keeps the whole system monitored by multiple controllers, eliminating the losses caused due to negligence by the operator. This technology can be widely used in dispersion-type large-scale buildings. So, network fire alarm control technology will become the development direction in the future. Based on the implementation of fire detection alarm systems by a network, automatic fire detection alarm systems as a whole and the detection parts of the system or network sub-system, should consequently, develop toward miniaturization, with the development and improvement of microprocessors, integrated circuit technology, nanotechnology, and information processing technology. In addition, the installation and use and management of automatic fire detection alarm systems, becomes simple, inexpensive and accessible.

(2) Advanced fire detection technology

Because a fire is an extremely complicated physical and chemical process and has a strong correlation with the environment, different environments and different combustion materials make combustion products diverse. Therefore, unit detecting technology using a single parameter to detect fires is difficult to meet the complexity and characteristics of fires. The fire characteristic signal uneven response sensitivity of single parameter detector cause its detection ability to be limited. It can only according to the different

place and the possibility of fire type to choose detector. An accidental improper selection will cause errors and omissions. As many environments are complex, it is not reasonable to use a single parameter detector. Capturing multiple environment parameters can make it more effective to distinguish the fire and non-fire signal, thereby improving the reliability of the alarm. Multi-sensor data fusion technology simulates the biology function which comprehensively processes complex problems, and converts the various appreciable information into valuable data for understanding environmental changes. Data fusion technology is the integration process of processing, control, and decision-making for a variety of information sources. Currently, it is the photoelectric smoke temperature compound detector which is relatively mature. Both Gent and Thom in the UK, Panasonic in Japan, and System Sensor in the United States have launched their own smoke temperature compound detectors. In Germany, the utilization of photoelectric smoke, ionization smoke, and a mild CO compound probe has reached a very high level. At present, the China Nanjing Fire Group has successfully developed a SH9432S intelligent photoelectric smoke sensing compound detector. The production of the compound detector provided the impetus for the development of fire detection direction. The multi-sensor data fusion technology applied in a fire detection alarm has the following advantages: it increases the survival ability of the system; it extends the coverage space; it extends the time range; it improves the credibility; it reduces the ambiguity of information; it improves the detection performance; it improves the spatial resolution, and it increases the measurement space dimension. Compound fire detection has become the mainstream one of fire detection methods.

(3) Intelligent fire alarm algorithm

Setting experience threshold values, through binary logic, early fire detector is the on-off type, of which the output is the fire or no fire signal. The threshold can be a fire's parameters or its functional form. The traditional threshold method for a complicated signal detection of fire state is too simple. Especially, the alarm threshold and the delay time of the alarm set are too simple. It is necessary to seek values closer to the real fire. The introduction of an addressable analog, means that fire detection technology, signal processing technology, and artificial intelligence technology have begun a broader cross-binding, and are leading the hardware and software combination of fire detection technology into a new stage of development. Since Y. Okayama (Japan) and S. Nakanishi (Japan) applied the neural network and fuzzy logic to fire detection technology, the performance of existing detection systems has been greatly improved. Fire detection systems no longer simply rely on switch signals, but now also rely on the analog reflecting the characteristics of the fire. The fuzzy logic idea conforms to the fire signal characteristics of uncertainty and randomness, while the self-learning ability of a neural network make the system adapt to environmental changes.

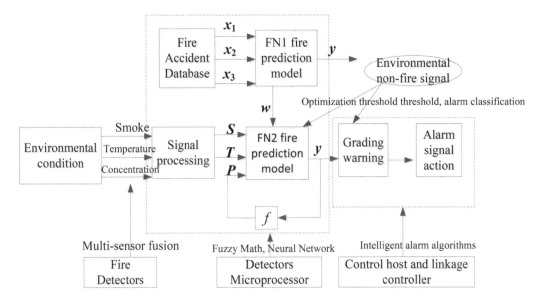

Figure 2. Fire detection alarm information processing model.

Their fault tolerance improves the reliability of the system.

Based on the multi-sensor detection, according to the established detection alarm algorithm model, a software program of detection algorithm should be written, which would be ported to the hardware system. This could be used in fire detection experiments of different fire types, in experimental data analysis, and in alarm grading system settings. An alarm grading system is mainly divided into primary, secondary, tertiary and quaternary levels. The primary level alarm signal is mastered by the regional controller; the secondary level is mastered by a central alarm controller, (whether or not to take fire-fighting action is artificially controlled); the tertiary level is mastered by the control centre, which directly employs the fire-fighting equipment, and the quaternary level is mastered by the control centre, who will start the highest level of the alarm signal networking alarm. This alarm logic will be able to overcome the contradiction between sensitivity and a false alarm which exists in traditional alarm logic, thereby having broad development prospects.

Combined with the fire detection algorithm based on a fuzzy neural network and the fire alarm grading system considering environmental parameters, an intelligent fire detection alarm system algorithm is put forward. By introducing the environment parameters into the detection alarm algorithm, detectors can change the permissions system of the alarm system according to the environment parameters (non-fire).That means that the detection alarm system has a wider scope of application, and reduces problems in detector project selection. Micro-processing technology is developing very rapidly; low-cost,

low-power, high-performance small microcontrollers abound. This provides adequate material protection for the development of intelligent detectors. Intelligent fire detection alarm systems are a major development in the field of modern fire direction.

4 A NEW FIRE DETECTION ALARM INFORMATION PROCESSING MODEL

Based on the research about fire safety warnings by my tutor-led task force, by analysing the application status of detection alarm systems, combined with multi-input composite signal detection technology, fuzzy logic signal processing technology, and the adaptive features of artificial neural networks, an intelligent fire detection alarm information processing model which may provide a reference for the core structure of fire detection alarm systems in the future, using multi-sensor information fusion based on a fuzzy neural network, was put forward.

A trend continuing algorithm combined with trends and duration judgment can be used in the intelligent fire detection alarm information processing model. By used the calculating the duration for the change of the fire trends which is over the trends threshold and the continuing of over the trends threshold exceeds the duration threshold as the two threshold signals to determine whether the fire signal. The cumulative function is as follows:

$$k(n) = \begin{cases} \left[k(n-1) + 1 \right] u(\tau(n-1) - s_c), s_c > 0 \\ \left[k(n-1) + 1 \right] u(s_c - \tau(n-1)), s_c < 0 \end{cases} \quad (1)$$

where $\tau(n)$ is trend value, s_c is the trend alarm threshold. The output can be expressed as when there is a trend continuous:

$$y(n) = \begin{cases} \left[y(n-1) + (\tau(n) - S_c) \right] u \left(k(n) - N_t \right), \\ \left[y(n-1) + (\tau(n) - S_c) \right] u \left(N_t - k(n) \right), \end{cases} \quad (2)$$

where N_t is the duration of trends. It ensures that only the duration of trends in the N_t above, only the calculated output, when the durations of trends in the N_t low, the outputs are 0, so that the glitch does not make the system produce false alarms.

There will be fire detection alarm information processing model which is a combination of the above algorithms and neural network training:

In the Figure 2, FN1 is a fire detection model which is obtained based on empirical data, FN2 is the identification model for fire warning, T is the output signal of the temperature sensor, S is the output signal of the smoke sensor, and the CO concentration signal can also be used in the model. The fire parameter signals were completed in the sensor of the detector. Quantify the rule of thumb based on empirical data, the weights and thresholds index after training of FN1 which to learn based on the rules of experience data delivered to FN2, the detection processing model with feedback error signal based on smoke, temperature and CO concentration composite signal which implemented in the microprocessor of the detector was take form. After fuzzy reasoning, the FN2 output values reach the fire alarm signal output value (analog).Then according to the set up alarm grading system, the FN2 alarm grading warning. This part can be obtained by MATLAB simulation. It passes through the detector fuzzy neural network trained data information transmission to the fire alarm controller, then the alarm classification for the data is calculated, which finally determines the fire rating. According to the determination of the level of the fire system, the system will launch the corresponding fire control system and linkage control device.

5 CONCLUSION

The fire detection alarm system is the core to building fire protection systems, and there will be more and more research topics carried out with broad prospects for development. Through the research of the present situation and the development of fire detection alarm systems, this paper mainly completed the following studies:

Firstly, from the angle of system research and analysis of the current fire detection alarm systems, and in summary, the alarm controller range is primarily limited at this stage, and the contradiction between the sensitivity and accuracy requirements render the system vulnerable to similar environmental characteristics and fire signals from disturbance causes the system to false positives, false negatives.

Secondly, in summary, the fire detection alarm system is currently moving in the direction of development, which is sensitive towards the response speed, multi-sensor data fusion, a structured complex logic, intelligent detection and alarm algorithms, the volume of miniaturization, the alarm signal classification, and system networks. A future fire detection alarm information processing model has been put forward.

Thirdly, the fire detection alarm system has an uncompleted database, and during the collection of reliability data on fire detection alarm systems, there exists a lot of difficulties; there ought to strengthen the fire based data collection and research.

REFERENCES

Chen T., Yuan H.Y. & Fan W. C. 2001. The developing fire detection technology. *Fire Safety Science*, 10(2): 109–113.

Dong W. 2009. *Design EMU fire alarm system*. Dalian University of Technology.

Du J.H. & Zhang R.C. 2004. Research status and development trend of fire detectors. *Fire Technique and Products Information*, 7: 10–15.

Guo J., Wang R.L. & Li M. 2004. Current status and development trends of fire detection technology. *Journal of Liaoning Technical University*, 23(2): 209–213.

He Y. K. & Guan Z. H. Future development of fire detection technology [J]. *Journal of Liaoning institute of technology*, 2004, 24(6): 11–17.

Hu Z. G. 2013. Design and Research on Fire Detection and Alarm System based on Multi-sensor Information Fusion Technology. *Xi'an University of Architecture and Technology*.

Li Y. H. 2007. The application status research and development trend of fire automatic alarm technology. *Modern Business Trade Industry*, 19(7): 195–196.

Li L. 2001. Hot-points and tendency of fire detection technology in China. *Fire Safety Science*, 10(2): 116–120.

Liu J., Zhao W.D. & Zhao D. 2006. Based on CAN Bus fire alarm system design [J].*Intelligent Buildings*, 5: 54–56.

Liu J. & Zhao W.D. 2007. Based on ZigBee technology, fire alarm system design. *Microcontroller and Embedded Systems*, 1: 56–58.

Okayama Y. 1991. A primitive study of a fire detection method controlled by artificial neural net. *Fire Safety Journal*, 17(6): 535–553.

Tommy H. & Per B. 2007. Heimo Tuovinen. Reconstruction of an arson hospital fire. *Fire and Materials*, 31(2): 225–240.

Rose P. 2000. Multi-criteria fire detection systems using a probabilistic neural network. *Sensors and Actuators B: Chemical*, 69(3): 325–335.

Nakanishi S. 1995. Intelligent fire warning system using fuzzy technology. *Proceedings 10th Internationale Konferenz ueber Automatische Brandentdeckung*. Duisburg Germany: 203–212.

Wang S. & Dou Z. 1997. Implementations of fire detection signal processing algorithm. *Fire Technique and Products Information*, 10: 22–23.

Wang S. & Dou Z. 1998. Fire detection and signal processing. Wuhan: *Huazhong University Press*: 154–161.

Xing Z.X., Chen L. & Chu D. Z. 2012. The research of fire detection alarming system and its reliability. *Journal of Safety Science and Technology*, 8(3): 151–154.

Xu Q., Zhan F.R. & Su G.F. 2002. Development and discussion on the fire smoke detection technology. *Fire Safety Science*, 11(2): 113–118.

Zhao D., Zhao W.D. & Liu J. 2006. Automatic alarm and linkage control system integration. Intelligent Buildings and Urban Information, 112(3): 103–105.

Zhao D., Zhao W.D. & Liu J. 2007. Application of Information Technology in Intelligent Security Monitoring System. *China Public Security*, 7: 132–135.

Zhao W.D., Li W.G., Ding W.T. et al. 2013. Fire environment monitoring device for fire fighters based on MSP430F169 chip. *Journal of Railway Science and Engineering*, 10(3): 103–107.

Electronics, Information Technology and Intellectualization – Song & Kwak (Eds)
© 2015 Taylor & Francis Group, London, ISBN 978-1-138-02741-1

Exploration and practice for the agricultural value of unmanned helicopters

Tingyang Meng
University of Michigan – Shanghai Jiao Tong University Joint Institute, Shanghai, China

ABSTRACT: A Commercial agricultural unmanned helicopter has many defects such as being extremely expensive, requiring maintenance, and having a single function. In this research, the unmanned helicopter uses pure electric power and is separated flying platform from basic task load. It can perform a variety of tasks such as aerial photography and pesticide spraying. Exploration and practice show that it has many advantages such as being low-cost, requiring simple maintenance, operating with stability and reliability, and its functions are easy to expand according to need.

1 INTRODUCTION

In recent years, the application of unmanned aircraft has become very extensive, especially the unmanned helicopter applied to agricultural use. Because the helicopter has many advantages such as low requirements for landing sites, a vertical take-off and landing, and an ability to fly at ultra-low speed and altitude, it is suitable for agricultural production; instead of manually spraying pesticide the downwash of the spray attachment made by the helicopter rotor, ensures a uniform application. However, because the current commercial agricultural unmanned helicopter is driven by a gasoline engine, is very complex to use and maintain, has poor flight stability, pollutes easily, is expensive, has a single function etc., this paper devotes its research to overcoming all the deficiencies above by using pure electric power and an independent design between the platform and the basic task.

2 DESIGN AND CONSTRUCTION OF AN UNMANNED HELICOPTER PLATFORM

2.1 Power system and mechanical structure

The electrical unmanned helicopter has the layout of a single rotor and a single tail rotor. Its power is supplied by a 44.4v Lithium battery, and this powers a DC brushless motor which has a maximum continuous power of 3300W. The main rotor head uses three steering CCPM (Collective-Cyclic Pitch Mixing) at an angle of 120°, and the tail rotor is driven by a belt. This produces small vibration, is easy maintenance, less noisy etc. The task load mount frame is arranged on the lower part, and it can mount different task loads to perform a variety of tasks, and it is easy to design a new mount task and expand its functions in the future.

2.2 Control system

The electrical unmanned helicopter mainly uses radio remote control to operate. Its control system is composed of a wireless transmitter, wireless receiver, three axis gyro, brushless electronic governor, tilted plate rudder and tail servo control, open source ArduPilot Mega (APM), etc. The control actions of the operator are converted to radio signals and received by the receiver. The throttle signal of the receiver directly controls the electronic governor, and the electronic governor drives the motor, reaching and stabilizing at a specified speed. The receiver aileron, elevator pitch, direction, and sensitivity connect with the three axis gyro and input control signal. The three axis gyro tests the three axis angular velocity of the helicopter, and controls the total pitch, period pitch and tail rotor pitch by rudder, which then controls the aircraft flight attitude. Basic flight states include forwards, backwards, shift to left or right, spin to left or right, rise and fall, still hovering on the vertical, as well as the superposition state. The APM receives the control signal from the receiver, transmits to the three axis gyro by processed, and then controls the aircraft flight in automatic flight mode. Here, we use two of the six flight modes of APM: the manual mode (manual) and the augmentation model (stabilized). In the stability augmentation mode, the plane will automatically restore the level posture and situ hover if there is no input control. It can realize precise hovering and automatic driving if the plane sets up a GPS module and uses the locating compass functions of a GPS.

2.3 Liquid spraying system R

The liquid spraying system and basic flight platform are independent of each other. The spraying system can be mounted on the mount bracket of the flight platform if needs be. The spraying control circuit connects to the receiver by standard plugs and is controlled. Usually, there are three methods of driven liquid spraying: single pressure, sustained pressure and non-pressure.

The experiment shows that the non-pressure method has lower feasibility. So we only have attempted the single pressure and sustained pressure. During the actual test, the spraying system of the sustained pressure method has a stable spraying rate, high operational reliability, small volume, not to occupy more payload. We think it is more suitable for agriculture.

2.4 Aerial photography system

The aerial photography system and basic flight platform are independent of each other. It can be mounted onto the mount bracket of a flight platform if needs be. The aerial photography control circuit connects to the receiver and APM by standard plugs and is controlled by them, and it can take single photos, multiple burst and shoot videos. It consists of two main parts, a stabilized augmentation platform and a digital camera. The function of the stabilized augmentation platform is to reduce impact shaking and jittering from the flight platform when the camera works. Here, we attempted to two forms of stabilized augmentation platform: one is referred to as the Stan Nikon stabilized augmentation platform used in photography, and known as a 'Stan Nikon platform', the other is an activity mechanism which is driven by a high speed server and has two freedom degrees. It is called the 'conventional platform'.

3 RESEARCH AND PRACTICE

3.1 Exploration and practice of pesticide spraying

In order to realize a fast and wide range of pesticide spraying, the aircraft needs to fly at a super low altitude in the operation area, whilst continuing to climb up and down, hovering and flattening, sometimes flying around obstacles (woods and high-voltage cables etc.), with frequent take-offs and landings on a simple runway because it is not only loading Co. each time but also working in field. Therefore, spraying demands that an aircraft fly with stability at very low altitude, and spraying should be uniform and reliable, fast and efficient.

It should plan the flight path first for spraying in the specified range, and then undertake the spray operation. For those regions that have a regular boundary, it can adopt a 'zigzag' reciprocating scanning spraying operation. The spacing between the two routes is approximately equal to the width the aircraft can cover whilst spraying in a single route. The height should be approximately 1 metre above the top of the plants. In order to maximise the utilization spraying range of the aircraft, the spraying device is always vertical to the spray route, i.e. the direction is always parallel to the route direction. The width of the spraying device in this paper is two metres. Considering the spraying

Figure 1. Working diagram of pesticide spraying.

effect and, combining it with actual field test results, the width of the spraying route between two spaces is 2.5m.

In the actual test (Figure 1), the level flight speed is approximately 5m/s, the height is 1m, and it can spray about 1 acre per minute. The spraying speed is more than 50 times artificial. In theory, in one day it can spray over 300 acres if the working group (one plane) consists of two men.

The frame and transmission system will be redesigned later. The main rotor can have a larger diameter and lower speed in order to obtain a higher efficiency. Another problem is with the control. At present, the aircraft cannot leave artificial control while it is flying. The main reason is that there are often various obstacles which are predicted by the flight control system in the spraying target area, such as poles, wires etc., and the current helicopter flight control system cannot detect obstacles enabling the aircraft to move around them by itself. But, it can automatically follow the path of flight planning if it has an installed GPS, and it will have certain advantages in a large area which has no barriers.

3.2 Exploration and practice of aerial photography

The design requirements of aerial photography include that it should be of high resolution; the photos should have enough resolution for identifying the crop growth or damage, it should be of high efficiency, that is, the photos should be obtained in a relatively short period of time, and they should be normative aerial photos, and the picture size and angle should be more appropriate to facilitate subsequent processing; it should be of high reliability, that is, the aerial process should be reliable and respond to time demands and ensure safety.

It is an adopted grid aerial region and low shooting and then modified split [1]. Within the grid target area, each cell corresponds to a photo, with the aircraft at a fixed altitude. Firstly, the flight height and mesh size are determined according to the camera parameters. Assuming the resolution is x million pixels and the camera angle is $\alpha°$, in order to achieve centimetre level resolution, the calculation formula is as follows (unit: m):

The ratio of length to width is 4:3,

$$flyingheight = \frac{\sqrt{\frac{10000x}{12}} \times 2.5}{\tan\frac{\alpha}{2}}, gridlenth = \sqrt{\frac{10000x}{12}} \times 4, gridwidth = \sqrt{\frac{10000x}{12}} \times 3$$

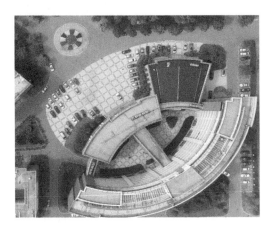

Figure 2. Actual aerial photo.

Figure 3. Google satellite image.

The ratio of length to width is 16:9,

$$flyingheight = \frac{\sqrt{\frac{10000x}{144}} \times 12.5}{\tan\frac{a}{2}}, gridlenth = \sqrt{\frac{10000x}{144}} \times 16, gridwidth = \sqrt{\frac{10000x}{144}} \times 9$$

Secondly, there is a certain difference between the aerial image and the actual ground scenery, and this is because of the variation of the aircraft flight attitude and flight height and the camera attitude, and it should be modified. Rough correction can be achieved by calibrating the square basic grid 5 m × 5 m. Accurate correction can be achieved by 3D projective transformation to transform photographic images into projection images that have the same flight altitude and camera lens at vertical downward (called standard projection image). In actual shooting, overlapping between square grids is usually taken as two times the flight error.

Figure 4. Correction mosaic map.

The aerial photograph test has been done at the Minhang campus of Shanghai Jiao Tong University. The test is the actual shooting a slightly worse result than in theory, but it achieved the basic requirements. The main problem is that the augmentation effect of the platform is not enough and the flying photo attitude is too high (over 100 m), so this impacts on the shooting quality. The maps shown below are part of the actual aerial photograph map (Figure 2): Google satellite image (Figure 3), part of revised image stitching (Figure 4) [2]. The actual aerial effect is satisfactory.

After the actual shooting, there were some shortages found from the instant images taken by the aircraft camera. We will use CNC aluminium and a carbon fibre plate to make components of the platform to resolve the problem of overall rigidity, and install an inertial measurement module to get attitude feedback, to control the platform posture jointly, and to resolve the problem of control mode. This requires optimizing the control algorithm and rewriting the control program [3-4]. Another way of trying to achieve this is by the console server attempting to use a stepper motor or a brushless DC motor.

4 PRACTICAL APPLICATION AND PROSPECTS

4.1 *Application of aerial photography technology in agriculture*

The panoramic photo is a very important research datum in many aspects such as disaster assessment, detection of seedling growth, water conservancy planning, etc. At present, the main methods used to obtain panoramic photos are satellite remote sensing, aerial photographs, and unmanned aerial photographs [5]. In shooting speed and accuracy, the satellite remote sensing is the fastest and has the maximum range, but it has low resolution, a high cost, low flexibility, and its use is regulated. The aerial photograph is an in-between method and it has speed, a large range, high resolution, generally flexibility, but with limited usage. The unmanned aerial photograph is a suitable method which has high speed, a small range, high resolution, and high flexibility and is very economical, so it is very suitable for small ranges and unexpected environmental events, such as water, crop pests and diseases etc.

4.2 Application of an unmanned helicopter spraying pesticides in agriculture

Looking at investigations into domestic farm unmanned helicopters, why is the existing agricultural unmanned helicopter used for spraying pesticides not popular, and why does it not get a wide range of applications? There are two main reasons. One is that the price is very expensive, another is that the technical support is lacking, and this is a very important reason. However, the unmanned helicopter is only 300$, and the maintenance is very simple, but the operation still needs some experience.

4.3 Prospect of an agricultural unmanned helicopter platform

Compared with multi-rotor aircraft, the single rotor helicopter has many incomparable advantages. The single rotor has higher lift efficiency than multi-rotor aircraft. So a single rotor helicopter load is greater and has a longer life. According to the practice and experiment, the basic parameters of the improved type have been determined: the main rotor length should be 800 mm~1000 mm, the battery is a 12 core lithium battery, the current is 5ah~20ah, and the motor power is 3 kW with 80 amps governor. At present, pesticide spraying companies on behalf of pesticides spraying have appeared gradually, and this can be regarded as a promotion way. So the maximum threshold is its usability problem which the unmanned helicopter implements agricultural task.

5 CONCLUSIONS

This paper focuses on two task loads of an unmanned helicopter, pesticide spraying and aerial photography

in agriculture. We put forward the design schemes according to actual needs, and gave the design and production of the corresponding device. We compared the advantages and disadvantages of each device with different designs by studying experimental flights, and recording and describing the actual hardware devices and the test results, and we obtained satisfactory experiment results. However there are still many problems to be solved, and there are some predetermined targets to be implemented in the future.

ACKNOWLEDGEMENTS

I would like to thank my teacher Professor Ma Chengbin and other students of the project team for their guidance. This paper is supported by "Fifth 'Shanghai Jiao Tong University Student Innovation Plan' Project" and "Chinese National College Students Innovative Practice Project in 2012" (project number: IPP5204).

REFERENCES

Fan, Caizhi et al. 2008. Attitude Control System Design and Experiment Research of a Small-scale Unmanned Helicopter. *Manufacturing Automation*, (11)30:64–67.

Feng, Guoyu et al. 2007. Geometric Rectification of Aerial Image-based Camera Projection Mode. *Modern Electronic Technique*. (23):7–9, 12.

Gu, Donglei et al. 2005. Design of Flying Control System for an Autonomous Helicopter. *Journal of Nanjing University of Aeronautics & Astronautics*, (4)37:476–478.

Luo, Xiaobo. 2008. *Research on Automated Mosaic Method of Aerial Image. National University of Defense Technology*, Changsha. CHN.

Zou, Changhui et al. 2011. Future Prospects in Utilization of Photo-taking Unmanned Aerial Vehicle in Low Altitude of RS System in Plateau Mountain Area of Guizhou. *Journal of Guizhou Normal University (Natural Science)*. 2(29):24–28.

Electronics, Information Technology and Intellectualization – Song & Kwak (Eds)
© 2015 Taylor & Francis Group, London, ISBN 978-1-138-02741-1

MRI image segmentation using information granules and MRFs

Shuangyun Xie & Yanan Dang
College of Information Engineering, Taiyuan University of Technology, Taiyuan, Shanxi, China

Fang Wang
Key Lab of Advanced Transducers and Intelligent control Systems, Ministry of Education and Shanxi Province,
Taiyuan University of Technology, Ministry of Education, Taiyuan, China

ABSTRACT: In order to obtain an accurate segmentation of spinal marrow in an MRI image, a Markov random field image segmentation algorithm based on an information granule is proposed. The information granule is established at first, then the region of the image is also created for the epitaxy of the granule, and the region features are extracted both internally and externally to be the connotation of the granule. The internal region features are denoted by second-order moment and the average value of pixels intra-regionally, and the external features by a region adjacency matrix. By combining the region features and information granule with the MRF, the image segmentation is finished, so a wrong segmentation caused by single pixel is avoided. Finally, the rule of maintaining big granule is used to correct the segmentation result. Compared with the result of the Markov random field, the effectiveness and veracity of this method are verified by the simulation experiment.

Keywords: information granule; feature extraction; MRF

1 INTRODUCTION

The accurate segmentation and quantization of medical structure in medical images is an important issue of medical structure analysis, and also a precondition of visualization, medical diagnosis and medical plan formulation (Hyeok et al. 2009). Because of the advantages of multi-parameter imaging, arbitrary cross-section imaging, and high contrast soft tissue and so on, Magnetic Resonance Imaging (MRI) is becoming more and more popular in clinical application (Joshua 2014, Wyatt 2003), compared with other imaging techniques. With the influence of noise, field shifting effects and organization difference, MRI is characterised by being fuzzy and non-uniform, leading to non-unified segmentation methods. The Markov Random Field (MRF) has been widely used for image segmentation (Yang et al. 2012), due to its excellent description of mutual information between neighbouring pixels and easy combination with other image processing methods. Yet wrong segmentation caused by a single pixel will be produced if MRF image segmentation, the method that treats a single pixel as the minimum unit, is applied to MRI images directly.

In this paper, the theory of granular computing (Lin 1997) is introduced, the low-rise pixel characteristics are shielded by the establishment of region based on different granularities, and an MRI image segmentation algorithm, based on the information granule, is proposed integrating the MRF. It is called MRF based on information granule (IGMRF). The

simulation result indicates that the algorithm can reach the result of segmentation quickly and efficiently, providing assistance for medical analysis.

2 GRANULATION OF IMAGE

2.1 Information granule

An information granule comes with a simple sub-problem or module decomposed in accordance with its feature and performance, whilst handling much complex information. The nature of granular computing is the expression and disposition of the granule. "Big granule" means large object. "Small granule" means small object. Problems will be resolved in the different granularity.

$G = (IG, EG)$ is used to describe the granule; here IG is called the connotation of G, and EG is the epitaxy of G. Connotation is the exposed knowledge of the granule in a specific context. The vector below shows the connotation of the granule:

$$IG = (ig_1, ig_2, \ldots, ig_n) \qquad (1)$$

where, $ig_1, ig_2, \ldots, ig_n$ represents each element of IG respectively. EG is a set of objects contained by a granule. In this paper, EG is denoted by the image region, which makes up with all the pixels in the region. Connotation is the feature of the region, including texture, spatial features, and so on.

2.2 Epitaxy of the granule

Epitaxy *EG* is the pixel region contained by a granule. In order to obtain *EG*, a series of methods to pre-process have been applied:

Apply SARD filtering algorithm (Yang et al. 2012) to the original image I, which becomes I'. Finish gradient based on I', and convert into gradient image, ∇I by *Sobel*. Label minimum pixel in ∇I, and use patulous minima transform (H-minima) from Soille (Soille 1991), remove local minima, of which the value is less than threshold value H, by setting the threshold of H. Apply compulsive minimum transform (Soille 1991) based on morphology to $\nabla I'$ to get $\nabla I''$, which is equal to $\nabla I'$ except compelling the pixels of "1" to minimum. Apply watershed to $\nabla I''$ to get I_{ws}, the region partition image.

After region partition, every region in has been numbered. Each region denotes an epitaxy of the information granule.

2.3 Connotation of the granule

Connotation, the general and common characteristics of all the pixels in a granule under a certain condition, will be obtained by feature extraction, in both intra-regional and inter-regional ways.

Because organizations of different types present big differences in the whole pixel intensity as an image, the average pixel intensity in the region has been chosen to show the important intra-regional feature of the spatial distribution properties of pixel intensity, in order to represent the regularity of the region from a macroscopic view, and to neglect the intra-regional local irregularity.

Calculate second-order moment of corresponding pixels in each region based on gray level co-occurrence matrix (GLCM) (Seba 2013), as follows:

$$f_{sec} = \sum_{i=0}^{L-1}\sum_{j=0}^{L-1} P^2(i,j) \qquad (2)$$

where $P(i,j), (i,j = 0, 1, 2, \ldots, L-1)$ is the gray level co-occurrence matrix; L denotes the grayscale of the image; i, j respectively denotes the gray value. The step length of GLCM is set to 1, and the angle is set to four directions, i.e., $0°, 45°, 90°$ and $135°$. The corresponding four second-order moments were calculated in order to overcome the effects of the angle by calculating the average value of four matrixes, a value that can reflect the uniformity of greyscale distribution and the roughness of texture intra-regionally.

The regional distribution is reflected by the inter-regional feature, including adjacent regions and adjacent region numbers. The model of the region adjacency matrix (RAM) is proposed to express the context of regions and extract the inter-regional features. RAM is made up with A and V. A denotes the adjacency relation matrix, and V denotes the relation statistical vector. Here, $A = \{a_{mn}\}$ belongs to a matrix of $s \times s$ size, $m, n \in [1, s]$, and s is the number of

Figure 1. Schematic of region distribution.

regions. $V = \{v_1, v_2, \ldots, v_s\}$, v_i denotes the numbers of adjacent region.

$$v_i = \sum_{j=1}^{s} a_{ij}, i \in [1, s] \qquad (3)$$

$$a_{mn} = \begin{cases} 1 & \text{region } m \text{ adjacent to } n \\ 0 & \text{otherwise} \end{cases} \qquad (4)$$

For example, the region distribution of an image is shown as Figure 1, in which the number means the serial number of regions. A and V are shown as follows:

$$A = \begin{bmatrix} 1 & 1 & 0 & 1 & 1 & 0 \\ 1 & 1 & 1 & 0 & 1 & 1 \\ 0 & 1 & 1 & 0 & 0 & 0 \\ 1 & 0 & 0 & 1 & 1 & 1 \\ 1 & 1 & 0 & 1 & 1 & 1 \\ 0 & 1 & 0 & 1 & 1 & 1 \end{bmatrix}$$

$$V = (4, 5, 2, 4, 5, 4)$$

Above all, the connotation of the information granule is $IG = (ig_1, ig_2, ig_3, ig_4)$. Here $ig_1 =$ intra-regional mean pixel intensity of epitaxy; $ig_2 = f_{sec}$; $ig_3 = A$; $ig_4 = V$. The dimensions and orders of magnitude among those indicators of connotation are not the same, so it is necessary to deal with them with normalization.

3 IMAGE SEGMENTATION BASED ON IGMRF

3.1 MRF image segmentation

If a random field X in a planar rectangular lattice meets, then:

$$P(\mathbf{X}_{ij} = x_{ij} \mid \mathbf{X}_{kl} = x_{kl}, (k,l) \neq (i,j))$$
$$= P(\mathbf{X}_{ij} = x_{ij} \mid \mathbf{X}_{kl} = x_{kl}, (k,l) \in \eta_{ij}) \qquad (5)$$

also $P(X = x) > 0$, then X is called MRF neighboured with η. Where $x =$ random field; $x_{ij} =$ fulfilling of x; η_{ij} denotes the neighbourhood system of pixel (i,j). Treating the image as a random process in lattice, equation (5) describes the spatial dependence among pixels: one pixel can be decided by the neighbouring pixels.

In the Bayesian framework, image segmentation can be finished by maximum posteriori estimation (MAP) (Barker 1998), as follows:

$$\hat{\mathbf{X}} = \arg\max_{\mathbf{X}} P(\mathbf{X} \mid y) = \arg\max_{\mathbf{X}} P(y \mid \mathbf{X}) P(\mathbf{X}) \qquad (6)$$

where feature model $P(X \mid y)$ = posterior probability of X, segmentation sign result, when y, image feature, is known already. $P(y \mid X)$ = conditional probability distribution of X which is known as marker results. $P(X)$ denotes the prior probability of X in a spatial model.

3.2 IGMRF image segmentation

The model of IGMRF is developed by combining a granulated image with MRF to finish the updating and merging of the granules. IGMRF uses the granule as the minimum operation unit. By replacing the features of a single pixel with ig_1 and ig_2, and replacing the pixel neighbourhood system with the RAM, IGMRF completes excellent segmentation with the ability of shielding interference caused by the low-rise single pixels. Assuming that the feature model of IGMRF is Gaussian distribution, according to equation (6), the minimum objective function is as follows:

$$\arg\min\left\{\sum\left\{\frac{1}{2}\log\left(2\pi\sigma_{l_r}^2\right)+\frac{\left(u\left(ig_1,ig_2\right)-\mu_{l_r}\right)^2}{2\sigma_{l_r}^2}\right\}\right.$$
$$\left.+\beta\sum_{(m,n)\in Q(\mathbf{A},\mathbf{V})}U\left(l_m,l_n\right)\right\} \quad (7)$$

where l_r = category labels of regions; $u(ig_1, ig_2)$ = the integration of ig_1 and ig_2; μ_{l_r} = mean pixel intensity of category l_r; $\sigma_{l_r}^2$ = variance of pixel intensity of category l_r; β = weight ratio of the feature model and the spatial context model; (m, n) = adjacent element pair of A, and the corresponding elements of V must be nonzero at the same time; $Q(A, V)$ = set of all the pairs; $U(l_m, l_n)$ = energy between region m and n. The definition is as follows:

$$U\left(l_m,l_n\right)=\begin{cases}1 & l_m\neq l_n\\0 & \text{otherwise}\end{cases} \quad (8)$$

The optimal value of objective function is calculated by the iterated conditional mode (ICM) (Fwu et al. 1998) in the paper.

3.3 Correction of wrong segmentation

In the experiments of MRI image segmentation with IGMRF, some problems have been found as follows:

(1) Several big granules containing a lot of pixels come into being in the process of granulating.
(2) The big granules have the same status with other granules, as the minimum unit of IGMRF.
(3) Once there is a wrong segmentation of a big granule, it can cause disturbance which cannot be ignored as it will affect the final result.

On this basis, the rule of maintaining a big granule has been proposed as follows:

If $\omega > \omega^*$, g_{wn} was judged to be a big granule.

$$f(g_{wn})=\begin{cases}\text{initial category value} & \omega\geq\omega^*\\\text{iteration value} & \omega<\omega^*\end{cases} \quad (9)$$

where $\{g_1, g_2,\dots, g_{wn},\dots, g_n\}$ is the particle-size partition of the image, in which the granules arrange themselves in the ascending order of the pixels contained, $\omega \in [0, 1]$; ω^* = correction coefficient; $f(g_{wn})$ = category value of g_{wn}. After classifying the pixels in the image into three categories according to gray value, the initial category value of the pixels in a big granule reflects a high accuracy.

3.4 The flow of the IGMRF algorithm

The total flow of the IGMRF algorithm is as follows:

(1) Apply SARD, *Sobel*, H-minima, patulous minima transform, compulsive minimum transform, and watershed transform to make a region partition. The regions are the epitaxy of the information granule.
(2) Obtain the connotation of the granule by feature extraction, including intra-regional features such as the mean pixel intensity, ig_1, the second-order moment of GLCM, ig_2, the serial numbers of regions, inter-regional features, and the RAM.
(3) Combine the defined information granule with MRF to create the model of IGMRF.
(4) Calculate the optimal value of the objective function by ICM.
(5) Set the gray value of each pixel according to the result of granule merging.

4 EXPERIMENT AND ANALYSIS

This paper selects a local spinal marrow MRI image and the same image with Gaussian noise as the simulation objects. They are in the size of 242×101 pixels. The window size of GLCM is 5×5. Let β equal 3.0, and the maximum times of ICM equal 30.

The original images are shown in Figure 2. Figure 3 shows the segmentation results of MRF. Figure 4 shows the segmentation results of IGMRF. Figures 2–4 show the original image and the results on the left, and all of them with Gaussian noise on the right.

It is obvious that MRF does not perform very well in cutting out intact vertebrae and other organizations. Several spots appear in the region of the vertebrae. This phenomenon is more serious on the image with Gaussian noise. The results of IGMRF are not only accurate, but also have a lot of advantages compared with MRF, such as distinct edge profiles, less spots in the vertebrae, obvious regionality, and a higher capacity of noise suppression.

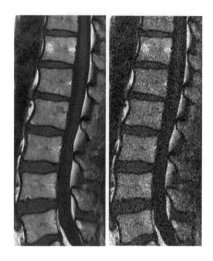

Figure 2.   Original image.

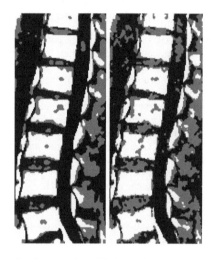

Figure 3.   Segmentation of the image based on MRF.

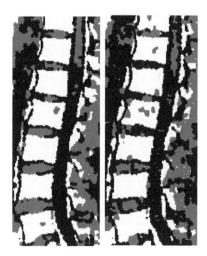

Figure 4.   Segmentation of the image based on IGMRF.

5   CONCLUSION

This paper segments MRI images using information granules and MRFs. There are two key points: one is the construction of the granule, including connotation and epitaxy, obtained respectively by feature extraction and region division, the other is the establishment of an IGMRF model, which is developed from the combination of the information granule and the MRF. The simulation experiment proves that IGMRF is feasible. In the future, IGMRF can combine with genetic algorithms or something else to obtain exquisite image segmentation, and play a positive role in medical diagnoses.

ACKNOWLEDGMENTS

This work is supported by the Natural Science Foundation Project of Shanxi Province (2012021030-1). The corresponding author is Ms Fang Wang, email: wangfang@tyut.edu.cn.

REFERENCES

Bae Min Hyeok, Pan Rong, Wu Teresa & Badea Alexandra. 2009. Automated segmentation of mouse brain images using extended MRF. *NeuroImage* 46 (3): 717–25.

Barker S A. 1998. Image Segmentation Using Markov Random Field Models. Ph. D Dissertation. Cambridge, UK: University of Cambridge, Department of Engineering.

Fwu J, Djuric P. 1996. Unsupervised vector image segmentation by a tree structure-ICM algorithm. *IEEE Transactions on Medical Image* 15(6): 871–881.

Joshua E. Johnson, Terence E. McIff, Phil Lee, E. Bruce Toby & Kenneth J. Fischer. 2014. Validation of radiocarpal joint contact models based on images from a clinical MRI scanner. *Computer Methods in Biomechanics and Biomedical Engineering* 17 (4): 378–387.

Lin T Y. 1997. Granule Computing, Announcement of the BISC Special Interest Group on Granular Computing.

Seba Suasn & M. Hanmandlu. 2013. A non-extensive entropy feature and its application to texture classification. *Neurocomputing* 120: 214–225.

Soille, P. 1991. *Morphological Image Analysis: and Applications*. Berlin, Germany: Springer Verlag.

Wyatt Paul P & Noble J Alison. 2003. MAP MRF joint segmentation and registration of medical images. *Medical Image Analysis* 7(4):.539–52.

Yang Xuezhi & David A. Clausi. 2012. Evaluating SAR Sea Ice Image Segmentation Using Edge-Preserving Region-Based MRFs. *IEEE Journal of Selected Topics in Applied Earth Observations and Remote Sensing* 5(5): 1383–1393.

Electronics, Information Technology and Intellectualization – Song & Kwak (Eds)
© 2015 Taylor & Francis Group, London, ISBN 978-1-138-02741-1

A ν-support vector regression for the inversion of sound speed profiles

Y.D. Xu & J.L. Li
Department of Information Science and Electronic Engineering, Zhejiang University, Hangzhou, China

ABSTRACT: Inversion of the Sound Speed Profile (SSP) with acoustic measurements is a strongly nonlinear problem. A method based on a ν-Support Vector Regression (ν-SVR) is proposed, where the ν-SVR nonlinear regression is employed to build the approximated inverse functions for the unknown SSP with the measured acoustic field data from a vertical linear array. Empirical Orthogonal Functions (EOFs) are introduced to reduce the parameters in the inversion algorithms. Simulation results show the proposed method correctly inverses the time-evolving SSP. It outperforms the standard Particle Filter (PF) and the ensemble Kalman Filter (EnKF), especially when the SSP changes fast.

1 INTRODUCTION

The Ocean Acoustic Tomography (OAT) introduced by Munk and Wunsch (1979) at the end of the 1970s is a powerful measurement tool to monitor large regions in ocean environments. In all the environment parameters, Sound Speed Profile (SSP) is one of the most important parameters that greatly affect the sound propagation (Carrière et al. 2009, Ballard & Becker 2010). The ensemble Kalman Filter (EnKF) is one of the most frequently used methods in SSP inversion problems (Carrière et al. 2009). One limitation in the EnKF is the underlying assumption of Gaussian forecast and observation errors (Guingla et al. 2013). In order to tackle it, the standard Particle Filter (PF) has been developed (Yardim et al. 2010, Guingla et al. 2013). However, this method has received criticism because of the problem of particle degeneracy (Li & Zhou 2013).

Recently, the Support Vector Machine (SVM) developed by Vapnik has gained popularity in classification and function estimation problems (Vapnik 1999, Jo et al. 2009, Wei 2009, Nasien et al. 2010, Lim & Chang 2012, Liu et al. 2014, Suganyadevi & Babulal 2014). ε-Support Vector Regression (ε-SVR) is an extension of the SVM theory for regression problems by introducing ε-insensitivity loss function proposed by Vapnik (Vapnik 1999, Shi et al. 2014). However, its parameters are hard to set. The ν-SVR is a different implementation of the SVR which allows easy setting of the SVR parameters (Schölkopf et al. 2000, Pérez-Cruz & Artes-Rodriguez 2001, Hu et al. 2014).

Inversion of the Sound Speed Profile (SSP) with acoustic measurements is a strongly nonlinear problem. ν-SVR is suitable to deal with the SSP inversion problem due to its advantage of being able to control the complexity and generalization ability of the model, and is the very method to deal with SSP inversion problems on limited samples and large dimensions.

This paper is organized as follows. Firstly, we briefly recall the basic ingredients of the ν-SVR algorithm and give a summary of the steps implementing the SSP inversion scheme. Secondly, Shallow Water 2006 (SW06) experimental results show the validity of our proposed method with comparison to the standard PF and the EnKF methods. Finally, the conclusions and suggestions for future work are included.

2 V-SUPPORT VECTOR REGRESSION

The given set of training samples $\{(\mathbf{x}_i, y_i), i = 1, 2, \ldots, l\}$, in which $\mathbf{x}_i$ is the input vector and y_i is the corresponding target value. With a ν-SVR nonlinear regression, the training data $\mathbf{x}_i$ is mapped by Φ into a high-dimensional feature space where a linear function

$$\hat{f}(\mathbf{x}_i) = \mathbf{w} \cdot \Phi(\mathbf{x}_i) + b \tag{1}$$

is performed, where b is a bias and w is a weight vector of the same dimension as the feature space. The regression problem is given as:

$$\min_{w,b} \quad \frac{1}{2} \mathbf{w}^T \mathbf{w} + C \left(\nu \varepsilon + \frac{1}{l} \sum_{i=1}^{l} (\xi_i + \xi_i^*) \right)$$

$$\text{s.t.} \quad y_i - \mathbf{w} \Phi(\mathbf{x}_i) - b \leq \varepsilon + \xi_i^* \tag{2}$$
$$\mathbf{w} \Phi(\mathbf{x}_i) + b - y_i \leq \varepsilon + \xi_i$$
$$\xi_i \geq 0, \xi_i^* \geq 0, \varepsilon \geq 0, i = 1, 2, \cdots, l$$

At each point x_i, an error of ε is allowed. Everything above ε is captured in slack variables ξ_i and ξ_i^*, which are penalized in the objective function via a regularization constant C, chosen a priori. The size of ε is traded off against model complexity and slack variables via a constant $\nu \geq 0$. ($*$) is a shorthand implying both the variables with and without asterisks.

Making use of the Lagrange multiplier method, we introduce multipliers α_i^*, η_i^*, $\beta \geq 0$. This regression problem shown in (2) is formulated as follows:

$$L = \frac{1}{2}\|\mathbf{w}\|^2 + Cv\varepsilon + \frac{C}{l}\sum_{i=1}^{l}(\xi_i + \xi_i^*) - \beta\varepsilon$$

$$-\sum_{i=1}^{l}(\eta_i\xi_i + \eta_i^*\xi_i^*) - \sum_{i=1}^{l}\alpha_i\left[\xi_i + \varepsilon + y_i - \mathbf{w}\,\Phi(\mathbf{x}_i) + b\right] \qquad (3)$$

$$-\sum_{i=1}^{l}\alpha_i^*\left[\xi_i^* + \varepsilon - y_i + \mathbf{w}\,\Phi(\mathbf{x}_i) + b\right]$$

To minimize the expression (2), we have to find the saddle point of L, that is, minimize over the primal variables $\mathbf{w}$, ε, b, ξ_i^* and maximize over the dual variables α_i^*, η_i^*, β. Setting the derivatives with respect to the primal variables equal to zero yields four equations:

$$\begin{cases} \dfrac{\partial L}{\partial \mathbf{w}} = 0 \rightarrow \mathbf{w} = \sum_{i=1}^{l}\left(\alpha_i^* - \alpha_i\right)\Phi(\mathbf{x}_i) \\[2mm] \dfrac{\partial L}{\partial \varepsilon} = 0 \rightarrow C \cdot v - \sum_{i=1}^{l}(\alpha_i + \alpha_i^*) - \beta = 0 \\[2mm] \dfrac{\partial L}{\partial b} = 0 \rightarrow \sum_{i=1}^{l}\left(\alpha_i - \alpha_i^*\right) = 0 \\[2mm] \dfrac{\partial L}{\partial \xi_i^{(*)}} = 0 \rightarrow \dfrac{C}{l} - \alpha_i^{(*)} - \eta_i^{(*)} = 0 \end{cases} \qquad (4)$$

Rewriting the constrains and noting that η_i^*, $\beta \geq 0$ do not appear in the dual, we arrive at the v-SVR optimization problem: for $v \geq 0$, $C \geq 0$, maximize

$$W(\alpha_i, \alpha_i^*) = \sum_{i=1}^{l} u_i(\alpha_i - \alpha_i^*)$$

$$-\frac{1}{2}\sum_{i=1}^{l}\sum_{j=1}^{l}(\alpha_i - \alpha_i^*)(\alpha_j - \alpha_j^*) \times \left\langle \Phi(\mathbf{x}_i)\Phi(\mathbf{x}_j)\right\rangle \qquad (5)$$

subject to

$$\sum_{i=1}^{l}(\alpha_i - \alpha_i^*) = 0 \qquad (6)$$

$$\alpha_i, \alpha_i^* \in \left[0, \frac{C}{l}\right] \qquad (7)$$

$$\sum_{i=1}^{l}(\alpha_i + \alpha_i^*) \leq C \cdot v \qquad (8)$$

It is found that parameter $0 \leq v \leq 1$ and v is an upper bound on the fraction of errors and a low bound of the fraction of support vectors, thereby simplifying the selection range of the parameter's combination compared to ε-SVR.

The regression problem then takes the form:

$$\hat{f}(\mathbf{x}) = \sum_{i=1}^{l}(\alpha_i - \alpha_i^*) \times \left\langle \Phi(\mathbf{x}_i)\Phi(\mathbf{x}_j)\right\rangle + b \qquad (9)$$

Mercer's theorem is applied to the inner product kernel:

$$K(\mathbf{x}_i, \mathbf{x}) = \left\langle \Phi(\mathbf{x}_i)\Phi(\mathbf{x}_j)\right\rangle \qquad (10)$$

The regression problem is thus given by:

$$\hat{f}(\mathbf{x}) = \sum_{i=1}^{l}(\alpha_i - \alpha_i^*)K(\mathbf{x}_i, \mathbf{x}) + b \qquad (11)$$

Due to the nature of these constraints, typically only a subset of the solution values $(\alpha_i - \alpha_i^*)$ are nonzero, and the associated data values are called the support vectors.

The kernel functions of v-SVR commonly include linear, polynomial, Radial Basis Function (RBF) and sigmoid. In general, the RBF kernel is a reasonable first choice of kernel function. Therefore, the RBF kernel has been considered in this paper. The RBF kernel is:

$$K(\mathbf{x}_i, \mathbf{x}) = \exp\left(-\|\mathbf{x}_i - \mathbf{x}\|^2 / \sigma^2\right) \qquad (12)$$

Our proposed inversion scheme based on the v-SVR algorithm above involves several steps as outlined below:

- Set the acoustic model, collect the acoustic field data and determine the prior bounds for the unknown SSP
- Generate the simulated input/output training set. Here, the output is the set of SSP which are selected in the prior searching space, and the inputs are the corresponding pressure amplitudes calculated by the forward acoustic model at each sensor of a vertical hydrophone.
- Train the v-SVR model for each SSP.
- Apply the trained v-SVR models to the experimental data.

3 EXPERIMENTAL RESULT

3.1 Experimental environment

In this section, we will show the result of SSP inversion using the v-SVR algorithm. The data sets of sound speed come from Shallow Water 2006 (SW06) (Tang et al. 2007, Ballard & Becker 2010). SW06 is a series of experiments conducted from June to September 2006 in the vicinity of the New Jersey continental shelfbreak. As the synchronous acoustic pressure data sets are inaccessible, we used normal mode code KRAKEN to calculate the acoustic pressure data. We used the data from August 26 to August 27 to do the simulation experiment and the sampling interval was 20 minutes. The simulation environment, which comes from SW06, is shown in Figure 1 and the simulation parameters are listed in Table 1.

The total number of input/output training pairs is 1000. To design a v-SVR, a commonly used function, the Radio Basic Function (RBF), is selected as the kernel function. After conducting the grid searching and five-fold cross-validation technique (Hsu et al., 2003) for training data, we obtained $C = 100$ and $g = 0.001$, where $g = 1/\sigma^2$.

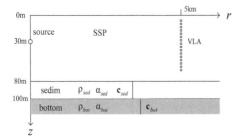

Figure 1. The simulation environment.

Table 1. Simulation parameters.

Source depth	30 m	Source frequency	224 Hz
Source range	5 km	Array SNR	20 dB
No. of hydrophones	16	Track length	48 hrs
Array start, Δz	8.2 m, 3.75 m	Track interval	20 min

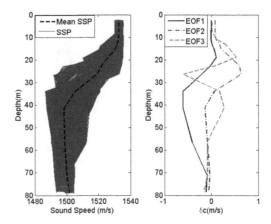

Figure 2. (a) SSP samples and mean SSP. (b) First three EOFs.

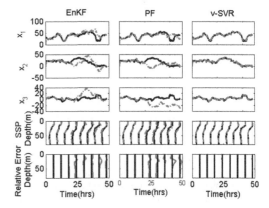

Figure 3. Tracking results of the EnKF, PF and ν-SVR. The solid lines are the true trajectories and the dashed-dotted lines denote the tracking results.

3.2 *EOFs representation of SSP*

In order to reduce the search space, Empirical Orthogonal Functions (EOFs) are introduced. EOFs have already been proved a low-dimensional scheme to describe the SSP and are very efficient in reducing the number of variables to estimate (LeBlanc & Middleton 1980, Yardim et al. 2009). In fact, EOFs are eigenvectors of an SSP data covariance matrix which can be estimated from an historical SSP database. Performing an eigendecomposition of the SSP data covariance matrix, we can obtain the required EOFs $\{\delta c_l(z)\}$, $l = 1, 2, \cdots, D$. Therefore, any SSP can be approximately expressed in terms of EOFs:

$$c_i(z) \approx \overline{c}(z) + \sum_{k=1}^{K} \alpha_k \delta c_k(z) \qquad (13)$$

where $\overline{c}(z)$ is the mean SSP and α_k is the kth coefficient of EOFs computed as:

$$\alpha_k = [c_i(z) - \overline{c}(z)]^T \delta c_k(z) \qquad (14)$$

In this way, any SSP can be described in k variables.

Figure 2 shows the SSP samples from SW06 data from August 26 to August 27

3.3 *Results and comparison*

The results of the three EOF coefficients inversion are presented in Figure 3, where the results of SSP tracking every 12 hours are given. We find that the PF and EnKF algorithm can track the time-evolving EOF coefficients at the very beginning of the period. However, in the fast changing region of the EOF coefficients, both of the two methods failed to inverse the EOFs. The ν-SVR algorithm can track the time-evolving very well and outperforms the PF and the EnKF, which is quite obvious from the relative errors shown in the last row.

The errors in the SSP estimation can be evaluated by the depth-integrated Root Mean Square Error (RMSE) metric. It essentially calculates the difference between the true sound speed and the estimated sound speed and then integrates across these values in depth to provide a single number defined by:

$$\Delta c_{RMS} = \sqrt{\frac{1}{D} \sum_{z=z_1}^{z=z_D} [c_{true}(z) - c_{est}(z)]^2} \qquad (15)$$

By examining the depth-integrated RMSE evolution, the divergence in the filters can be seen more obviously. The three different methods' results of RMSE are shown in Figure 4.

From Figure 4, we can conclude that at the first 20 hours, the three methods can all inverse SSPs, and when the SSPs change relative fast, especially from 20 hours to 45 hours, the PF and EnKF failed but the ν-SVR still have the high performance.

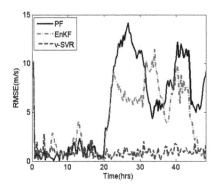

Figure 4. RMSE of PF, EnKF and ν-SVR.

4 SUMMARY

In this paper, a new method of SSP inversion based on ν-SVR was proposed. We investigated the use of ν-SVR to estimate the inverse functions between SSP and the measured acoustic data. Visual results and RMSE were compared among the proposed method, the PF method, and the EnKF method. Comparison results show that the proposed method shows advantages in performance over both the PF and EnKF methods. In the fast change domain especially, the proposed methods dealt with the inversion problems whilst the other two methods failed. One limitation in the EnKF application was the underlying assumption of Gaussian forecast and observation errors and the PF had the problem of particle degeneracy. These drawbacks led the two methods' failure in the inversion problem under the poor environment. ν-SVR is suitable to deal with the SSP inversion problem for its advantage of being able to control the complexity and generalization ability of the model and is, therefore, the very method to deal with SSP inversion problems on limited samples and large dimensions. As the proposed method has not considered new measurement data, our future work would consist of investigating a self-adaptive ν-SVR algorithm for the SSP inversion.

REFERENCES

Ballard, M. S. and K. M. Becker (2010). "Inversion for range-dependent water column sound speed profiles on the New Jersey shelf using a linearized perturbative method." *The Journal of the Acoustical Society of America* **127** (6): 3411–3421.

Carrière, O. and J. Hermand, et al. (2009). "Inversion for time-evolving sound-speed field in a shallow ocean by ensemble Kalman filtering." *Oceanic Engineering, IEEE Journal of* **34** (4): 586–602.

Guingla, P. and A. Douglas, et al. (2013). "Improving particle filters in rainfall – runoff models: Application of the resample – move step and the ensemble Gaussian particle filter." *Water Resources Research* **49** (7): 4005–4021.

Hsu, C. and C. Chang, et al. (2003). A practical guide to support vector classification.

Jo, Q. and J. Chang, et al. (2009). "Statistical model-based voice activity detection using support vector machine." *Signal Processing, IET* **3** (3): 205–210.

LeBlanc, L. R. and F. H. Middleton (1980). "An underwater acoustic sound velocity data model." *The Journal of the Acoustical Society of America* **67** (6): 2055–2062.

Li, J. and H. Zhou (2013). "Tracking of time-evolving sound speed profiles in shallow water using an ensemble Kalman-particle filter." *The Journal of the Acoustical Society of America* **133** (3): 1377–1386.

Lim, C. and J. Chang (2012). "Enhancing support vector machine-based speech/music classification using conditional maximum a posteriori criterion." *IET signal processing* **6** (4): 335–340.

Liu, X. and D. Liu, et al. (2014). "Optimal Support Vector Regression Algorithms for Multifunctional Sensor Signal Reconstruction." *TELKOMNIKA Indonesian Journal of Electrical Engineering* **12** (4): 2762–2768.

Munk, W. and C. Wunsch (1979). "Ocean acoustic tomography: A scheme for large scale monitoring." *Deep Sea Research Part A. Oceanographic Research Papers* **26** (2): 123–161.

Nasien, D. and S. S. Yuhaniz, et al. (2010). Statistical learning theory and support vector machines. *Computer Research and Development, 2010 Second International Conference on, IEEE.*

Shi, Y. Z. and M. Xu, et al. (2014). "Knowledge Based ε-Support Vector Regression Method." *Applied Mechanics and Materials* **462**: 472–475.

Suganyadevi, M. V. and C. K. Babulal (2014). "Support Vector Regression Model for the prediction of Loadability Margin of a Power System." *Applied Soft Computing.*

Tang, D. and J. N. Moum, et al. (2007). "Shallow Water' 06: A joint acoustic propagation/nonlinear internal wave physics experiment".

Vapnik, V. N. (1999). "An overview of statistical learning theory." *Neural Networks, IEEE Transactions on* **10** (5): 988–999.

Wei, G. (2009). Geoacoustic Inversion Based on Support Vector Machine. *Natural Computation, 2009. ICNC'09. Fifth International Conference on, IEEE.*

Yardim, C. and P. Gerstoft, et al. (2009). "Tracking of geoacoustic parameters using Kalman and particle filters." *J. Acoust. Soc. Am* **125** (2): 746–760.

Yardim, C. and P. Gerstoft, et al. (2010). "Geoacoustic and source tracking using particle filtering: Experimental results." *The Journal of the Acoustical Society of America* **128** (1): 75–87.

Electronics, Information Technology and Intellectualization – Song & Kwak (Eds)
© 2015 Taylor & Francis Group, London, ISBN 978-1-138-02741-1

Metallogenic prediction based on GIS through automatic database building and extraction of regional gravity and magnetic structure

S.S. Ye & S.C. Wang
Institute of Mineral Resources Prediction with Synthetic Information, Jilin University, Changchun, China

Y.X. Ye
Key Laboratory of Symbolic Computation and Knowledge Engineering of Ministry of Education, Jilin University, Changchun, China
National Engineering Research Center of Geophysics Exploration Instruments, Changchun, Jilin University, Changchun, China

G.X. Zhao & F. Teng
Tianjin Institute of Geology and Mineral Resources, Tianjin, China

ABSTRACT: The application of intelligent geophysical interpreting, technology of mapping and database building based on GIS have been discussed under the guidance of the theory and the method of the mineral resources prediction of synthetic information in this paper. The automatic extraction of regional gravity and magnetic structures to build databases has been realized. With the prediction of the iron mine as a case, the whole process of the application of synthetic information in minerals' prediction of GIS technology has been stated. Through the realization of the data process of regional gravity and magnetic data, and of line-ring structure extraction, geophysical spatial information and map databases have been automatically built. Through a combination of geological and geophysical interpretation, the spatial information and map databases of geophysical ore-finding (iron mine) have been built. Through model units, an ore-finding model of comprehensive information has been built. Through geological units, geophysical-geological spatial information and map databases of synthetic information of predictive models have been built. According to the requirement of different mathematical models for mineral prediction of synthetic information for the data models, the spatial information in the synthetic information databases of predictive models has been extracted, and the geological variables have been converted, and then converted mutually into a digital geological multivariate statistical variable matrix. Different statistic prediction models of the multivariate statistical methods have been established to complete the whole process of metallogenic prediction by the computer.

Keywords: Gravity and Magnetic Structures, Synthetic Information, Metallogenic Prediction, GIS Technology, Spatial Information and Map databases

1 INTRODUCTION

A very large standard basic database has been established on the basis of massive earth science data which has been accumulated by China in the past sixty years. According to incomplete statistics, including twelve major categories and more than fifty kinds of databases of national data resources (MLR, 2009a) which have already been completed and are currently being built, they mainly include: the spatial database of 1:5,000,000–1:50,000 digital geological maps; the 1:1,000,000–1:200,000 aeromagnetic survey database; the aerial remote sensing images database; the regional gravity survey database; the 1:500,000–1:200,000 regional geochemistry database (39 elements); the 1:200,000–1:50,000 small-scale

mineral geology database; the heavy concentrate database, etc. As the contradiction between the needs of the rapid development of the national economy and shortages of mineral resources, many of China's standard basic databases will be used to predict and evaluate mineral resources (Wang et al. 2007; Porwal & Kreuzer 2010).

The theory and the method of the mineral resources prediction of synthetic information have now become the general theory and the method of China's mineral resources prediction (Wang 2010; MLR 2009b). Prediction of mineral resources with synthetic information is a series of systems of thoughts and methods with integrity theory and methods in predicting minerals (Wang et al. 2003). Its guiding principles are the ore-forming theory and mathematical geology. With the geological bodies as units, the relation between geology, geochemistry, geophysics, and remote sensing from the perspective of the geological evolution, is studied. According to the thought of systematic theory, ore-finding models of synthetic information are established. The ore-finding model of synthetic information is to use the mineral resources body as a unit to apply this assemblage in order to build the statistical model. The bodies of mineral resources have different classes (such as ore field, ore deposit, ore body and etc.). Different classes of ore-finding models are statistical models rather than deterministic models. Synthetic information ore-finding models are based on geological information to study geological, geophysical, geochemical, and remote sensing information, and to study the transformation regularity of information between them to apply indirect information to look for concealed mineral resources bodies and blind mineral resources bodies, and to achieve the purpose of prospecting (Wang 2010).

The authors have researched mineral resource prediction by applying the method of synthetic information for a long time. We have combined geology, geophysics, geochemical, and remote sensing closely with computer technology, developed "the prediction expert system of the metallogenic series of comprehensive information" (Wang et al. 1999), and then combined the geographic information system and finally successfully developed the "Mineral Resources Prediction System with Synthetic Information" (Ye et al. 2003, 2004, 2005, 2007). Thus quantification and intelligence of comprehensive information mineral resources prediction based on GIS have been basically realized.

2 DATABASE–BUILDING OF GEOPHYSICAL PROSPECTING INTERPRETATION BASED ON GIS

2.1 Regional gravity and magnetic data process and expert interpretation

Geophysics applied in comprehensive information mineral prediction is used mainly to study the gravity and magnetic ore-controlling structure form. Basement structures which gravity and magnetism reflect have the characteristics of regional structures. The cross place of line-ring structures of gravity and magnetism to reflect basement structure form is mineralization concentration regions (Wang et al., 2003).

2.1.1 Regional gravity and magnetic data processing

Regional gravity and magnetic data is usually processed by potential field transformation in frequency domain to separate the deep source anomaly field and the shallow local anomaly field. The frequency spectrum of the regional gravity and magnetic fields has obvious differences of features of high-low frequency. The features of high-low frequency are separately

reflected in the objective existence of geological bodies. As upward continuation is higher, information can reflect the existence of deep, large geological bodies, such as old basement, large concealed magmatic bodies, etc. Conversely, the existence of shallow, small geological bodies can be reflected. The fault structures and ring structures of existence can be interpreted and inferred according to the differences between density or magnetism of the geological bodies.

2.1.2 Workflow of regional geophysical prospecting interpretation

Experts who are looking for minerals, from the aim of prediction of mineral resources, interpret maps of gravity and magnetic potential field conversion of different continuation altitudes, to infer line-ring structures on corresponding different depth planes and to construct the framework maps (Ye et al. 1996) (Figure 1). The basic principle and workflow of regional geophysical interpretation are to use a horizontal first derivative of various directions (EW, NE, SN, NW) on the plane of different altitudes to explain and to infer linear structures of gravity and magnetism on each plane. With the vertical second derivative on the planes of different altitudes, ring structures of gravity and magnetism on each plane can be interpreted and inferred. Through the vertical correlation of line-ring structures on each plane of different altitudes, spatial distribution regularity of mineral resource bodies can be studied from a three-dimensional perspective.

2.2 The intelligent geophysical interpretation and mapping and establishment of databases based on GIS

2.2.1 Establishing the unified projection coordinate system of the different scale data

Synthetic mineral prediction in the GIS environment must be performed in the uniform space coordinates system. Based on a 1:200,000 standard geological map in the scope of the study area: the area in east Wu Zhu-E Shengaobi of Inner Mongolia: the east longitude $116°00' \sim 119°00'$ and along the north latitude $45°20' \sim 46°40'$, combined with the data of regional physical prospecting, chemical prospecting and minerals, etc., the unified projection coordinate system of the research area, with the east longitude $111°00'$ as the longitudinal coordinates, has been built together in this paper. Figure 2 is used to construct the sketch map of distribution about the study area. The sketch map relates to five maps of 1:200,000 mappable unit scope with different scales, which are about the data of geology and physical prospecting.

2.2.2 Intelligent interpretation system of geophysical prospecting

According to the whole process of geophysical data processing and interpretation of geophysical protecting experts (including knowledge of interpretation and experience), under the GIS environment, the intelligent geophysical interpretation system (Ye et al. 2004,

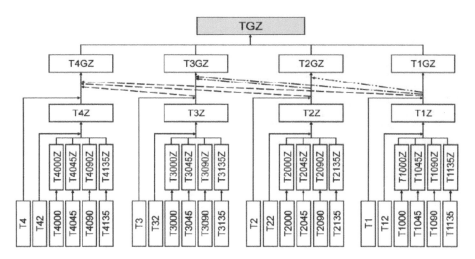

Figure 1. Workflow of aeromagnetic interpretation.
Instructions from bottom to top layers of Figure 1:
1. Horizontal directions (taking an east-west direction as 0 direction, moving counterclockwise: 0°, 45°, 90°,135°), derivative maps: T1000 T1045 T1090 T1135-first altitude of upward continuation, T2000 T2045 T2090 T2135-second altitude of upward continuation, T4000 T4045 T4090 T4135-upward continuation of fourth altitude; T12 T22 T32 T42-vertical second derivative maps of different altitudes (first, second, third, and fourth) of upward continuation individually; Tl T2 T3 T4-anomalies maps of different altitudes (first, second, third, and fourth) of upward continuation individually.
2. Horizontal directions (0°,45°,90°,135°), derivative axis maps: T1000Z T1045Z T1090Z T1135Z-first altitude upward continuation, T2000Z T2045Z T2090Z T2135Z-second altitude upward continuation, T3000Z T3045Z T3090Z T3135Z-third altitude of upward continuation, T4000Z T4045Z T4090Z T4135Z-upward continuation of fourth altitude.
3. T1Z T2Z T3Z T4Z-linear structure maps of different altitudes (first, second, third, and fourth) of upward continuation individually.
4. T1GZ T2GZ T3GZ T4GZ-structure framework maps of different altitudes (first, second, third, and fourth) of upward continuation individually.
5. TGZ-framework map of Aeromagnetic Interpretation.

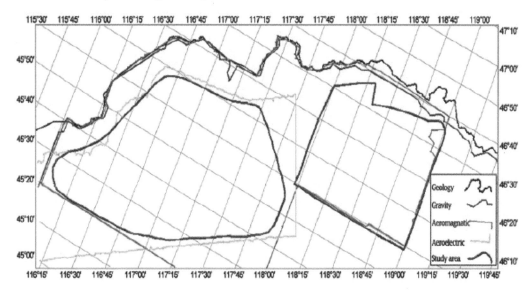

Figure 2. The map of distribution range of geological and physical prospecting data based on the GIS study area.

2006) from data input to data output of the final gravity and magnetic linear structure maps, the ring structure maps, the framework maps, and the integration of maps and the databases, has been designed and developed by the authors. Among them, the left part below in Figure 3 is the main menu about regional gravity and magnetic interpretation of geophysical interpretation systems, and about forward and inversion algorithm

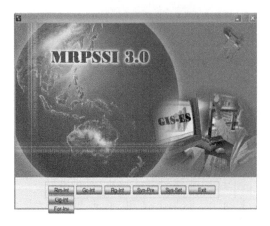

Figure 3. Functional modules of geophysical interpretation under mineral resources prediction system with synthetic information 3.0 version.
Instructions:
MRPSSI3.0: mineral resources prediction system with synthetic information 3.0 version; Rm-Int: regional magnetic interpretation; Gg-Int: regional gravity interpretation; For-Inv: calculus of forward and inverting; Gc-Int: geochemical interpretation; Rg-Int: regional geological interpretation; Syn-Pre: synthetic prediction;

methods of gravity, magnetism, and electricity. Figure 4 is the main interface of subsystem of magnetic interpretation.

2.2.3 *Building the geophysical prospecting spatial information and maps database by automatic interpretation*

On account of the gravity or magnetic data input by a geophysical prospecting interpretation system of GIS, the potential field conversion (Pf-Tra) is carried out to establish the series of figure files of regional gravity or magnetic data processing and interpreting results (see the list of files under "Pft-Que" of "Md-Man" menu in Fig. 4, and twenty-four

Potential field conversion maps on the bottom of Fig. 1). On this basis, sixteen derivative axes of four altitudinal planes of different continuation is extracted (Fig. 1). Spatial level relating between derivative axes in every plane is carried out, and the linear structures (T1Z T2Z T3Z T4Z in Fig. 1) in each plane are built. The spatial vertical correlation of liner structures in each plane is carried out, and spatial information and the graphic databases of the structural framework maps of gravity or magnetism are automatically established, such as structural strike of gravity or magnetic

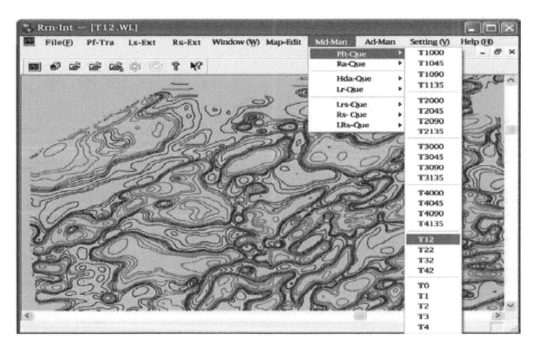

Figure 4. Main interface of regional magnetic interpretation.
Instructions:
Pf-Tra: potential field transformation; Ls-Ext: linear mstructure extraction; Rs-Ext: ring structure extraction; Md-Man: maps database management; Ad-Man:attribute database management; Pft-Que: potential field transformation query; Ra-Que: residual anomaly query; Hda-Que: horizontal derivative axis query; Lr-Que: level relating query; Lrs-Que: linear structures query; Rs-Que: ring structures query; LRs-Que: line-ring interpreted structures query; T0: the map of anomalies of original magnetic data reduction to the pole. Sys-Set: system settings.

interpretation, structural dip, and so forth (Fig. 5 and Fig. 6).

3 COMBINING GEOPHYSICAL INTERPRETATION OF GEOLOGY AND ESTABLISHING DATABASES

3.1 Workflow of the Geophysical Interpretation Combined with Geological Bodies

The aim of the interpretation design of geophysical prospecting spatial analysis combined with geology is

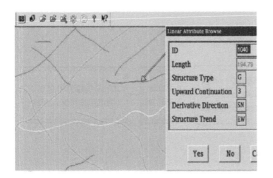

Figure 5. Maps database of gravity linear structures of the third (8km) of altitude upward continuation.
Instructions:
G: structure type of gravity interpretation.

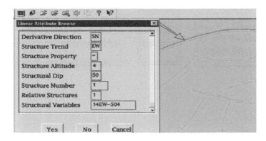

Figure 6. Maps database of aeromagnetic structures of the fourth altitude (10 km) of upward continuation
Instructions:
T4EW-S04: combining information (T4: aeromagnetic fourth altitude of upward continuation; EW: EW-trending -: structure property of negative derivative axis; S0: structural dip as south; 4: structure of four altitudinal plane.)

to achieve the distribution range of buried and half-hidden old basement, stratum and rock bodies delimited by geophysical information interactive delineation according to the exposed and semi- exposed rock and stratum. Geological information of spatial analysis comes from the spatial information and map databases of geological mineralization of regional geological interpretation of the geological interpretation system (Ye et al. 2012). The main part of geophysical information is the vertical second derivative in the geophysical spatial information and map databases, and residual anomaly. Figures 7, 8, 9, and 10 are the workflow combined with the geological-geophysical interpretation. Among them, the spatial database includes spatial information and map databases.

3.2 Establishing the spatial information and map databases of geophysical ore-finding

Zuo Renguang et al., (Zuo & Xia 2007) of China University of Geosciences have done the application researches for the GIS-based ore deposits information extraction assistant.

Geophysical ore-finding spatial information and map databases are established by interpretation of geophysical prospecting spatial information and map databases combined with regional geology in this paper: Firstly, it is necessary to complete geological interpretation, and to establish spatial information and map databases of geological mineralization of mineral resources bodies (Ye et al. 2012): Based on the digital geologic map, space is located with the map of mineral deposit (Fig. 11). Ore-controlling factors of known typical ore deposits (points) are analysed by analysing mineralization information of the ore-bearing strata and rocks to determine the ore mineral resources bodies, to integrate, and to establish spatial information and map databases of geological mineralization of mineral resources bodies. Secondly, is the geophysical interpretation combined with geology: The geophysical ore-finding spatial information and maps database are established by technique of spatial overlay analysis. This technique is used to integrate the information and figures which reflect the features of geophysical prospecting of mineral resources bodies.For example, in magnetic iron ore prediction, Figures 12, 13, 14, 15, and 16 show the establishment of the geophysical prospecting marks etc., which reflect delimitation of ore-controlling rock bodies.

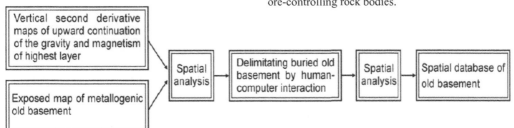

Figure 7. Delimitating old basement through the vertical second derivative maps of the highest top of upward continuation of gravity and magnetism.

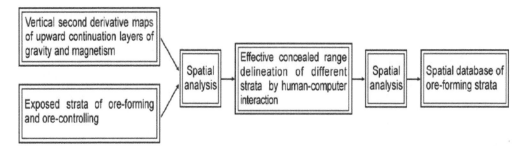

Figure 8. Flow chart for delimitating strata of ore-forming.

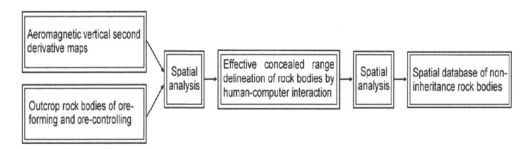

Figure 9. Flow chart for delimitating non-inheritance rock bodies.

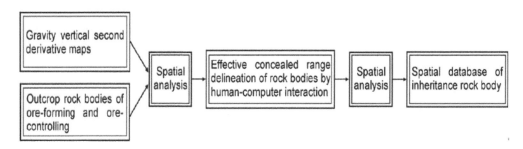

Figure 10. Flow chart for delimitating inheritance rock bodies.

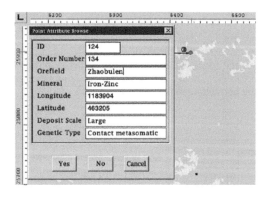

Figure 11. Spatial analysis of geological mineralization information about typical ore deposits and granite bodies of the Yanshanian period.

relative low density gravity anomaly ▨ relative high density gravity anomaly ▦ Yanshan period granite (γJ)

Figure 12. Granite bodies of the Yanshanian period as gravity low density anomaly bodies.

negative residual gravity anomaly   positive residual gravity anomaly   Yanshan period granite (γJ)

Figure 13.   Granite bodies of the Yanshanian period as gravity negative residual anomaly bodies.

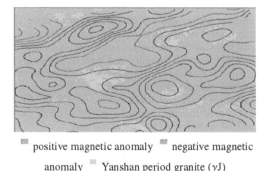

positive magnetic anomaly   negative magnetic anomaly   Yanshan period granite (γJ)

Figure 14.   Granite of Yanshan period is located at one side of negative value of the gravity gradient zone.

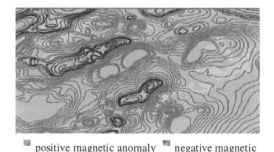

positive magnetic anomaly   negative magnetic anomaly   Yanshan period granite (γJ)

Figure 15.   Granite bodies of the Yanshan period as local positive magnetic anomaly.

4   APPLICATION OF GEOPHYSICAL INFORMATION IN MINERAL RESOURCE PREDICITION WITH SYNTHETIC INFORMATION

Zhao Pengda and Cheng Qiuming have completed the successful quantitative forecasting for deep exploration of Sn and Cu mineral deposits, in conjunction with advanced GIS technology (Zhao et al 2008). As everyone knows, geophysical prospecting methods are

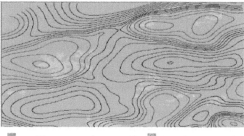

positive magnetic anomaly   negative magnetic anomaly   Yanshan period granite (γJ)

Figure 16.   Granite of the Yanshan period is located at one side of positive value of the gradient zone of the third altitude (4 km) of aeromagnetic upward continuation.

effective methods to discover magnetic iron ore. Gravity and magnetic information can be used as direct marks of ore-finding.

4.1   Establishing the ore-finding model with synthetic information as the model unit

Mineral resources bodies are composed of large, moderate, and small deposits as a related whole that is inalienable. The prospecting model of synthetic information takes the mineral resource body as a unit. Through extraction and integration of geophysical ore-finding spatial information, the geophysical-geological ore-finding model of the best composition of prospecting marks is set up, and the ore-finding model with synthetic information is also set up.

In this paper, establishing the ore-finding model with synthetic information of a prognosis iron mine used as a model unit based on GIS, is used to carry out the spatial metallogenic analysis though model units with the spatial information and maps database of geophysical ore-finding overlain. Distribution characteristics of geophysical anomalies of the exposed and semi-exposed ore deposits are analysed. Through similarity analogies, the prospecting marks of geophysical-geological prospecting of hidden ore and semi-hidden ore are identified. The geophysical-geological ore-finding model with synthetic information of a mine field grade based on model units is established.

4.1.1   The model unit delimitated based on typical ore deposits

The introduction of GIS and application provide a very convenient means for building the prospecting model. In this paper, mine fields which are composed of one or a number of typical deposits are used as the necessary condition of mineralization of partition of the model unit. The spatial information and maps database of geophysical ore-finding are combined with ore-forming stratum of geological mineralization and ore-forming rock bodies (Ye et al. 2012) as the sufficient condition, as shown in Figures 11,

12, 13, 14, 15, and 16. That is overlain spatially with iron-polymetallic deposits (ore occurrence) of the map of mineral deposits to locate spatial location of the model unit. It is used to combine geophysical prospecting local anomalies to determine the delimited region of model units and the boundary conditions, and to delimit model units by human-computer interaction.

4.1.2 Establishment of the ore-finding model with synthetic information (geophysical-geological) based on the model units

With the model unit, extraction and integration of spatial information from the geophysical ore-finding spatial information and maps database, and the spatial information and the maps database of geological mineralization (Ye et al. 2012) are carried out. The main synthetic geophysical- geological prospecting marks of the ore-finding model with synthetic information (iron ores) are:

1) D_2 t: the Middle Devonian Tal Baggett group
2) D_3 a: Upper Devonian series Angell tone group
3) O_{1-2} t-b: the Middle-Lower Ordovician and layer of Copper Mountain group and Duobaoshan group
4) γJ: Early Yanshan (Jurassic Period) granite (porphyritic granite; medium-grained biotite granite)
5) Local positive intensive aeromagnetic anomalies
6) Local low-density gravitational anomaly
7) Negative residual gravity anomaly
8) the side of bend at gravity gradient belt of negative value
9) ore-controlling of aeromagnetic NE-trending and NW-trending linear structures
10) ore-controlling of the cross place of linear structures of aeromagnetic upward continuation of the different altitudes
11) ore-controlling of gravity NE-trending and EW-trending linear structures
12) ore-controlling of the intersected section of gravity structures, as well as the composition of various prospecting marks.

4.2 Establishing the spatial information and map databases of synthetic information of the prediction model

The predictive model mainly refers to the mathematic model of statistic prediction and the mathematic model of the prediction of mineral resources (Wang 2010). Because geological data from the actual observation has a uniqueness of disunity, the more effective method is to set up the predictive models as geological units. Through the unit division of geological bodies and mineral resources bodies, the data will be normalized, standardized, and then, processed (Ye et al. 2012).

4.2.1 Establishment of the unified geological unit, the model unit, and the predictive unit

The geological unit, also known as the geological statistics unit, includes the model unit of mineral

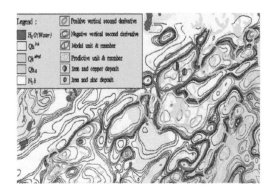

Figure 17. One of the markers of the units delimited (vertical second derivative of aeromagnetic upward continuation 1500 m).

resource bodies and the predictive unit of mineralization anomaly bodies. Namely, the model unit is established with a geological unit which contains the ore field, and ore deposits (ore nodes). The predictive unit is established with the geological unit which contains ore mineralization anomalies. Therefore, the geological body, which is the concrete object of the research of mineral resources prognosis, has been realized according to the geological unit.

In this paper, the unit division emphasizes understanding geological bodies and mineral resources bodies with complete integrity. With the geological bodies and mineral resources bodies divided unifiedly as a unit, namely, with ore deposits (ore nodes) contained as a model unit, and with the mineralization anomaly as a predictive unit, the regional ore-controlling characteristics of the two units have a broad comparison with the geological unit of ore field grade.

The predictive unit is based on prospecting marks of the geophysical-geological, ore-finding model. With the similar condition of the model unit, the mode of human-computer interaction of GIS technology is adopted. Therefore, the prediction units of iron deposits (Fig. 17 & Fig. 18) are delimited and found in hidden, and half-hidden areas which contain geological anomalies. There are twenty-one geological units delimited for the prediction of iron mines in this study area.

4.2.2 Establishment of the geophysical-geological spatial information and maps database of the prediction model

The ore-finding spatial information and the maps database with synthetic information (iron ores) of geological statistics units is established: The map of the distribution of geological units (Fig. 17 & Fig. 18) of the predictive unit combined with the model unit, is used as base maps database of spatial information integration of the prediction model. The spatial information and the maps database of geophysical and geological mineralization are used as the extraction object of ore-finding synthetic information. The ore-finding spatial information and the maps database are

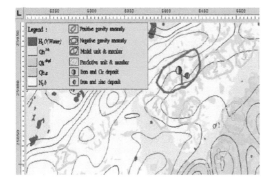

Figure 18. Another of the markers of the unit's delimited (side of bend at gravity gradient belt of negative value).

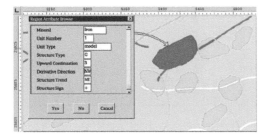

Figure 19. One of maps of the synthetic spatial information database of geophysical iron-mine-finding (the linear structure of third height (8 km) of gravity upward continuation).
Instructions:
G: the structure type of gravity interpretation;
+: derivative axis of positive anomaly of the first derivative of horizon

established by the spacial overlayer analysis of GIS. The chart entity of the map is the geological unit. The the spacial overlayer analysis of GIS which is adopted in this process included the regional-regional discriminant analysis, region-point discriminant analysis, and region-lines discriminant analysis (Fig. 19 & Fig. 20). Finally, the spatial information and maps of predictive models of synthetic information of iron deposits are built.

4.3 *Establishing the prediction model—location and quantitative prediction models of iron deposits*

Based on the synthetic spatial information database of the predictive model as the geological unit, the geological and mathematical model of translation is studied to realize extraction of spatial information in the synthetic information database of the prediction model and switch of geological variables. A matrix of multivariate geological statistical variables of the digital form is transformed to establish different statistical prediction models of multivariate statistical methods. Based on digital characteristics of mineral resources as model units and observation values of geological variables, statistical and mathematical models of location

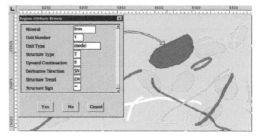

Figure 20. Another map of the synthetic spatial information database of geophysical iron-mine-finding (the linear structure of third altitude (4km) of aeromagnetic upward continuation).
Instructions:
T: the structure type of aeromagnetic structure
-: derivative axis of negative anomaly of the first derivative of horizon

and quantitative prediction of mineral resources are established individually, which are used for statistical prediction of ore-forming probability, and the potential resources amount of the prediction units.

In this paper, based on the established synthetic spatial information database of prediction models, original variables used in the prediction of iron-polymetallic deposits are extracted, integrated, and constituted according to the different goals of prediction. There are two methods of variable assignment. One is to state qualitative assignments. The existence of some kind of information in the unit is "1", and non-existent or unclear information is "0". The other is a measuring assignment of the quantitative geological body.

For example, the units contain the individual number of aut-crop rock bodies, and measuring value of the area, etc. (Table 1 & Table 2).

5 THE RESULTS OF PREDICTION AND DISCUSSION

5.1 *The results of iron deposit prediction*

In view of the characteristics of a few iron ore model units in the study area, and of the prognosis data of the geological units of a few known large, medium-sized deposits etc., the different statistical prediction models of multivariate statistical methods are established, according to requirements for the data models, which come from the different mathematic models of mineral resources prediction of synthetic information. The study area is divided altogether into twenty-one geological statistics units of predicting iron ore.

The different grade targets are comprehensively appraised: One grade is nine; two grades are six; three grades are six. Among them, there are six large ore deposit target areas, and six midsize iron deposit target areas (Fig. 21). The partitioning of the grades of location prediction in the iron ore target area is from the statistical results (Table 3) of the principal component method of the product matrix of characteristic

Table 1. Variables Table of Extraction and Switch of the Prediction Model of Geological Units (Iron Ores).

Unit No.	Unit Area	γJ Area	γJ Numbers	$\gamma\pi J$ Area	$\gamma\pi J$ Numbers	D_2t Area	D_3a Area	D_3a Numbers	$O_{1-2}t\text{-}b$ Area	$O_{1-2}t\text{-}b$ Numbers
1	463.088	64.3046	2	0	0	76.9655	0	0	0	0
2	752.438	214.278	3	0	0	458.365	0	0	0	0
2	1530.6	279.373	6	0	0	289.875	4.75528	1	0	0
4	690.521	36.6339	2	0	0	49.0893	0	0	0	0
5	559.078	146.814	1	0	0	104.039	0	0	0	0
6	731.722	68.2254	1	0	0	0	187.408	6	0	0
7	422.495	0	0	0	0	65.332	0	0	0	0
8	523.709	0	0	0	0	0	89.8563	2	0	0
9	521.157	0	0	0	0	0	42.7953	3	187.846	1
10	365.357	0	0	0	0	0	0	0	0	0
11	622.375	23.1315	1	0	0	0	0	0	88.1399	4
12	549.567	106.289	2	0	0	0	0	0	0	0
14	228.263	58.3974	1	0	0	0	0	0	0	0
13	386.247	86.4285	1	0	0	0	0	0	116.092	2
15	723.515	0	0	71.0814	1	0	0	0	0	0
16	775.241	0	0	44.4421	4	0	0	0	0	0
17	751.84	0	0	47.9395	1	0	11.9425	4	0	0
18	936.39	0	0	48.0851	2	0	0	0	0	0
19	1411.74	0	0	43.8666	3	0	0	0	0	0
20	2054.74	425.827	1	0	0	0	0.895927	1	0	0
21	1677.34	98.9841	5	0	0	0	0	0	0	0

Table 2. Variables Table of Extraction and Switch of the Prediction Model of Geological Units (Iron Ores), continued.

Unit No.	T1Z LOA	T2Z Cross	T3Z→ EW	T3Z→ NE	T3Z→ SN	T3Z→ NW	T3Z+	T3Z−	T3Z Cross	T3Z items	T3Z Density	T3Z/Area
1	16.91463	1	1	0	1	0	1	1	1	2	0.00431884	0.10128
2	72.84635	1	1	0	0	0	0	1	0	1	0.00132901	0.032993
3	115.0097	1	1	1	0	0	0	2	1	2	0.00130668	0.078004
4	89.14193	1	0	1	0	1	1	1	1	2	0.00289636	0.043368
5	30.30181	1	0	0	0	0	0	0	0	0	0	0
6	84.78691	1	0	1	1	0	1	1	1	2	0.00273328	0.085368
7	19.4474	0	1	2	0	0	2	1	1	3	0.00710067	0.120348
8	58.7689	1	0	1	1	1	2	1	1	3	0.00572837	0.157657
9	80.283	1	1	1	0	0	1	1	1	2	0.00383762	0.042723
10	54.78489	1	0	2	0	0	1	1	0	2	0.00547409	0.127477
11	77.95681	1	0	2	0	1	2	1	1	3	0.00482024	0.127309
12	53.77442	0	1	2	0	0	2	1	1	3	0.00545884	0.127234
14	51.22116	1	0	0	0	0	0	0	0	0	0	0
13	61.02453	1	1	0	0	0	1	0	0	1	0.00258901	0.057738
15	83.38283	1	0	0	0	0	0	0	0	0	0	0
16	79.38255	1	1	0	0	0	0	1	0	1	0.00128992	0.052665
17	42.28383	1	1	0	0	1	1	1	1	2	0.00266014	0.064929
18	76.52853	1	0	1	0	1	0	2	1	2	0.00213586	0.067504
19	174.932	1	1	1	1	0	2	1	1	3	0.00212503	0.069639
20	217.7377	1	0	2	2	1	3	2	1	5	0.0024334	0.035789
21	235.2998	1	1	1	2	1	4	1	1	5	0.00298092	0.083686

Table 1, Table 2 Instructions:

T1Z LOA: tectonic total length of aeromagnetic upward continuation first altitude (1 km)

T2Z Cross: the structure intersection of aeromagnetic upward continuation second altitude (2 km)

T3Z→EW: east west structural strike items of aeromagnetic upward continuation third altitude (4 km)

T3Z→NE: north east structural strike items of aeromagnetic upward continuation third altitude (4 km)

T3Z→SN: south north structural strike items of aeromagnetic upward continuation third altitude (4 km)

T3Z→NW: north west structural strike items of aeromagnetic upward continuation third altitude (4 km)

T3Z+: structure property of aeromagnetic upward continuation third altitude (4 km) as positive

T3Z−: structure property of aeromagnetic upward continuation third altitude (4 km) as negative

T3Z Cross: the structure intersection of aeromagnetic upward continuation third altitude (4 km)

T3Z Items: the structures numbers of aeromagnetic upward continuation third altitude (4 km)

T3Z Density: the structures total items of Aeromagnetic upward continuation third altitude (4 km) to divide with its unit area

T3Z/Area: the structures total length of Aeromagnetic upward continuation third altitude (4 km) to divide with its unit area

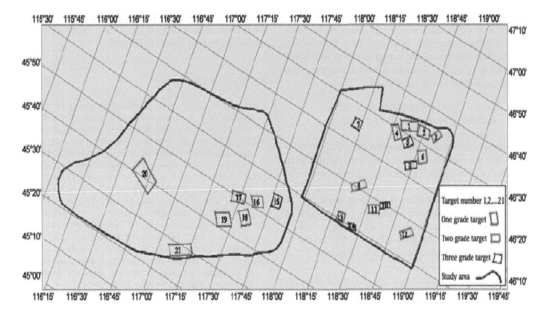

Figure 21. The distributing map of the iron mine target area of prediction of synthetic information.

Table 3. The Ranking Units of Characteristic Analysis.

Unit number	Unit type	Statistical score	Unit grade
1	model	4.3984	
10	model	3.126	
13	model	3.6675	
		Connection degree	
5	predictive	1.8738	III
15	predictive	2.3526	
14	predictive	2.5249	
7	predictive	2.7932	
8	predictive	2.9608	
12	predictive	3.4259	II
3	predictive	3.4386	
16	predictive	3.5638	
6	predictive	3.5855	
9	predictive	3.5982	
20	predictive	3.6118	
18	predictive	3.6716	I
4	predictive	3.7015	
11	predictive	3.7543	
17	predictive	3.7969	
19	predictive	3.9174	
2	predictive	3.9644	
21	predictive	4.4432	

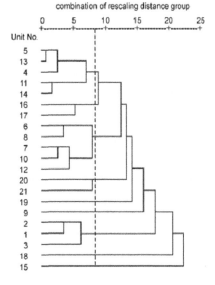

Figure 22. Dendrogram of predicting iron deposits using average linkage (between groups).

analysis. The scale of the potential resources amount in the iron deposit target area is derived from statistical results (Fig. 22) of cluster analysis.

5.2 Results and discussion

There are several works for the quantitative estimation mineral resources (Porwal & Kreuzer 2010; Carranza 2011). The following conclusions can be drawn from the above-mentioned in our work.

5.2.1 Automatic mapping and building database of spatial information based on GIS

Integrated technology of geophysical prospecting spatial information based on GIS is the comprehensive technology combined with geophysical data processing, interpretation and inference, the GIS spatial analysis mapping, and building the databases. The integrated technology implements the whole process of the regional geophysical prospecting interpretation, the geophysical interpretation of combination with the

213

regional geology, and automatic mapping and building the maps database, which has changed the state of the previous mapping separated from establishing the attribute database, and completed the automatic integration of spatial information and mapping and building database at the same time.

5.2.2 Three-dimensional ore-finding marks according to the different upward continuation altitudes

The spatial information and map databases of interpretation and integration of different upward continuation altitudes of geophysical prospecting, are relative to the tow-dimensional map of geology and mining which can be transformed into the ore-finding marks of the three-dimensional mineral resources bodies, and of the three-dimensional prediction of geologic bodies, in order to provide the possibility for prediction of hidden deposits and deeply concealed deposits.

5.3 Using the statistical unit as geological bodies is more accorded with geological facts

It is recognized that it is reasonable to use a geological unit as the division of a statistical unit of prediction. The ore-controlling geological meaning is clear, which has been widely applied. The geological units are partitioned with synthetic ore-finding marks of ore-finding models with synthetic information to determinate the delimited region and boundary conditions of units, which can objectively reflect ore-forming and ore-controlling conditions of geological bodies to conform to the geological facts.

5.4 Information of spatial modelling with integrity

This technology, in the process of extraction of metallogenic information, is unlike the traditional method that requires many people to be involved, which results in too long a period of work, with errors which can hardly be avoided. The integration technology of spatial information in this study is the synthetic application of GIS spatial analysis between geological units and the maps database and database management technology. Extraction of information of geological mineralization and geophysical ore-controlling is required to integrate the information for the automatic vector spatial overlay analysis between different digital maps, and to establish the maps metallogenic information which can be excavated by the integration of spatial information quickly, comprehensively, and deeply. Therefore, there are a series of systems and integrity in this kind of information.

ACKNOWLEDGMENTS

This work was supported in part by the project of National Natural Science Fund under Grant Nos. 41172294; Provincial Department of cooperation projects, Sub-Project: Predicted Study with Mineral Resource Prediction of Synthetic Information on Key Metallogenic Belt at East Wu Zhu in Inner-Mongolia under Grant Nos.1212010781028.

REFERENCES

Carranza, E. 2011. From predictive mapping of mineral prospectivity to quantitative estimation of number of undiscovered prospects. *Resource Geology* 61:30–51.

MLR. 2009a. Overall schema of inspecting national mining right. Ministry of Land Resources, China, http://www.mlr.gov.cn/xwdt/zytz/200906/P020090603580308614626.doc

MLR. 2009b. Overall schema of resource evaluation of national mineral potency. Ministry of Land and Resources, China, http://www.mlr.gov.cn/xwdt/zytz/200906/P020090603580307879093.doc

Porwal, A. & Kreuzer, O. 2010. Introduction to the special issue: ineral prospectivity analysis and quantitative resource estimation. *Ore Geology Reviews* 38:121–127.

Wang, S. 2010. The new development of theory and method of synthetic information mineral resources prognosis. *Geological Bulletin of China* 29:1399–1403.

Wang, S., Ye, S. & Yang, Y. 1999. The prediction expert system of the metallogenic series of comprehensive information. Changchun: Press of Changchun.

Wang, S., Ye, S. & Zhou, D. 2003. Theory and method of mineral resource prediction based on synthetic information. *Journal of China University of Geosciences* 14:269–273.

Wang, X., Ju, F. & Xi, Q. 2007. Research on spatial data mining techniques based on mineral belts foundation database. In Porceedings of the IAMG '07: Geomachematics and GIS analysis of resource, environment and hazards (pp. 343–346.

Ye, S., Chen, T. & Ma, S. 1996. A study on irregular graph inference mechanism of an expert system in geoscience. *Progress in Geology of China* pp: 1084–1088.

Ye, S., Qiao, J. & Ye, Y. 2012. Integrate of geological interpretation spatial information of mineral resource prediction based on GIS. *Journal of Jilin University (Earth Science Edition)* 42: 1214–1222.

Ye, S., Wang, S. & Li, D. 2003. Application of GIS in mineral resource prediction of synthetic information. *Journal of China University of Geosciences* 14: 234–241.

Ye, S., Wang, S. & Liu, W. 2004. The basic principles of GIS and application development. Changchun: Press of Jilin University.

Ye, S. 2007. Applied study on the prediction with the prediction system of comprehensive information mineral resources in the concentration region of polymetallic deposits at the southeast section of Daxinganling Mountains in Inner-Mongolia. Changchun: Press of Jilin University.

Ye, S., Zhou, D. & Dong, Y. 2005. Mineral resources prediction system of synthetic information. In The 4th ISPRS workshop on dynamic and multi-dimensional GIS 174–178.

Ye, S., Zhou, D. & Sun, F. 2006. The spatial information integration research of gravity and magnetic regional field based on GIS. *Progress in Geophysics* 21: 84–92

Zhao, P., Cheng, Q. & Xia, Q. 2008. Quantitative prediction for deep mineral exploration. *Journal of China University of Geosciences* 19:309–318

Zuo, R. & Xia, Q. 2007. Development and application of ore deposits information extraction assistant mapping system based on MapGIS. *Journal of China University of Geosciences*, 18:266–268.

Electronics, Information Technology and Intellectualization – Song & Kwak (Eds)
© 2015 Taylor & Francis Group, London, ISBN 978-1-138-02741-1

The research for informatization online music enterprise and its integration strategies

Delin Sun & Yu Li
Jiangxi Provincial Key Lab for High-Performance Computing, Jiangxi Normal University, Nanchang, China

Ying Yu & Tingting Cai
Jiangxi Normal University, Nanchang, China

ABSTRACT: This paper aims to explore the integration connotation of informatization and online music enterprise, the informatization's effect on the development of online music enterprise, and the integration strategies of the two elements.

1 INTRODUCTION

Online music, a form of music full of vitality, is cultivated by informatization. Naturally, the integration of informatization and online music foster the online music enterprise. However, the informatization and online music's effect on online music enterprise is not reversible. Online music enterprises can feedback as well and give a new intention and extension to the online music. The integration of informatization and online music will expand the field for research and bring more value to it.

2 THE INTEGRATION CONNOTATION OF INFORMATIZATION AND ONLINE MUSIC ENTERPRISE

The influence of informatization on online music is manifested as a communication means that is continually improving with the unceasing progress of science and technology. Modern media, a kind of material instrument based on the Internet, can propagate and carry information like image, text, audio frequency and video which is exchanged between transmitter and receiver. Nowadays, the media industry is gradually moving into a digitization direction. It is widely applied in electronic techniques and digitized network technology. Digitized communication technology is the outcome of the integration of communication technology and digital technology. It is widely applied in the traditional media industry and a new form of digitized media is generated with it.

The development of modern media technology injects fresh vitality into online music which manifests itself through the display of new features:

A. Individualization. Advanced digital techniques enable people to compose and transcribe distinctive music composition however they like. Therefore, users are offered more opportunities to express themselves.

B. Superficial. Because of the development of digital techniques, large amounts of musical works can be appreciated by audiences through assistance means such as, visual sense. This fact impels more and more audiences into becoming unwilling to understand music further. As a result, music becomes superficial.

C. Universalness. Music users around the world are able to download and appreciate music that they like or need from the Internet anywhere, anytime.

Music is one of the major contents of communication that are propagated through modern communication technology. In the days when communication technology was backward, music was merely inherited and propagated by oral tradition and handwritten transcripts. Now, by applying this communication and recording modes, music can be easily distorted and lost during the inheritance process. Fortunately, with advanced modern media technology, music can be well preserved and appreciated by people all over the world. Music can be spread around the world in an instant through broadcasting, the wireless network and the Internet. Music needs media technology to propagate, whilst media relies on music to enrich its contents. The two elements are inseparable in a network of mutuality.

Thanks to modern media, musicians from all over the world are enabled to communicate with each other and learn from one another's musical culture, therefore, bringing a boost to music. With a wider transmission platform, music now welcomes a new era.

3 THE PROBLEMS IN THE INTEGRATION OF INFORMATIZATION AND NETWORK MUSIC ENTERPRISES

3.1 The lack of relevant industry standards in network music

The mess in the order of the network music industry is partly invoked by the lack of industry standards, which mainly presents itself in the following three aspects:

3.1.1 The music products are lacking in standards

Only the empowered side always gets the paper authorization. Therefore, because the record company cannot offer the complete data such as the sound file, the lyrics and the album cover, the enterprises which can offer original services are unable to effectively meet the agreement, thereby squandering time and energy.

3.1.2 The information of music products is in disorder

At present, no normative standard has been adopted for music products in the network platform.

3.1.3 The mixed foreign language translation offered by the service providers and the description disparity of products' information caused information differences between every net

In addition, the songwriter copyrights and singing rights are always separated. The content copyright used by operators to run music services is now mainly used to audit the neighbouring rights. As for the scattered songwriter copyright, it takes the claiming way while there is no according for the proportion of claiming.

3.1.4 The ownership of music products is mixed

Because the copyright of network music is not clear or the ownership of it is mixed, the situation is that the empowered side is often not always able to check whether the true copyright belongs to the certifier or not. Sometimes the empowered products will be indicted for copyright infringement by a third side, causing the financial losses on the channel's side. For example, for the same song, the same performer and the same album, three different websites may all declare ownership to it. Then we require the record company to label all three websites, but the outcome is that there is no ownership for any of the music products. The ambiguous ownership for music products has disturbed the market order.

3.2 Customers' consumption habit of paying should be improved

Piracy is still the biggest obstacle to the development of the online music industry. The key to improving the habit of consuming music products lies in fostering the awareness of copyrighted music. Since the Internet emerged in China, a group of Internet companies from the early days have applied the Napster model when they engage in music products. Most Chinese netizens have been accustomed to free Internet resources, and piracy is the means of obtaining them. As it is hard to change the free consumption habit of Chinese netizens in a short time, companies are likely to meet a big barrier when carrying out the pay-for-consumption model. At present, a great number of online music companies provide consumers with free resources or information, which makes the business model of carrying our licensed online music so single that it is hard to make a great profit.

3.3 The system of distribution and negotiation within the industry needs to be built

The authors and the suppliers differ widely in their settlement systems and copyright price. They argue a lot. Therefore they must negotiate on a profit distribution system and improve the communication system.

4 INFORMATIZATION'S EFFECT ON THE DEVELOPMENT OF ONLINE MUSIC ENTERPRISE

4.1 Social networks have revolutionized the music industry

Since the online music service was launched, online music manufacturers have attempted to develop its social contact function. There is no denying that music is the most sociable media. Online music can be more penetrative and popular when it takes advantage of music's sociality. Meanwhile, related sales brought by social contact can expand the industry chain of online music, add more economic benefits and create new marketing channels and profit patterns. For instance, Spotify and Facebook's corporation helped Spotify to immediately become Facebook's default music service supplier, which could then be applied by more than 800 million customers. Consequently, Spotify gained nearly one million new customers in the first month after its corporation with Facebook. Like.fm, another social music-sharing website, allows its clients to associate with the account numbers of Facebook, Twitter and fm.

While users listen to music through websites or software, the information like music chapter, affiliated website and music player can be synchronized to Like.fm and SNS anytime in order to interact with other users.

Apart from the modes mentioned before, music's application in Sina microblog and microblog social games reconfirms the developing trend of music social contact as well. Online music's penetration into animation and games enriches the communication channels and service modes of music, which makes music-sharing act as one application endowed with a vital function for sociality.

4.2 *The creative application brought by the mobile internet*

At present, with more maturity in the mobile network environment, the quick popularization of different kinds of mobile terminal and the need for listening to high-quality music anytime and anywhere, will undoubtedly bring huge commercial opportunities. Several pioneering enterprises have broken into this situation by using Smart phones to speed up the layout of the handheld music consumer market. The wireless terminals of the built-in intelligent operating systems change with every passing day, and various applications of wireless music emerge one after another. The two points make wireless music become the most capable detailed field to grow among the online music market. From the major layout of purchase in 2011, in which the equity of Duomi Music and then that of Kugou was purchased by A8 Music, we can conclude that the field in mobile internet music has been paid much attention.

On the other hand, the specificity of the mobile network environment has gradually brought administrative changes in the needs of digital music and operating habits. There is a requirement for more time and convenience to constantly stimulate the music network operators to create new things. For example, in the past, a simple content obtainment which relied on searching for targeted keywords has developed into today's intelligent recommendation. Also, the simple personal collection for appreciation has changed into the sharing of music by Weibo, or communication of preference by entering the community. At the same time, the product realization has also changed a lot. At the beginning, the most present way of music is functional player, now gradually changed into online player. The streaming media technique and cloud skipper music play even make music liberate from the constraints of the physical medium.

4.3 *Open platform has changed the industry development pattern*

The open platform has become the main orientation of network industry development. Of course, network music is not excluded; Kugou Music, QQ Music and other such traditional music producers successively opened the platform in 2011. Jingdong Mall's plan was to put forward the digital music services in the later part of 2012, forming a more complete digital content service platform. The China mobile wireless music base will open API by 1:3 and offer industry solutions. Developers can call an original music library to open API, then develop all kinds of music applications.

The cooperation between the music industry and the network platform can gather all kinds of high-quality applications and content resources. It can knock through every field and platform. Not only can it meet administers' requirements of searching for music, but it can also make it more effective for enterprises to look for targeted administers. The future industry development tends to dock all kinds of high-quality resources with the needs of the administrator by any terminals and application open platforms. Then it can integrate the network music industry, the mobile network and the traditional industries, achieving its commercial value more quickly.

5 INFORMATIZATION ONLINE MUSIC ENTERPRISE AND ITS INTEGRATION STRATEGIES

5.1 *Improve laws and regulations*

Currently, some national regulations are lacking in feasibility, and the responsibilities of government departments, as well as the details of implementation, are not temporarily clear and definite. Many companies, therefore, hope that the government will further perfect the laws and regulations. From 2000, "Notice on the Online Business of Audio-visual Products" licences started to be issued by the government, provided that a company who wanted to run an online business of audio-visual products was approved by the administrative department of the market. However, many shops without licences are still running this kind of business. The government, therefore, should make further improvements on the laws and regulations about the protection of music products.

5.2 *Strengthening technical support*

We should rely on science and technology as a powerful tool for protecting music products. Technological tools are mainly used to control the assessment of the musical works. According to the access permission, the companies can control the number of those who do not get the access permission, and they can also cut down the quantity of those people trying to copy and use the musical works through unlawful ways. To some extent, technical assistance can help business owners protect their interests, and can increase their profits. On the one hand, companies' costs may increase because of technological innovation, on the other hand, the spread of pirated music will decrease. Measures for the protection of music products can greatly improve people's music enthusiasm, and their pursuit for artistic innovation will be encouraged.

5.3 *Set up the mechanism of charge*

In order to improve the interest distribution mechanism and promote the development of the network market, a better system of payment needs to be established. The implementation of a pay system needs the support of laws and regulations.

In the early stages of the pay model, MP3 network operators may provide free downloads, However, they could then provide a good service for VIP customers in the later period, such as experiencing a better music sound and unlimited downloading of music, etc. In this way, they can expand the group of music users, and their influence.

6 COMPLIMENTARY CLOSE

Chinese music is developing in diversification, and the trend in the development of network music is unstoppable. All in all, our country's online music industry is good, with regards to the business model, as the carriers constantly penetrate upstream and downstream, CP and the SP, terminal manufacturers are constantly expanding their market space and with the increase in the wireless music user's demand, the whole network of the music industry pattern is seeking the optimal group in the constant collision model, leading to an increase in the innovative applications and strengthening the industry penetration gradually.

7 THANK YOU

In this paper, one of the achievements about special tasks for the Ministry of Education, Humanities and Social Science research project (Engineering science and technology talent training research) ≪E-commerce and e-business business excellence engineers training mode research≫ (Project grant no.12GDGC013).

REFERENCES

You-Quan Ouyang (2007). The general theory of cultural industry, Hunan people's publishing house 6, 26–28.

Sunrui (2004). The history of Chinese pop music. China Federation of Literary and Art Circles 8, 156–161.

Shewugong (2013.Telecom operators to operate online music business analysis. Industrial Applications3, 122–126.

Fanfenglong (2010). China's online music business development model. Business and Economics Review 5, 22–27.

Songqianqian, Zhuxuan (2010). China's online music industry profit model analysis. Knowledge economy 6, 155–161.

Linli (2012). The online music market in US. Business perspective 6, 223–228.

Author index